The Immune Response to Implanted Materials and Devices

Bruna Corradetti
Editor

The Immune Response to Implanted Materials and Devices

The Impact of the Immune System on the Success of an Implant

Springer

Editor
Bruna Corradetti
Department of Life and Environmental Sciences
Marche Polytechnic University
Ancona, Italy

ISBN 978-3-319-45431-3 ISBN 978-3-319-45433-7 (eBook)
DOI 10.1007/978-3-319-45433-7

Library of Congress Control Number: 2016948732

Printed on acid-free paper

This Springer imprint is published by Springer Nature
The registered company is Springer International Publishing AG
The registered company address is: Gewerbestrasse 11, 6330 Cham, Switzerland

Foreword

In the powerfully emerging world of smart, or functional materials, I cannot imagine a class with greater potential impact on healthcare and societal benefits than biomaterials with an ability to modulate inflammatory response—precisely the subject focus of this exceptionally timely monograph edited by Dr. Bruna Corradetti.

All materials for use in healthcare elicit an inflammatory response, bar none; but exactly as inflammation can be a fundamental step in a healing process, or a formidable foe, if frustrated into a chronic manifestation, this biological response to a material interface can be essentially helpful, or profoundly detrimental. Materials technology, and our understanding of the many facets of inflammation, has finally reached a point of sufficient maturity and convergence, to make it possible, for biomaterials to be designed so as to elicit a beneficial, or at least a functionally neutral response from the biology with which they contact.

The downstream vision from this exciting vantage point potentially portends transformational breakthroughs in multiple domains of healthcare, ranging from lifelong orthopedic implants, to indwelling molecular sensors, brain-machine interfaces, regenerative biomaterial-cell combinations for applications in pancreatic and hepatic medicine, central and peripheral nervous system repair, T-cell transplantation and novel therapeutic systems. They comprise both, drug-delivery implants and systemic administration constructs, with the ability to preferentially concentrate at inflammatory sites, sense their biological surrounding, and respond accordingly to optimize therapeutic benefit and minimize adverse effects.

I express my enthusiastic support for Dr. Corradetti's efforts in realizing this extraordinary collection of contribution from world-leading experts, to place the convergence of inflammatory modulation and biomaterials on a firmer footing, for decades of scientific work in this nascent era. It has been an honor to serve in an editorial advisory capacity for this volume, and a great added privilege to be able to do so in concert with two exceptionally distinguished scientists as Dr. Anthony Atala and Ali Khademhosseini. My gratitude goes to them and to the authors for their outstanding contributions.

With all of this, I wish you all happy readings and a pathway of rewarding research, enhanced by the contents of this important monograph.

Sincerely,
Dr. Mauro Ferrari

Preface

This textbook is intended to be a resource for biomaterial scientists and biomedical engineers, in both industry and academia, interested in the development of smart strategies able to exploit the self-healing properties of the body and achieve functional tissue restoration. Nowadays, many textbooks and journals discuss the broad spectra of material properties that can be customized for any specific applications but only few of them characterize in detail the host response, as the driving factor in determining the success of an implant.

Thanks to the perspectives offered by experts in the field of regenerative medicine, tissue engineering, surgery, immunology, nanomedicine, and transplantation, this textbook will guide the readers throughout the fascinating cascade of events activated in the body following the implant of biomaterials and devices. In Chap. 1 Dr. Badylak provides an overview of the host response to various categories of biomaterials for regenerative medicine applications, from a host-centric and a biomaterial-centric perspective. In Chap. 2 Dr. Anderson discusses the humoral and cellular events occurring at the implant site immediately following implantation. In Chap. 3, Dr. Giachelli presents the current understanding of macrophages, their functions in physiological processes and dysfunction in response to the foreign body, as well as approaches to guide them towards resolution of the foreign body-elicited inflammatory response. Dr. Dobrovolskaia proposes in Chap. 4 regulatory challenges, translational considerations, and literature case studies pertinent to the immunological safety of nanotechnology-based devices. Dr. Sant and Dr. Goldsmith provide a discussion about the effects of natural vs. synthetic biomaterials, as well as the role of the biomechanical environment on tissue fibrosis, in Chaps. 5 and 9, respectively. Highlights about the role of the biomechanical and physicochemical properties in osteo-immunomodulation and the effect of surface topographical modification on the cellular and molecular mechanisms associated with osseointegration are reported in Chaps. 6 and 8, by Dr. Xiao and Dr. Ivanovski. In Chap. 7, Dr. Li describes challenges and opportunities in targeting key elements of the innate immune system in favor of transplant survival. In Chap. 10, Dr. Sabek reviews possible solutions for the challenges encountered in the pancreatic islet transplantation field, while in Chap. 11 Dr. Tacke discusses current strategies to target macrophages in liver diseases and cancer. Novel concepts of T-cell immunomodulation for their clinical translation are presented by Dr. Hildebrandt in Chap. 12 to allow the transfer of the knowledge gained to implanted materials and devices.

It has been a particular privilege for me to collaborate with each of the authors participating in this project, and I feel grateful for their inspired work and for the time they devoted to make this volume possible. I wish to express my public gratitude to Dr. Anthony Atala, Dr. Ali Khademhosseini, and Dr. Mauro Ferrari for serving as Editorial Advisors for this book, for their constant support, outstanding suggestions, and visionary ideas. It has been an honor working with you.

My greatest hope is that this book will stimulate further discussions and investigations on the powerful role of the host response in regenerative processes allowing for the development of cutting-edge approaches able to exploit it and achieve functional tissue healing.

Bruna Corradetti

Ancona, Italy

Contents

Contributors

Sara Khalid Al-Ageel Department of Pathology and Laboratory Medicine, King Faisal Specialist Hospital & Research Center, Riyadh, Saudi Arabia

TUMCells Interdisciplinary Center for Cellular Therapies, TUM School of Medicine, Technische Universität München, München, Germany

James M. Anderson Departments of Pathology, Macromolecular Science And Biomedical Engineering, Case Western Reserve University, Cleveland, OH, USA

Stephen F. Badylak McGowan Institute for Regenerative Medicine, University of Pittsburgh, Pittsburgh, PA, USA

Matthias Bartneck Department of Medicine III, RWTH University Hospital Aachen, Aachen, Germany

Wayne Carver Department of Cell Biology and Anatomy, University of South Carolina, School of Medicine, Columbia, SC, USA

Zetao Chen Institute of Health & Biomedical Innovation, Queensland University of Technology, Brisbane, QLD, Australia

State Key Laboratory of High Performance Ceramics and Superfine Microstructure, Chinese Academy of Sciences, Shanghai Institute of Ceramics, Shanghai, China

Marina A. Dobrovolskaia Nanotechnology Characterization Laboratory, Cancer Research Technology Program, Leidos Biomedical Research Inc., Frederick National Laboratory for Cancer Research, National Cancer Institute at Frederick, Frederick, MD, USA

Karen V. Eaton Department of Bioengineering, University of Washington, Seattle, WA, USA

Amanda M. Esch Department of Molecular and Medical Genetics, Oregon Health & Science University, School of Medicine, Portland, OR, USA

Mauro Ferrari Department of Nanomedicine, Houston Methodist Research Institute, Houston, TX, USA

Weill Cornell Medical College, New York, NY, USA

Vennece Fowlkes Department of Mathematics and Science, Hagerstown Community College, Hagerstown, MD, USA

Cecilia M. Giachelli Department of Bioengineering, University of Washington, Seattle, WA, USA

Edie C. Goldsmith Department of Cell Biology and Anatomy, University of South Carolina, School of Medicine, Columbia, SC, USA

Sandra Grass Medizinische Klinik III, Klinikum rechts der Isar, Technische Universität München, München, Germany

Stephen M. Hamlet Menzies Health Institute Queensland, Griffith University, Nathan, QLD, Australia

Martin Hildebrandt TUMCells Interdisciplinary Center for Cellular Therapies, TUM School of Medicine, Technische Universität München, München, Germany

Saso Ivanovski Menzies Health Institute Queensland, Griffith University, Nathan, QLD, Australia

School of Dentistry and Oral Health, Griffith University, Southport, QLD, Australia

Melissa F. Jackson Department of Bioengineering, University of Washington, Seattle, WA, USA

Sirui Jiang Departments of Pathology, Macromolecular Science And Biomedical Engineering, Case Western Reserve University, Cleveland, OH, USA

Peixiang Lan Immunobiology and Transplant Science Center, Houston Methodist Hospital and Houston Methodist Research Institute, Houston, TX, USA

Xian C. Li Immunobiology and Transplant Science Center, Houston Methodist Hospital and Houston Methodist Research Institute, Houston, TX, USA

Ricardo Londono McGowan Institute for Regenerative Medicine, University of Pittsburgh, Pittsburgh, PA, USA

Susan A. Lund Department of Bioengineering, University of Washington, Seattle, WA, USA

Omaima M. Sabek Department Cell and Molecular Biology, Weill Cornell Medical College, New York, NY, USA

Department of Surgery, Houston Methodist Hospital, Houston, TX, USA

Shilpa Sant Department of Pharmaceutical Sciences, School of Pharmacy, University of Pittsburgh, Pittsburgh, PA, USA

Department of Bioengineering, Swanson School of Engineering, University of Pittsburgh, Pittsburgh, PA, USA

McGowan Institute for Regenerative Medicine, University of Pittsburgh, Pittsburgh, PA, USA

Kishor Sarkar Department of Pharmaceutical Sciences, School of Pharmacy, University of Pittsburgh, Pittsburgh, PA, USA

Michelle E. Scarritt McGowan Institute for Regenerative Medicine, University of Pittsburgh, Pittsburgh, PA, USA

Marta Scatena Department of Bioengineering, University of Washington, Seattle, WA, USA

Frank Tacke Department of Medicine III, RWTH University Hospital Aachen, Aachen, Germany

Chengtie Wu State Key Laboratory of High Performance Ceramics and Superfine Microstructure, Shanghai Institute of Ceramics, Chinese Academy of Sciences, Shanghai, China

The Australia-China Centre for Tissue Engineering and Regenerative Medicine, Queensland University of Technology, Brisbane, QLD, Australia

Yin Xiao Institute of Health and Biomedical Innovation, Queensland University of Technology, Brisbane, QLD, Australia

State Key Laboratory of High Performance Ceramics and Superfine Microstructure, Shanghai Institute of Ceramics, Academy of Sciences, Shanghai, China

Yingfei Xue Department of Pharmaceutical Sciences, School of Pharmacy, University of Pittsburgh, Pittsburgh, PA, USA

Yue Zhao Immunobiology and Transplant Science Center, Houston Methodist Hospital and Houston Methodist Research Institute, Houston, TX, USA

Abbreviations

ADA	Adenosine deaminase
aGvHD	Acute graft vs. host disease
ALP	Alkaline phosphatase
AMR	Antibody-mediated rejection
APC	Antigen-presenting cells
AST	Arginine stimulation test
ATMP	Advanced therapy medicinal product
BMI	Body mass index
BMP	Bone morphogenetic protein
CaP	Calcium phosphate direct deposition
CARPA	Complement activation related pseudoallergy
CaSR	Calcium sensing receptor
CCL	CC chemokine ligand
CDP	Common DC progenitor
cGvHD	Chronic graft vs. host disease
CID	Chemical inducer of dimerization
cMoP	Common myeloid progenitor
CSF	Colony-stimulating factor
CSFR	Colony-stimulating factor receptor
CXCL	Chemokine (C-X-C motif) ligand
CXCR	Chemokine receptor
DAF	Decay accelerating factor
DAMP	Damage-associated molecular pattern
DC	Dendritic cell
DDA	Degree of deacetylation
DKK-1	Dickkopf-1
DPP	Dipeptidyl peptidase
ECad	Epithelial cadherin
ECM	Extracellular matrix
EDRF	Endothelial-derived relaxing factor
EGF	Epidermal growth factor

egf-α	Tumor necrosis factor
EPCs	Endothelial progenitor cells
ER	Endoplasmic reticulum
ETS	E26 transformation-specific
FACS	Fluorescence-activated cell sorting
FBC	Foreign body capsule
FBGCs	Foreign body giant cells
FBR	Foreign body reaction
FDA	Food and Drug Administration
FG	Fasting glucose
FGF	Fibroblast growth factor
FXIIA	Activated Hageman factor
GDSC	Glutaraldehyde cross-linked collagen
GlcN	D-Glucuronic and D-glucosamine
GM-CSF	Granulocyte-macrophage colony-stimulating factor
GvHD	Graft vs. host disease
H1/H2	Histamine receptor
HA	Hyaluronic acid
HDSC	Hexamethylenediisocyanate
HETE	Hydroxyeicosatetranoic acid
HIF	Hypoxia-inducible factors
HLA	Human leukocyte antigen
HMGB	High-mobility group box chromosomal protein
HRG	Histidine-rich glycoprotein
HSA	Human serum albumin
HSC	Hepatic stellate cells
HSCT	Hematopoietic stem cell transplantation
HUVECs	Human umbilical vein endothelial cells
IAT	Islet auto-transplantation
IBMIR	Instant blood mediated immune reaction
ICOS	Inducible costimulatory
IDE	Investigational device exemption
IFG	Impaired fasting glucose tolerance
IFN	Interferon
IGF	Insulin growth factor
IgG	Immunoglobulin G
IL	Interleukin
IL-R	Interleukin receptor
ILC	Innate lymphoid cells
IND	Investigational new drug
iNOS	Inducible nitric oxide synthase
IVGTT	Intravenous injection of glucose tolerance test
KC	Kupffer cells
KIR	Killer cell immunoglobulin-like receptors
KLF	Kruppel-like factor

LPS	Lipopolysaccharide
LRR	Leucine-rich repeat motifs
LTB_4	Leukotriene B_4
LVAD	Left ventricular assist devices
M1	Classically activated macrophages or pro-inflammatory macrophages
M2	Alternatively activated macrophages or anti-inflammatory/pro-wound healing macrophages
MCP-1	Monocyte chemotactic protein 1
M-CSF	Macrophage colony-stimulating factor
MDP	Monocyte-macrophage DC progenitor
MDSC	Myeloid-derived suppressor cells
MHC	Major histocompatibility complex
MIP	Macrophage inflammatory protein
miR	microRNA
MMP	Matrix metalloprotease
modSLA	Sandblasted hydrophilic nano-rough surface
MoMF	Monocyte-derived macrophage(s)
MPS	Mononuclear phagocyte system
MSCs	Mesenchymal stromal cells
MSFM	Memphis serum-free media
MWCNT	Multi-walled carbon nanotubes
NBD-PE	1,2-Dioleoyl-sn-glycero-3-phosphoethanolamine-N-(7-nitro-2-1,3-benzoxadiazol-4-yl)
NF-kB	Nuclear *factor* kappa
NGF	Neuronal growth factor
NK	Natural killer
NLR	NOD-like receptors
NO	Nitric oxide
OGTT	Oral glucose tolerance test
OPG	Osteoprotegerin
OSM	Oncostatin M
PAMAM	Polyamidoamine
PAMP	Pathogen-associated molecular patterns
PBMA	Poly(butylmethacrylate)
PCA	Procoagulant activity
PCBMA	Poly(carboxybetaine methacrylate)
PCL	Poly(ε-caprolactone)
PDGF	Platelet-derived growth factor
PDMS	Polydimethylsiloxane
PDO	Polydioxanone
PEG	Polyethylene glycol
PGA	Polyglycolide
PI/IRI	Proinsulin to immunoreactive insulin
PIBCA	Polyisobutyl
PIHCA	Polyisohexylcyanoacrylate

PLA	Polylactide
PLGA	Poly(lactic-co-glycolic acid)
PLGA-PLL	Poly(lactic-co-glycolic acid)-poly-L-lysine
PMB	Poly(2-methacryloyloxyethyl phosphorylcholine(MPC)-co-n-butylmethacrylate(BMA)s)
PMNs	Polymorphonuclear leukocytes
POPC	1-Palmitoyl-2-oleoyl phosphatidylcholine
PPAR	Peroxisome proliferator-activated receptor
PRR	Pattern recognition receptor
PTFE	Polytetrafluoroethylene
PU	Polyurethane
PVA	Polyvinyl alcohol
PVA-SPION	Poly(vinyl alcohol)-coated superparamagnetic iron oxide nanoparticles
QD	Quantum dots
RANKL	Receptor activator of nuclear factor kappa-B ligand
RBC	Red blood cells (erythrocytes)
RES	Reticuloendothelial system
RGD	Arginine-glycine-aspartic acid
RLR	RIG-like receptors
ROS	Reactive oxygen species
SIBS	Poly(styrene–isobutylene–styrene) copolymer
SLA/Sr	Sandblasted micro-rough surface containing strontium
SLA	Sandblasted micro-rough surface
SOST	Sclerostin
SRBC	Sheep red blood cells
STZ	Streptozotocin
T1DM	Type 1 diabetes mellitus
TAM	Tumor-associated macrophages
T-cells	Thymocytes
TGF	Transforming growth factor
TIMP	Tissue inhibitor of metalloprotease
TLR	Toll-like receptors
TNF	Tumor necrosis factor
t-PA	Tissue-type plasminogen activator
T-regs	T-regulatory cells
VEGF	Vascular endothelial growth factor
VEGFR	VEGF receptor
VFH	Vinylidene fluoride-hexafluoropropylene copolymer
Zr-SLA & Zr-modSLA	Zirconium alloy SLA and modSLA surfaces

Chapter 1
Host Response to Implanted Materials and Devices: An Overview

Michelle E. Scarritt, Ricardo Londono, and Stephen F. Badylak

Abstract The host response to implanted materials and devices is influenced not only by the design of the material itself, but also by the local and systemic environment of the host. Much of the early response follows the well-described cascade of events of wound healing from hemostasis to scar formation. An implanted material can positively or negatively modulate this cascade of events, culminating in a constructive remodeling response, a persistent inflammatory response, a foreign body response with encapsulation, or an adaptive immune response. An overview of these events, as well as the influence of biologic versus synthetic materials, is discussed in this chapter.

Keywords Host response • Immune response • Hemostasis • Scar • Leukocyte • Macrophage • Constructive remodeling • Extracellular matrix

1.1 Introduction

The host response to an implanted biomaterial is determined by factors related to both the material itself and the host into which the material is placed. The long-term functional outcome, that is, the ability of the material to perform its intended function, is ultimately determined by the host response.

The evolution of and advances in biomaterials during the past 30–40 years, including the raw materials, device configuration, and manufacturing methods, have focused upon material properties such as degradability, pore size, surface functionality, and mechanical properties, among others. With the exception of studies related to the foreign body reaction (FBR) to nondegradable (e.g., permanent) implants, relatively little attention has been given to the host innate and acquired immune response elicited by these materials following implantation. The present chapter provides an overview of the host response to various categories of biomaterials from both a host-centric and a biomaterials-centric perspective.

M.E. Scarritt • R. Londono • S.F. Badylak (✉)
McGowan Institute for Regenerative Medicine, University of Pittsburgh, 450 Technology Drive, Suite 300, Pittsburgh, PA 15219, USA
e-mail: badylaks@upmc.edu

B. Corradetti (ed.), *The Immune Response to Implanted Materials and Devices*,
DOI 10.1007/978-3-319-45433-7_1

The immediate events following implantation include the adsorption of plasma proteins on the surface of the implant followed by all biologic processes associated with acute inflammation. These processes include the innate immune response to the biomaterial itself, and the response to the unavoidable tissue injury associated with the surgical procedure. Simultaneously, activation of the initial steps of the adaptive immune system occurs with downstream sequelae that either positively or negatively affects implant integration. An overview of the continuum of events associated with the innate and adaptive immune response is depicted in Figs. 1.1 and 1.2.

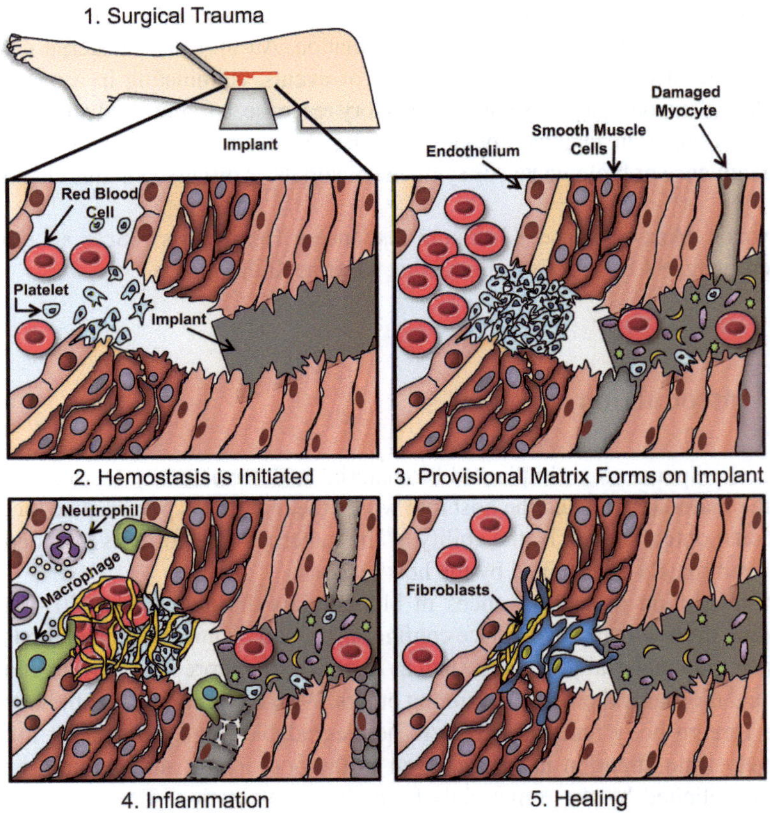

Fig. 1.1 Continuum of events following implantation of a material or device. (1) The surgical procedure inevitably damages the tissue at the implantation site. (2) Vascular damage initiates the coagulation cascade leading to the formation of a platelet-fibrin-red blood cell clot. Vascular damage also facilitates blood-implant interaction. (3) Proteins from the blood and interstitial fluid dynamically adsorb to the implant (Vroman effect). (4) A mileu of cytokines and chemokines are released by activated cells at the implant/injury site. Neutrophils, followed by monocytes and macrophages, are recruited to clear debris. Persistence of leukocytes/macrophages leads to chronic inflammation. (5) Healing is initiated and includes formation of granulation tissue, angiogenesis, and remodeling

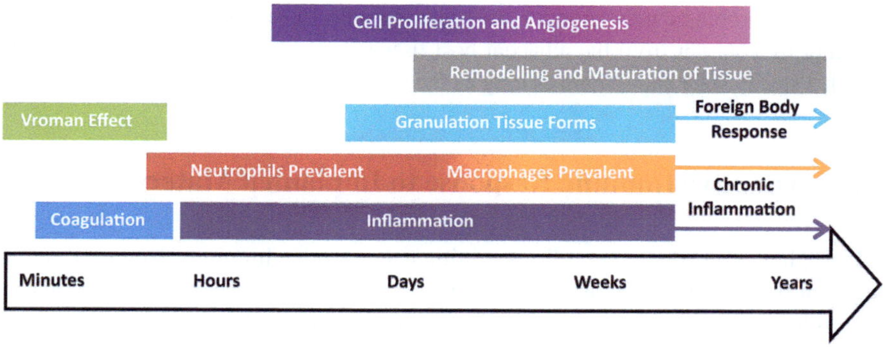

Fig. 1.2 Timeline of the host response following implantation of a material or device. The events that encompass the host reaction to an implant can be grouped according to broad response times. The Vroman effect and coagulation cascade occur within minutes of surgery, while immune cells infiltrate within hours and can persist for years after implantation

1.2 Innate Immune Response to Implanted Materials

Tissue injury following surgical implantation of any biomaterial is associated with well-described processes that include hemostasis, inflammation, and the formation of scar tissue.

Hemostasis—the process of blood clotting—occurs rapidly following injury. Injury to endothelial cells exposes the underlying vascular basement membrane causing platelets to adhere, activate, and initiate the coagulation cascade [1, 2]. As a result, a fibrin-platelet clot forms to prevent or slow further hemorrhage.

The acute inflammatory response is initiated by cytokines and chemokines released from damaged cells [3]. Acute inflammation is marked by an influx of neutrophils followed within 24–48 h by mononuclear macrophages [4]. Activated neutrophils and macrophages have a phagocytic function that includes release of proteolytic enzymes which degrade cellular debris and the extracellular matrix (ECM). In addition to clearing cellular debris, these phagocytes engulf and destroy any bacteria and foreign substances and present antigen peptide fragments to thymocytes (T-cells). The acute inflammatory response normally subsides within 3–5 days. Persistence of polymorphonuclear leukocytes (e.g. neutrophils) is an indication of a chronic-active inflammatory response typically associated with infection or implant toxicity.

The formation of granulation tissue occurs in the later stages of the innate immune response and largely involves the proliferation of fibroblasts and endothelial cells. Fibroblasts create and remodel the extracellular matrix of the granulation tissue by synthesizing and secreting collagen, proteoglycans, and related molecules, while endothelial cells sprout and organize into new blood vessels to supply nutrients to, and remove waste from, the granulation tissue [5].

The presence of multinucleate giant cells at the interface with the implant is an indication of a FBR to the implanted material or device. Foreign body giant cells (FBGCs) form when monocytes and macrophages fuse in an attempt to engulf materials or debris greater than 50–100 μm in size [6]. In the later stages of granulation tissue formation, activated fibroblasts may produce a fibrous capsule to surround the implant in an attempt to isolate it from the adjacent host tissue. This fibrous capsule will typically reach a steady state and remain as long as the implant is present.

1.3 Adaptive Immune Response to Implanted Materials

Macrophages and dendritic cells may initiate an adaptive immune response through antigen presentation. Dendritic cells may also be drawn to the implant site by recognition of foreign substances. The foreign constituent is typically a pathogen, in which case dendritic cells internalize, process, and present antigens to T-cells via major histocompatibility complex (MHC) molecules. However, particles, ions, or degradation products from implanted materials or devices may also be recognized as foreign by macrophages and dendritic cells [7, 8]. Implantation of a material from an allogeneic or xenogeneic source, especially one that contains cells or cell debris, can exacerbate the host response due to the presence of non-self, foreign epitopes, which also elicit a T-cell mediated response. When a T-cell recognizes an antigen bound by a dendritic cell or macrophage, the T-cell becomes activated.

Subsets of activated CD4+ T-cells, termed helper T-cells, secrete cytokines that regulate inflammation. These helper T-cells can be activated to display pro-inflammatory (Th1) or anti-inflammatory (Th2) secretory profiles [9]. A Th1-mediated immune response is commonly associated with a pro-inflammatory response to xenogenic materials, materials with cytotoxic degradation products, and/or nondegradable synthetic materials, while Th2 responses typically support greater tolerance of the implant [10, 11]. A Th2-like secretory response has been implicated in the gradual development of a FBR [12, 13]. Th2 cells also engage in cross-talk with macrophages and are associated with a regulatory/anti-inflammatory macrophage phenotype (often termed type 2 macrophages (M2), which are discussed in more detail below) [14, 15]. Implantation of a biologic scaffold material derived from porcine small intestinal submucosa (SIS matrix) elicited a Th2 cytokine expression profile, constructive remodeling, and eventual graft acceptance in a mouse model of abdominal wall defect [16, 17]. Clinical implantation of SIS matrix similarly led to a Th2 cytokine secretory profile with no signs of rejection in follow-ups out to 2 years [11]. In a recent study, dendritic cell activation by adhesion to albumin/serum-coated tissue culture plates was associated with a Th2 secretory profile, whereas activation by adhesion to collagen and vitronectin corresponded to a Th1 profile [18]. Thus, this report suggests that the provisional matrix formed by protein adsorption to implanted materials may also influence the adaptive immune response.

1.4 Macrophages and Constructive Remodeling

Macrophages respond to all implanted materials including synthetic materials such as metals, ceramics, and cements as well as naturally occurring materials such as collagen and ECM scaffolds [19]. Macrophages are critical to the fate of an implant.

As previously discussed, macrophages can initiate the adaptive immune response through antigen presentation; however, macrophages are also necessary for debris clearance, resolution of the pro-inflammatory response, and tissue regeneration via constructive remodeling. Constructive remodeling is the process by which implanted materials induce, facilitate, or otherwise support the replacement of injured tissues with new, site-appropriate functional tissue [20]. Constructive remodeling typically occurs when the early innate immune response shifts from a pro-inflammatory environment toward a non-inflammatory, regulatory environment. Similar to helper T-cells, macrophages can be activated toward a pro-inflammatory (M1-like) phenotype or a regulatory (M2-like) constructive phenotype. When activated, pro-inflammatory macrophages produce cytokines and chemokines, such as IL-1β, IL-6, TNFα, and iNOS, and can induce a Th1 inflammatory response. Regulatory macrophages, however, mediate Th2 responses [21]. An M2-like phenotype has been shown to be associated with mitigation of the pro-inflammatory state, constructive remodeling, and immunoregulation [22, 23]. In a study illustrating the importance of macrophages in constructive remodeling, depletion of macrophages from the peripheral blood in a rat model prevented efficient degradation of an implanted biomaterial and thereby inhibited the constructive remodeling response [24]. Considering the importance of macrophages in other processes such as tissue development [25, 26], tissue homeostasis [27, 28], and true tissue regeneration in species such as the axolotyl [29], the macrophage can easily be considered an orchestrator of the host response.

1.5 Host Response to Biologic Versus Synthetic Biomaterials

The clinical outcome of biomaterial-mediated strategies for tissue repair depends, in part, upon a number of biomaterial-related factors including mechanical properties, composition, surface topography, ability to resist infection, and degradability, among others [30, 31]. However, the ultimate determinant of clinical outcome is the host response to the biomaterial itself.

Although the initial phases of the biomaterial-mediated tissue repair process (e.g. iatrogenic injury during implantation, hemostasis, and activation of the innate immune system) are similar regardless of the identity and characteristics of the implanted material, the later phases and clinical outcome of the tissue repair process vary greatly depending on the biomaterial. Differences are likely to be observed as early as the protein adsorption phase, as materials with different surface topography, molecular structure and charge distribution adsorb unique profiles of proteins to their surface. In turn, differences observed in the later phases of the biomaterial-mediated tissue repair process are more obvious, and include the ability of some materials to modulate the innate immune response, recruit stem cells to the injury site, or, at a minimum, provide a compatible microenvironment for such cells, and promote constructive tissue remodeling.

Shortly after implantation, hemostasis and the Vroman effect result in a temporary fibrin-rich matrix that bridges the gap between the implanted material and the adjacent host tissue [5, 32–34]. In the case of degradable biomaterials, this temporary matrix can serve as a bridge that facilitates cell migration and gradual infiltration into the biomaterial as the degradation process takes place. In the case of nondegradable materials, this temporary matrix serves as an interface between the biomaterial and the host.

As stated above, macrophages play a central role in the process of biomaterial-mediated tissue repair. Biomaterials, which tend to elicit a persistent M1 pro-inflammatory macrophage response, have been associated with clinical outcomes that include scar tissue formation, encapsulation, and seroma formation. In contrast, biomaterials associated with the presence of a predominantly M2 pro-remodeling macrophage phenotype after the M1 response promote clinical outcomes that include stem cell recruitment/proliferation and constructive tissue remodeling [23]. Hence, the macrophage response is an early predictor of the downstream outcome in the biomaterial-mediated tissue repair process. The biomaterial-related factors, which affect and modulate macrophage phenotype, have been the focus of recent studies, and likely will be central to the design of next generation biomaterials [30, 31].

1.5.1 Biologic Versus Synthetic Biomaterials

Synthetic biomaterials can typically be manufactured with great precision. Their mechanical properties can be finely tuned according to specific clinical applications, and their molecular composition can be reliably altered to match desired specifications. However, synthetic biomaterials—particularly nondegradable synthetic biomaterials—tend to produce a persistent pro-inflammatory response after implantation that includes the well-characterized foreign body reaction [35–41]. This inflammatory response usually reaches a steady state and eventually leads to a robust, organized fibrous tissue formation. In contrast, the properties of biomaterials derived from biologic sources are less amenable to fine tuning, modification, and precision manufacturing. Biomaterials derived from decellularized tissues, i.e., the extracellular matrix, vary in structure and composition depending upon the source tissue from which they are derived and the decellularization process used to produce these materials [42]. However, these materials have the ability to promote a pro-remodeling microenvironment including an M2 macrophage phenotype, and when used appropriately, can promote constructive remodeling [22, 43].

Recent investigations attempt to combine the highly tunable and desirable properties of synthetic materials with the ability to promote a "friendlier" host response and immunomodulatory properties of biologic materials [44]. A thorough and long-term characterization of the host response to such hybrid materials has yet to be conducted (Fig. 1.3).

1.5.2 Extracellular Matrix as Biologic Scaffold

The ability of biomaterials derived from the extracellular matrix to promote constructive tissue remodeling can be attributed to both their structure and composition [42]. The ECM is a complex milieu of both structural and functional bioactive molecules. The ECM was once thought to serve the sole purpose of providing form, structural support,

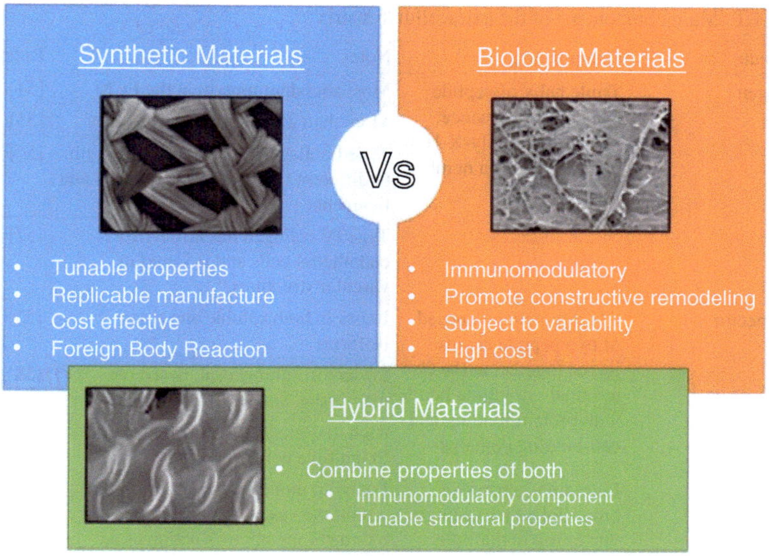

Fig. 1.3 Comparison of natural, synthetic, and hybrid biomaterials. Synthetic materials have highly tunable properties that can be adjusted with precision during manufacture depending on the intended applications. However, synthetic materials do not promote constructive tissue remodeling, and can produce foreign body response that leads to scarring and encapsulation. In contrast, biologic materials can promote constructive remodeling, but their mechanical properties and composition are subject to natural variability, and are less cost effective. Hybrid materials seek to combine the biocompatible properties of biologic materials, with the tunable mechanical properties of synthetic materials

and biomechanical properties to the different tissues. However, the ECM is now known to serve as a reservoir of information in the form of mechanical and biochemical cues that play key roles during development, homeostasis, and response to injury [45–47].

The ECM is secreted by the resident cells of each tissue and organ not only to provide structural support but also to facilitate communication between adjacent cells. In addition, the matrix itself engages in back and forth communication with the resident cells and each is responsive to the other. Hence, the extracellular matrix exists in a state of "dynamic reciprocity" with the local cells and the microenvironment [48].

The main components of the ECM include collagen, fibronectin, laminin, growth factors, cytokines, and glycosaminoglycans (Table 1.1). In addition, molecular fragments of these existing molecules, referred to as matricryptic peptides, in themselves possess biologic properties [46].

As the extracellular matrix undergoes structural and conformational alterations during degradation, exposure and/or release of these cryptic peptides into the microenvironment occurs (Table 1.2). The processes through which this is achieved include enzymatic cleavage, protein multimerization, adsorption of molecules to other ECM components, cell-mediated mechanical deformation, and ECM denaturation. Such properties as yet are not possible to design or manufacture in synthetic biomaterials. More importantly, since these processes are part of natural events, the host response when ECM materials are used as scaffold materials is markedly favorable.

Table 1.1 Main components of the extracellular matrix

Molecule	Composition	Notes	References
Collagen	Triple helix of peptide chains with sequence: Gly-Pro-X or Gly-X-Hyp where x can be a number of amino acids	Most abundant protein in the ECM	[54]
		More than 25 isoforms exist	[55]
		Type I collagen offers tensile strength to different tissues such as tendons and ligaments	[56]
		Type IV collagen has affinity for endothelial cells and is found in vascular structures	[57]
Fibronectin	Glycoprotein composed of two peptide chains joined together at the C terminal via sulfide bonds with protein-binding and cell-binding domains	Exists in both soluble and tissue isoforms	[55]
		Present in submucosal, basement membrane, and interstitial tissues	[58]
		Rich in domains that facilitate adhesion to multiple cell types via integrins	
Laminin	Laminin is a trimeric cross-linked polypeptide that exists in multiple configurations	Found in multiple tissues (particularly within the vasculature) within basement membranes acting as an adhesion molecule for different cell types and ECM	[59]
Glycosaminoglycans	Unbranched polysaccharides composed of repeating disaccharide units	Possesses the ability to retain water and bind to growth factors and cytokines sequestering them	[60]
Growth factors and cytokines	Small proteins (~5–20 kDa)	Growth factor and cytokine release from the ECM is a complex process that relies upon binding affinity, conformational changes, and degradation of the ECM during normal and pathologic processes	[61–63]
Matricryptic peptides	Molecular fragments of parent proteins	Structural and conformational changes in the ECM result in matricryptic peptide exposure, activation, and release into the microenvironment. These changes occur via enzymatic cleavage, protein multimerization, cell-mediated mechanical deformation, and ECM denaturation	[64–68]

Although the composition of the ECM has many common features across tissue types, differences do exist depending on the anatomic structure to which it belongs. For example, the extracellular matrix in tendons and ligaments needs to provide the necessary tensile strength to support and maintain the structure of the body, and hence, it is composed of mostly type I collagen [65]. Similarly, elastin is found in large amounts in compliant and elastic tissues such as the aorta [66]. Both type IV collagen and laminin are found in tissues with a basement membrane component such as urinary bladder and esophagus [67–69]. Therefore, although the molecular composition of the ECM is largely shared across tissues and species, the preferred source tissue from which each naturally occurring biomaterial is prepared for each clinical application has not been determined [20].

Table 1.2 Matricryptic peptides generated via ECM degradation

ECM parent molecule	Matricryptic peptide/site	Function	References
Collagen	C-terminal globular domain of collagen XVII (20 kd)	Angiogenesis inhibitor	[49]
	RGD fragment	Arteriolar vasoactivity	[50]
		Cell adhesion (αvβ3)	[51]
	(Pro-Pro/Hyp-Gly) collagen type I fragments	Cell migration	[52]
			[53]
	C-terminal telopeptide of collagen IIIα	Chemotaxis of progenitor cells	[46]
		Osteogenesis	[45]
	Peptide E1	Wound healing	[54]
	Peptide C2	Cell adhesion	[55]
Fibronectin	120-kd cell-binding domain	Cell migration	[56]
	40-kd gelatin-binding domain	Cell migration	[57]
	N- and C-terminal heparin binding fragments	Cell proliferation inhibition	[58]
	Type III repeat	Inflammatory pathway activation	[59]
	Fibronectin's III1 module	Cell growth and contractility	[60]
Laminin	RGD fragment	Cell adhesion	[61]
	alpha 5 beta 1 gamma1 fragment	Chemotaxis	[62]
		Inflammatory modulation	[63]
Elastin	VGVAPG sites	Cell migration	[64]

Tissue decellularization through which biomaterials composed of ECM are manufactured is typically a chemical, enzymatic, and mechanical process that aims to remove cellular material while preserving the structure and composition of the extracellular matrix. To date, several decellularization protocols have been developed, and the methods of tissue decellularization have been reviewed extensively [70, 71]. While biomaterials that have been properly decellularized have been shown to perform adequately in clinical applications, biomaterials that have been ineffectively decellularized tend to be associated with a persistent pro-inflammatory response and negative clinical outcomes [10, 43, 72]. Other factors that affect the host response to extracellular matrix-derived biomaterials include the age of the animal from which they are derived, post-processing modifications such as chemical crosslinking and solubilization, bacterial and endotoxin contamination, and methods of terminal sterilization [71, 73].

1.6 Host Response to Orthopedic Implants

Biomaterials used for orthopedic applications in the form of screws, plates, wires, rods, and external fixation devices include metals, plastics, and ceramics. Similar to biomaterials used in soft tissue and organ repair applications, the host response to the implanted construct consists of two main components. The first component is the acute inflammatory reaction in response to iatrogenic injury during implantation. The magnitude and type of this response will be determined by the size of the injury, anatomic location, surgical technique, underlying pathologic conditions in the local microenvironment, overall health of the patient, and pathogen contamination of the wound site. The second component of the host response is determined, for the most part, by biomaterial-related factors such as composition, surface chemistry, and physical properties, among many others.

One of the main differences between biomaterials used in soft tissue/organ repair applications and biomaterials used in orthopedic applications is that the primary function of the musculoskeletal system is to provide structural and mechanical support to the various anatomic structures in the body. Hence, while biomaterials designed for soft tissue/organ repair applications aim to promote deposition of site appropriate functional tissue, biomaterials for orthopedic applications aim primarily to provide structural support. By providing structural support, functional goals are achieved by definition. Although the structural properties of biomaterials employed in orthopedic applications are of particular importance, the host response to the implanted construct is equally important.

1.7 Clinical Considerations

1.7.1 Designing Materials and Devices for Immunomodulation

Mitigation of a chronic pro-inflammatory response to an implanted biomaterial is a desirable design goal. Such a response may begin by modulating protein adsorption, which forms the provisional matrix for cell attraction, adherence, and attachment to the implant. Identification of those signaling molecules that promote a permissive microenvironmental niche, and the development of biomaterials that either possess or attract such molecules, represents logical approaches to a host-centric biomaterial strategy. Surface topography, roughness, porosity, size, and shape of an implant have been shown to affect the intensity of the foreign body response, suggesting that biocompatibility of an implant can be improved simply through adjusting material geometry [74]. Modulation of surface chemistry of a material to limit/restrict protein adsorption and thereby control macrophage adhesion or activation has been suggested and indeed, macrophages display different phenotypic profiles when cultured on hydrophobic, hydrophilic, and ionic polymers [75]. However, as previously described, it is not necessarily the number of macrophages that respond to a biomaterial that determines outcome, but rather the phenotype of responding macrophages.

Surface modification of biomaterials with bioactive molecules such as growth factors, antibodies, drugs, or adhesion molecules has also been investigated. For example, incorporation of CD47, a marker for "self," on the surface of implants can reduce recognition of the material/device as "foreign" by immune cells [76, 77]. In addition to surface chemistry, the mechanical properties of the implant can contribute to the long-term outcome. Materials and devices can be designed to mimic the load bearing or elastic/contractile features of the tissue itself. Importantly, the mechanical properties of the implant can influence myofibroblasts—a contractile cell type that has been identified as the major source of collagen production in fibrosis [78, 79]. However, the mechanical properties of a biomaterial will be modified by the host response; therefore, although the mechanical properties of the material at the time of implantation can be controlled, the determinant of mechanical properties in the days, months, and years following implantation is a function of the host response to the material.

1.8 Conclusions

Both the host factors and biomaterial characteristics influence the host response and the long-term outcome of an implant. Depending on the clinical indication for biomaterial implantation, the ideal outcome likely requires a constructive remodeling response mediated by regulatory macrophages. Significant advances have been made in understanding the host response to biomaterials and have affected the way in which materials are designed and manufactured. Additional investigation of the innate and acquired immune response to biomaterials is needed for the next generation of biomaterials that do not simply augment or substitute the mechanical properties of injured or missing tissue, but faithfully reconstruct both the physiology and structural functions of target tissue. The topics covered in the various chapters of this textbook provide an excellent summary of what is known about the host response to biomaterials and devices as well as additional insights for future investigation.

References

1. Key N, Makris M, O'Shaughnessy D et al (2009) Practical hemostasis and thrombosis. Wiley-Blackwell, Oxford, p 328
2. Furie BC, Furie B (2005) Thrombus formation in vivo. J Clin Invest 115(12):3355–3362
3. Anderson J, Cramer S (2015) Perspectives on the inflammatory, healing, and foreign body responses to biomaterials and medical devices. In: Badylak SF (ed) Host response to biomaterials: the impact of host response on biomaterial selection. Elsevier, New York
4. Eming SA, Krieg JM, Davidson T (2007) Inflammation in wound repair: molecular and cellular mechanisms. J Invest Dermatol 127(3):514–525
5. Reinke JM, Sorg H (2012) Wound repair and regeneration. Eur Surg Res 49(1):35–43
6. Anderson JM, Rodriguez DT, Chang A (2008) Foreign body reaction to biomaterials. Semin Immunol 20(2):86–100
7. Konttinen YT, Pajarinen J, Takakubo Y et al (2014) Macrophage polarization and activation in response to implant debris: influence by "particle disease" and "ion disease". J Long Term Eff Med Implants 24(4):267–281
8. Park JE, Babensee J (2012) Differential functional effects of biomaterials on dendritic cell maturation. Acta Biomater 8(10):3606–3617
9. Romagnani S (2000) T-cell subsets (Th1 versus Th2). Ann Allergy Asthma Immunol 85(1):9–18, quiz 18, 21

10. Badylak SF, Gilbert TW (2008) Immune response to biologic scaffold materials. Semin Immunol 20(2):109–116
11. Ansaloni L, Cambrini P, Catena F et al (2007) Immune response to small intestinal submucosa (surgisis) implant in humans: preliminary observations. J Invest Surg 20(4):237–241
12. McNally AK, DeFife KM, Anderson JM (1996) Interleukin-4-induced macrophage fusion is prevented by inhibitors of mannose receptor activity. Am J Pathol 149(3):975–985
13. Higgins DM, Basaraba RJ, Hohnbaum AC et al (2009) Localized immunosuppressive environment in the foreign body response to implanted biomaterials. Am J Pathol 175(1):161–170
14. Gratchev A, Kzhyshkowska J, Utikal J et al (2005) Interleukin-4 and dexamethasone counterregulate extracellular matrix remodelling and phagocytosis in type-2 macrophages. Scand J Immunol 61(1):10–17
15. Mills CD, Ley K (2014) M1 and M2 macrophages: the chicken and the egg of immunity. J Innate Immun 6(6):716–726
16. Allman AJ, McPherson TB, Badylak SF et al (2001) Xenogeneic extracellular matrix grafts elicit a TH2-restricted immune response. Transplantation 71(11):1631–1640
17. Allman AJ, McPherson TB, Merrill LC et al (2002) The Th2-restricted immune response to xenogeneic small intestinal submucosa does not influence systemic protective immunity to viral and bacterial pathogens. Tissue Eng 8(1):53–62
18. Acharya AP, Dolgova NV, Clare-Salzler MJ et al (2008) Adhesive substrate-modulation of adaptive immune responses. Biomaterials 29(36):4736–4750
19. Fishman J, Wiles K, Wood K (2015) The acquired immune system response to biomaterials, including both naturally occurring and synthetic biomaterials. In: Badylak SF (ed) Host response to biomaterials: the impact of host response on biomaterial selection. Elsevier, New York
20. Badylak SF (2014) Decellularized allogeneic and xenogeneic tissue as a bioscaffold for regenerative medicine: factors that influence the host response. Ann Biomed Eng 42(7):1517–1527
21. Mantovani A, Sica A, Sozzani S et al (2004) The chemokine system in diverse forms of macrophage activation and polarization. Trends Immunol 25(12):677–686
22. Brown BN, Badylak SF (2014) Extracellular matrix as an inductive scaffold for functional tissue reconstruction. Transl Res 163(4):268–285
23. Brown BN, Londono R, Tottey S et al (2012) Macrophage phenotype as a predictor of constructive remodeling following the implantation of biologically derived surgical mesh materials. Acta Biomater 8(3):978–987
24. Valentin JE, Stewart-Akers AM, Gilbert TW et al (2009) Macrophage participation in the degradation and remodeling of extracellular matrix scaffolds. Tissue Eng Part A 15(7):1687–1694
25. Leid JM, Carrelha J, Boukarabila H et al (2016) Primitive embryonic macrophages are required for coronary development and maturation. Circ Res 118(10):1498–1511
26. Wynn TA, Chawla A, Pollard JW (2013) Macrophage biology in development, homeostasis and disease. Nature 496(7446):445–455
27. Bain CC, Mowat AM (2014) Macrophages in intestinal homeostasis and inflammation. Immunol Rev 260(1):102–117
28. Ginhoux S, Jung F (2014) Monocytes and macrophages: developmental pathways and tissue homeostasis. Nat Rev Immunol 14(6):392–404
29. Godwin JW, Pinto AR, Rosenthal NA (2013) Macrophages are required for adult salamander limb regeneration. Proc Natl Acad Sci U S A 110(23):9415–9420
30. Brown BN, Ratner BD, Goodman SB et al (2012) Macrophage polarization: an opportunity for improved outcomes in biomaterials and regenerative medicine. Biomaterials 33(15):3792–3802
31. Sridharan R, Cameron AR, Kelly DJ et al (2015) Biomaterial based modulation of macrophage polarization: a review and suggested design principles. Mater Today 18(6):13
32. Vroman L, Adams AL, Fischer GC et al (1980) Interaction of high molecular weight kininogen, factor XII, and fibrinogen in plasma at interfaces. Blood 55(1):156–159
33. Vogler EA (2012) Protein adsorption in three dimensions. Biomaterials 33(5):1201–1237
34. Boehler RM, Graham JG, Shea LD (2011) Tissue engineering tools for modulation of the immune response. Biotechniques 51(4):239–240, passim

35. Asawa Y, Sakamoto T, Komura M et al (2012) Early stage foreign body reaction against biodegradable polymer scaffolds affects tissue regeneration during the autologous transplantation of tissue-engineered cartilage in the canine model. Cell Transplant 21(7):1431–1442
36. Felländer-Tsai L, Reinholt FP, Turan I (1997) Complications with infection and foreign body reaction after silicone implant arthroplasty in the second metatarsophalangeal joint in an adolescent: a case report. J Foot Ankle Surg 36(6):452–456
37. Florin W, Mandel L (2012) Foreign body reaction to facial dermal fillers: case report. J Oral Maxillofac Surg 70(10):2352–2355
38. Hale CS, Patel RR, Meehan S (2011) Polyurethane foam: an underrecognized cause of foreign body granulomas. J Cutan Pathol 38(10):838–839
39. Khandwekar AP, Patil DP, Hardikar AA et al (2010) In vivo modulation of foreign body response on polyurethane by surface entrapment technique. J Biomed Mater Res A 95(2):413–423
40. Mamelak AJ, Katz TM, Goldberg LH et al (2009) Foreign body reaction to hyaluronic acid filler injection: in search of an etiology. Dermatol Surg 35(Suppl 2):1701–1703
41. Veleirinho B, Coelho DS, Dias PF et al (2014) Foreign body reaction associated with PET and PET/chitosan electrospun nanofibrous abdominal meshes. PLoS One 9(4), e95293
42. Badylak SF, Freytes DO, Gilbert TW (2009) Extracellular matrix as a biological scaffold material: structure and function. Acta Biomater 5:1–13
43. Brown BN, Valentin JE, Stewart-Akers AM et al (2009) Macrophage phenotype and remodeling outcomes in response to biologic scaffolds with and without a cellular component. Biomaterials 30(8):1482–1491
44. Wolf MT, Dearth CL, Ranallo CA et al (2014) Macrophage polarization in response to ECM coated polypropylene mesh. Biomaterials 35(25):6838–6849
45. Agrawal V, Kelly J, Tottey S et al (2011) An isolated cryptic peptide influences osteogenesis and bone remodeling in an adult mammalian model of digit amputation. Tissue Eng Part A 17(23-24):3033–3044
46. Agrawal V, Tottey S, Johnson SA et al (2011) Recruitment of progenitor cells by an extracellular matrix cryptic peptide in a mouse model of digit amputation. Tissue Eng Part A 17(19-20):2435–2443
47. Engler AJ, Sweeney HL, Discher DE et al (2007) Extracellular matrix elasticity directs stem cell differentiation. J Musculoskelet Neuronal Interact 7(4):335
48. Bissell MJ, Aggeler J (1987) Dynamic reciprocity: how do extracellular matrix and hormones direct gene expression? Prog Clin Biol Res 249:251–262
49. O'Reilly MS, Boehm T, Shing Y et al (1997) Endostatin: an endogenous inhibitor of angiogenesis and tumor growth. Cell 88(2):277–285
50. Davis GE (1992) Affinity of integrins for damaged extracellular matrix: alpha v beta 3 binds to denatured collagen type I through RGD sites. Biochem Biophys Res Commun 182(3):1025–1031
51. Mogford JE, Davis GE, Platts SH et al (1996) Vascular smooth muscle alpha v beta 3 integrin mediates arteriolar vasodilation in response to RGD peptides. Circ Res 79(4):821–826
52. Laskin DL, Kimura T, Sakakibara S et al (1986) Chemotactic activity of collagen-like polypeptides for human peripheral blood neutrophils. J Leukoc Biol 39(3):255–266
53. Albini A, Adelmann-Grill BC (1985) Collagenolytic cleavage products of collagen type I as chemoattractants for human dermal fibroblasts. Eur J Cell Biol 36(1):104–107
54. Banerjee P, Suguna C, Shanthi L (2015) Wound healing activity of a collagen-derived cryptic peptide. Amino Acids 47(2):317–328
55. Banerjee P, Mehta C, Shanthi A (2014) Investigation into the cyto-protective and wound healing properties of cryptic peptides from bovine Achilles tendon collagen. Chem Biol Interact 211:1–10
56. Clark RA, Wikner NE, Doherty DE et al (1988) Cryptic chemotactic activity of fibronectin for human monocytes resides in the 120-kDa fibroblastic cell-binding fragment. J Biol Chem 263(24):12115–12123
57. Schor SL, Ellis I, Dolman C et al (1996) Substratum-dependent stimulation of fibroblast migration by the gelatin-binding domain of fibronectin. J Cell Sci 109(Pt 10):2581–2590
58. Homandberg GA, Williams JE, Grant D et al (1985) Heparin-binding fragments of fibronectin are potent inhibitors of endothelial cell growth. Am J Pathol 120(3):327–332

59. You R, Zheng M, McKeown-Longo PJ (2010) The first type III repeat in fibronectin activates an inflammatory pathway in dermal fibroblasts. J Biol Chem 285(47):36255–36259
60. Hocking DC, Kowalski K (2002) A cryptic fragment from fibronectin's III1 module localizes to lipid rafts and stimulates cell growth and contractility. J Cell Biol 158(1):175–184
61. Aumailley M, Gerl M, Sonnenberg A et al (1990) Identification of the Arg-Gly-Asp sequence in laminin A chain as a latent cell-binding site being exposed in fragment P1. FEBS Lett 262(1):82–86
62. Adair-Kirk TL, Atkinson JJ, Broekelmann TJ et al (2003) A site on laminin alpha 5, AQARSAASKVKVSMKF, induces inflammatory cell production of matrix metalloproteinase-9 and chemotaxis. J Immunol 171(1):398–406
63. Adair-Kirk TL, Atkinson JJ, Kelley DG et al (2005) A chemotactic peptide from laminin alpha 5 functions as a regulator of inflammatory immune responses via TNF alpha-mediated signaling. J Immunol 174(3):1621–1629
64. Senior RM, Griffin GL, Mecham RP (1980) Chemotactic activity of elastin-derived peptides. J Clin Invest 66(4):859–862
65. Kannus P (2000) Structure of the tendon connective tissue. Scand J Med Sci Sports 10(6):312–320
66. Zou Y, Zhang Y (2012) Mechanical evaluation of decellularized porcine thoracic aorta. J Surg Res 175(2):359–368
67. Keane TJ, Londono R, Carey RM et al (2013) Preparation and characterization of a biologic scaffold from esophageal mucosa. Biomaterials 34(28):6729–6737
68. Faulk DM, Carruthers CA, Warner HJ et al (2014) The effect of detergents on the basement membrane complex of a biologic scaffold material. Acta Biomater 10(1):183–193
69. Marçal H, Ahmed T, Badylak SF et al (2012) A comprehensive protein expression profile of extracellular matrix biomaterial derived from porcine urinary bladder. Regen Med 7(2):159–166
70. Gilbert TW (2012) Strategies for tissue and organ decellularization. J Cell Biochem 113(7):2217–2222
71. Crapo PM, Gilbert TW, Badylak SF (2011) An overview of tissue and whole organ decellularization processes. Biomaterials 32(12):3233–3243
72. Keane TJ, Londono R, Turner NJ et al (2012) Consequences of ineffective decellularization of biologic scaffolds on the host response. Biomaterials 33:1771–1781
73. Londono R, Badylak SF (2015) Factors which affect the host response to biomaterials. In: Badylak SF (ed) Host response to biomaterials: the impact of host response on biomaterial selection. Elsevier, New York
74. Veiseh O, Doloff JC, Ma M et al (2015) Size- and shape-dependent foreign body immune response to materials implanted in rodents and non-human primates. Nat Mater 14:643–651
75. Jones JA, Chang DT, Meyerson H et al (2007) Proteomic analysis and quantification of cytokines and chemokines from biomaterial surface-adherent macrophages and foreign body giant cells. J Biomed Mater Res A 83(3):585–596
76. Stachelek SJ, Finley MJ, Alferiev IS et al (2011) The effect of CD47 modified polymer surfaces on inflammatory cell attachment and activation. Biomaterials 32(19):4317–4326
77. Finley MJ, Clark KA, Alferiev IS et al (2013) Intracellular signaling mechanisms associated with CD47 modified surfaces. Biomaterials 34(34):8640–8649
78. Hinz B, Phan SH, Thannickal VJ et al (2012) Recent developments in myofibroblast biology: paradigms for connective tissue remodeling. Am J Pathol 180(4):1340–1355
79. Klingberg F, Hinz ES, White B (2013) The myofibroblast matrix: implications for tissue repair and fibrosis. J Pathol 229(2):298–309

Chapter 2
Implications of the Acute and Chronic Inflammatory Response and the Foreign Body Reaction to the Immune Response of Implanted Biomaterials

James M. Anderson and Sirui Jiang

Abstract In vivo implantation of medical devices, prostheses, biomaterials, and tissue-engineered scaffolds initiates the innate immune response consisting of acute inflammation, chronic inflammation, and the foreign body reaction (FBR) within the first 2 weeks following implantation. This chapter focuses on these humoral and cellular events occurring at the implant site immediately following implantation. Following injury/implantation, blood-material interactions occur and the provisional matrix is formed. Acute inflammation consisting predominantly of polymorphonuclear leukocytes follows but resolves quickly, usually within the first week if not sooner, depending on the extent of injury at the implant site. Chronic inflammation consisting of monocytes, macrophages, and lymphocytes follows acute inflammation. This process also resolves quickly with biocompatible materials leaving monocytes and macrophages to interact at the interface of the implanted device or material. The FBR at the interface with biomaterials is composed of macrophages, which may fuse together to form foreign body giant cells (FBGCs). Outside the FBR at the biomaterial interface, fibrosis and fibrous encapsulation occur in the final stages of the healing response to the implanted biomaterial. Numerous challenges including lack of understanding of these responses in vivo currently limit projection to clinical application of the respective medical device, prosthesis, or biomaterial. The end-stage of the innate immune response consisting of the FBR at the interface with fibrous encapsulation has received extensive attention over the past decade. Numerous efforts have been made to downregulate the activity of macrophages and FBGCs at the interface and to decrease/eliminate the fibrous capsule formation. Ultimately, the success or failure of medical devices, implants, biomaterials, and tissue-engineered constructs is modulated by the interaction between their characteristics, patient conditions, and surgical technique.

J.M. Anderson (✉) • S. Jiang
Departments of Pathology, Macromolecular Science And Biomedical Engineering,
Case Western Reserve University, Cleveland, OH, USA
e-mail: jma6@case.edu

B. Corradetti (ed.), *The Immune Response to Implanted Materials and Devices*,
DOI 10.1007/978-3-319-45433-7_2

Keywords Acute inflammation • Chronic inflammation • Foreign body reaction • Innate • Adaptive • Immune response • Matrix formation

2.1 Introduction

Acute inflammation, chronic inflammation, and the foreign body reaction are generally considered to be early events occurring in the innate immune response, e.g., the first 2 weeks following implantation of a biomaterial or medical device. Moreover, the humoral and cellular components and their respective interactions in the innate and acquired immune responses must be appreciated within the context of the in vivo environment. This chapter focuses on the humoral and cellular events occurring at the implant site immediately following implantation. As such, the response to injury may predominate and mask specific interactions occurring at the biomaterial surface/tissue interface as they may relate to or impact the development of acquired immune responses.

The most commonly used term to describe appropriate host response to biomaterials used in a medical device is biocompatibility. A simplistic definition of biocompatibility is those materials, which do not induce an adverse tissue reaction. A more helpful definition of biocompatibility is the ability of a material to perform with an appropriate host response in a specific application [1]. This definition is helpful as it links material properties or characteristics with performance, e.g., biological requirements, with a specific application, a specific medical device, or biomaterial used as a medical device. The "appropriate host response" implies identification and characterization of tissue reactions and responses that could prove harmful to the host and/or lead to ultimate failure of the biomaterial, medical device, or prosthesis through biological mechanisms. Viewed from the opposite perspective, the "appropriate host response" implies identification and characterization of the tissue reactions and responses critical for the successful use of the biomaterial or medical device. Biocompatibility assessment is considered to be a measure of the magnitude and duration of the adverse alterations in homeostatic mechanisms that determine the host response [2]. Safety assessment or biocompatibility assessment of a biomaterial or medical device is generally considered to be synonymous.

Inflammation, wound-healing, and foreign body reaction are generally considered as parts of the tissue or cellular host responses to injury [3]. Table 2.1 lists the sequence/continuum of these events following injury. Overlap and simultaneous occurrence of these events should be considered; e.g., the foreign body reaction at the implant interface may be initiated with the onset of acute and chronic inflammation. From a biomaterials perspective, placing a biomaterial in the in vivo environment requires injection, insertion, or surgical implantation, all of which injure the tissues or organs involved.

The placement procedure initiates a response to injury by the tissue, organ, or body and mechanisms are activated to maintain homeostasis. Obviously, the extent of injury varies with the implantation procedure. The degrees to which the homeostatic mechanisms are perturbed and the extent to which pathophysiologic condi-

Table 2.1 Sequence/continuum of host reactions following implantation of medical devices

Injury
Blood-material interactions
Provisional matrix formation
Acute inflammation
Chronic inflammation
Granulation tissue
Foreign-body reaction
Fibrosis/fibrous capsule development

tions are created and undergo resolution are a measure of the host response to the biomaterial and may ultimately determine its biocompatibility. Although it is conceptually convenient to separate homeostatic mechanisms into blood-material or tissue-material interactions, it must be remembered that the various components or mechanisms involved in homeostasis are present in both blood and tissue, are inextricably linked, and are a part of the physiologic continuum. Furthermore, it must be noted that the host response is tissue-dependent, organ-dependent, and species-dependent. The similarities and differences between prostheses utilized between cells, tissues, organs, and species must be appreciated in the design of in vitro and in vivo experiments in the determination of biocompatibility and function.

2.2 Blood-Material Interactions/Provisional Matrix Formation

Immediately following injury, changes in vascular flow, caliber, and permeability occur. Fluid, proteins, and blood cells escape from the vascular system into the injured tissue in a process called exudation. The changes in the vascular system, which also include the hematologic alterations associated with acute inflammation, are followed by cellular events that characterize the inflammatory response.

Blood-material interactions and the inflammatory response are intimately linked, and in fact, early responses to injury involve mainly blood and vasculature. Regardless of the tissue or organ into which a biomaterial is implanted, the initial inflammatory response is activated by injury to vascularized connective tissue (Table 2.2).

Inflammation serves to contain, neutralize, dilute, or wall off the injurious agent or process. In addition, the inflammatory response initiates a series of events that may heal and reconstitute the implant site through replacement of the injured tissue by regeneration of native parenchymal cells, may form fibroblastic scar tissue, or a combination of these two processes. Since blood and its components are involved in the initial inflammatory responses, blood clot formation and/or thrombosis also occur. Blood coagulation and thrombosis are generally considered humoral responses and

Table 2.2 Cells and components of vascularized connective tissue

Intravascular (blood) cells
Erythrocytes (RBC)
Neutrophils (PMNs, polymorphonuclear leukocytes)
Monocytes
Eosinophils
Lymphocytes
Plasma cells
Basophils
Platelets
Connective tissue cells
Mast cells
Fibroblasts
Macrophages
Lymphocytes
Extracellular matrix components
Collagens
Elastin
Proteoglycans
Fibronectin
Laminin

are influenced by homeostatic mechanisms such as the extrinsic and intrinsic coagulation systems, the complement system, the fibrinolytic system, the kinin-generating system, and platelets. Thrombus or blood clot formation on the surface of a biomaterial is related to the well-known Vroman effect, in which a hierarchical and dynamic series of collision, adsorption, and exchange processes, determined by protein mobility and concentration, regulate early time-dependent changes in blood protein adsorption. From a wound-healing perspective, blood protein deposition on a biomaterial surface in tissue is described as provisional matrix formation. Blood interactions with biomaterials are generally considered under the category of hematocompatibility.

Injury to vascularized tissue in the implantation procedure leads to immediate development of the provisional matrix at the implant site. This provisional matrix consists of fibrin and inflammatory mediators, produced by activation of the coagulation and thrombosis and complement systems, respectively, activated platelets, inflammatory cells, and endothelial cells. These events occur early, within minutes to hours following implantation of a medical device and initiate the resolution, reorganization, and repair processes such as fibroblast recruitment. The provisional matrix provides both structural and biochemical components to the process of wound healing. The complex three-dimensional structure of the fibrin network with attached adhesive proteins provides a substrate for cell adhesion and migration. The presence of cytokines, chemokines, and growth factors within the provisional matrix provides a rich milieu of activating and inhibiting substances for cellular proliferative and

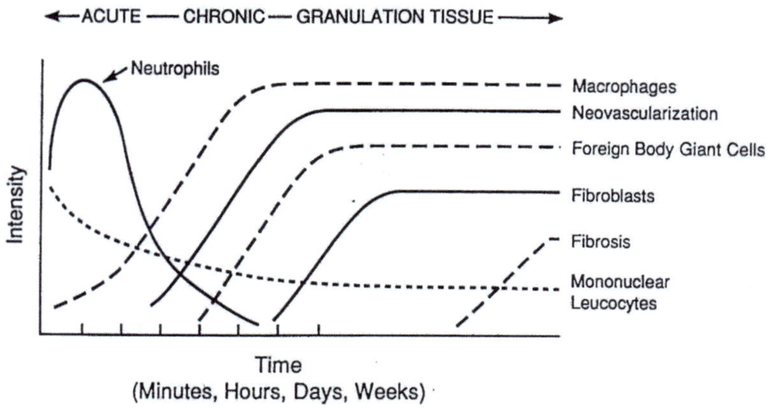

Fig. 2.1 The temporal variation in the acute inflammatory response, chronic inflammatory response, granulation tissue development, and foreign-body reaction to implanted biomaterials. The intensity and time variables are dependent upon the extent of injury created in the implantation and the size, shape, topography, and chemical and physical properties of the biomaterial

synthetic processes, mitogenesis, and chemoattraction. The provisional matrix may be viewed as a naturally derived, biodegradable, sustained release system in which these various bioactive molecules are released to orchestrate subsequent wound-healing processes. Although our understanding of the provisional matrix and its capabilities has increased, our knowledge of the key molecular regulators of the formation of the provisional matrix and subsequent wound-healing events is poor. In part, this lack of knowledge is because most studies have been conducted in vitro, and there is a paucity of in vivo studies that provide for a more complex perspective. However, attractive hypotheses have been presented regarding the presumed ability of adsorbed materials to modulate cellular behavior.

The predominant cell type present in the inflammatory response varies with time (Fig. 2.1). In general, neutrophils predominate during the first several days following injury and exposure to a biomaterial, and then are replaced by monocytes as the predominant cell type. Three factors account for this change in cell type: neutrophils are short lived and disintegrate and disappear after 24–48 h; neutrophil emigration from the vasculature to the tissues is of short duration; and chemotactic factors for neutrophil migration are activated early in the inflammatory response. Following emigration from the vasculature, monocytes differentiate into macrophages and these cells are very long-lived (up to months). Monocyte emigration may continue for day to weeks, depending on the extent of injury and type of implanted biomaterial. In addition, chemotactic factors for monocytes are produced over longer periods of time. In short-term (24 h) implants in humans, administration of both H1 and H2 histamine receptor antagonists greatly reduced the recruitment of macrophages/monocytes and neutrophils on polyethylene terephthalate surfaces [4]. These studies also demonstrated that plasma-coated implants accumulated significantly more phagocytes than did serum-coated implants.

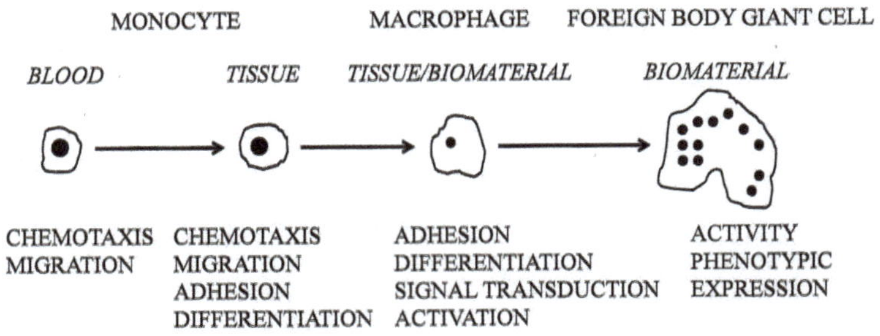

Fig. 2.2 In vivo transition from blood-borne monocyte to biomaterial adherent monocyte/macrophage to foreign body giant cell at the tissue/biomaterial interface

The temporal sequence of events following implantation of a biomaterial is illustrated in Fig. 2.2. The size, shape, and chemical and physical properties of the biomaterial may be responsible for variations in the intensity and duration of the inflammatory or wound-healing process, and thus the host response to a biomaterial.

2.3 Acute Inflammation

While injury initiates the inflammatory response, the chemicals released from plasma, cells, or injured tissues mediate the inflammatory response. Important chemical mediators of inflammation are presented in Table 2.3.

Several points must be noted to understand the inflammatory response and its relationship to biomaterials. First, although chemical mediators are classified on a structural or functional basis, complex interactions provide a system of checks and balances regarding their respective activities and functions. Second, chemical mediators are quickly inactivated or destroyed, suggesting that their action is predominantly local (e.g., at the implant site). Third, generally the lysosomal proteases and the oxygen-derived free radicals produce the most significant damage or injury. These chemical mediators are also important in the degradation of certain biomaterials [5–8].

Acute inflammation is of relatively short duration, lasting for minutes to hours to days, depending on the extent of injury and the type of implanted biomaterial. Its main characteristics are the exudation of fluid and plasma proteins (edema) and the emigration of leukocytes (predominantly neutrophils). Neutrophils (polymorp honuclear leukocytes, PMNs) and other motile white cells emigrate or move from the blood vessels into the perivascular tissues and the injury (implant) site. Leukocyte emigration is assisted by "adhesion molecules" present on leukocyte and endothelial surfaces. The surface expression of these adhesion molecules can be induced, enhanced, or altered by inflammatory agents and chemical mediators. White cell emigration is controlled, in part, by chemotaxis, which is the unidirectional migration of cells along a chemical gradient. A wide variety of exogenous and endogenous substances have been identified as chemotactic stimuli. Specific

Table 2.3 Important chemical mediators of inflammation derived from plasma, cells, or injured tissue

Mediators	Examples
Vasoactive agents	Histamines, serotonin, adenosine, endothelial-derived relaxing factor (EDRF), prostacyclin, endothelin, thromboxane α_2
Plasma proteases	
Kinin system	Bradykinin, kallikrein
Complement system	C3a, C5a, C3b, C5b-C9
Coagulation/fibrinolytic system	Fibrin degradation products, activated Hageman factor (FXIIA), tissue plasminogen activator (tPA)
Leukotrienes	Leukotriene B_4 (LTB_4), hydroxyeicosatetranoic acid (HETE)
Lysosomal proteases	Collagenase, elastase
Oxygen-derived free radicals	H_2O_2, superoxide anion
Platelet activating factors	Cell membrane lipids
Cytokines	Interleukin 1 (IL-1), tumor necrosis factor (TNF)
Growth factors	Platelet-derived growth factor (PDGF), fibroblast growth
	Factor (FGF), transforming growth factor TGF-α or (TGF-β), epithelial growth factor (EGF)

receptors for chemotactic agents on the cell membranes of leukocytes are important in the emigration or movement of leukocytes. These and other receptors also play a role in the transmigration of white cells across the endothelial lining of vessels and activation of leukocytes. Following localization of leukocytes at the injury (implant) site, phagocytosis and the release of proteolytic enzymes occur following activation of neutrophils and macrophages. The major role of the neutrophil in acute inflammation is to phagocytose microorganisms and foreign materials. Phagocytosis is seen as a three-step process in which the stimulus (e.g., damaged tissue, infectious agent, biomaterial) undergoes recognition and neutrophil attachment, engulfment, and killing or degradation. In regard to biomaterials, engulfment and degradation may or may not occur, depending on the properties of the biomaterial.

Although biomaterials are not generally phagocytosed by neutrophils or macrophages because of the disparity in size (e.g., the surface of the biomaterial is greater than the size of the cell), certain events in phagocytosis may occur. The process of recognition and attachment is expedited when the injurious agent is coated by naturally occurring serum factors called "opsonins." Two major opsonins are immunoglobulin G (IgG) and the complement-activated fragment, C3b. Both of these plasma-derived proteins are known to adsorb biomaterials, and neutrophils and macrophages have corresponding cell-membrane receptors for these opsonins. These receptors may also play a role in the activation of the attached neutrophil or macrophage. Other blood proteins such as fibrinogen, fibronectin, and vitronectin may also facilitate cell adhesion to biomaterial surfaces. Owing to the disparity in size between the biomaterial surface and the attached cell, frustrated phagocytosis may occur; a process that does not involve engulfment of the biomaterial but does cause the extracellular release of leukocyte products in an attempt to degrade the biomaterial.

Henson has shown that neutrophils adherent to complement-coated and immunoglobulin-coated nonphagocytosable surfaces may release enzymes by direct extrusion or exocytosis from the cell [9]. The amount of enzyme released during this process depends on the size of the polymer particle, with larger particles inducing greater amounts of enzyme release. This disparity suggests that the specific mode of cell activation depends, at least in part, upon the size of the implant and whether or not a material is in a phagocytosable form. For example, a powder, particulate, or nanomaterial may provoke a different degree of inflammatory response than the same material in a nonphagocytosable form such as film. In general, materials greater than 5 μm are not phagocytosed, while materials less than 5 μm can be phagocytosed by inflammatory cells, i.e., nanoparticles.

Acute inflammation normally resolves quickly, usually less than 1 week, depending on the extent of injury at the implant site. The presence of acute inflammation (i.e., PMNs) at the tissue/implant interface at time periods beyond 1 week (i.e., weeks, months, or years) suggests the presence of infection.

2.4 Chronic Inflammation

Chronic inflammation has a more heterogeneous histologic appearance than acute inflammation. In general, chronic inflammation is characterized by the presence of macrophages, monocytes, and lymphocytes, with the proliferation of blood vessels and connective tissue. Many factors can modify the course and histologic appearance of chronic inflammation.

Clinically, surgical pathologists commonly use the term chronic inflammation to describe the foreign body reaction. Caution is recommended in the use of this term as it demonstrates the breadth of histological findings that lead to the clinical diagnosis of chronic inflammation. Chronic inflammation that is predominantly composed of monocytes, macrophages, and lymphocytes is most commonly associated with toxicity or infection, whereas the foreign body reaction is most commonly composed of macrophages and foreign body giant cells.

Persistent inflammatory stimuli lead to chronic inflammation. While the chemical and physical properties of the biomaterial in themselves may lead to chronic inflammation, in situ motion of the implant or infection may also produce chronic inflammation. The chronic inflammatory response to biomaterials is usually of short duration and is confined to the implant site. The presence of mononuclear cells, including lymphocytes and plasma cells, is considered chronic inflammation, whereas the foreign-body reaction with the development of granulation tissue is considered the normal wound-healing response to implanted biomaterials (i.e., the normal foreign-body reaction). Chronic inflammation with the presence of collections of lymphocytes and monocytes at extended implant times (weeks, months, years) also may suggest the presence of a long-standing infection. The prolonged presence of acute and/or chronic inflammation also may be due to toxic leachables from a biomaterial [10].

The following example illustrates this point. In vivo subcutaneous implantation studies were conducted in rats and rabbits with naltrexone sustained release preparations that included placebo (polymer only) beads and naltrexone containing beads [11]. Histopathological tissue reactions were determined at days 3, 7, 14, 21, and 28. The only significant histological finding in both rats and rabbits at any time period was the persistent chronic inflammation that occurred focally around the naltrexone containing beads. The focal inflammatory cell density in both rats and rabbits was higher for the naltrexone beads than for the placebo beads at days 14, 21, and 28, respectively. This difference in inflammatory response between naltrexone beads and placebo beads increased with increasing time of implantation. Considering the resolution of the inflammatory response for the placebo beads with implantation time in both rats and rabbits, the more severe inflammatory reaction suggested that the naltrexone drug itself was the causative agent of the focal chronic inflammation present surrounding the naltrexone beads in the implant sites.

The important lesson from this case study is the necessary use of appropriate control materials. If no negative control, i.e., placebo polymer-only material, had been used, the polymer in the naltrexone containing beads also would have been considered a causative agent of the extended chronic inflammatory response. Similar chronic inflammatory responses have been identified with drugs, polymer plasticizers and other additives, fabrication and manufacturing aids, and sterilization residuals. Each case presents its own unique factors in a risk assessment process necessary for determining safety (biocompatibility) and benefit versus risk in clinical application.

Lymphocytes and plasma cells are involved principally in adaptive immune reactions and are key mediators of antibody production and delayed hypersensitivity responses. Although these cells may be present in nonimmunologic injuries and inflammation with biomaterials, their roles in such circumstances are largely unknown [12, 13]. Little is known regarding humoral (or adaptive) immune responses and cell-mediated immunity to synthetic biomaterials. The role of macrophages (cells of the innate humoral response) must be considered in the possible development of adaptive immune responses to synthetic biomaterials. Macrophages and dendritic cells process and present the antigen to immunocompetent cells and thus are key mediators in the development of adaptive immune reactions.

Monocytes and macrophages belong to the mononuclear phagocytic system (MPS), also known as the reticuloendothelial system (RES). These systems consist of cells in the bone marrow, peripheral blood, and specialized tissues. Table 2.4 lists the tissues that contain cells belonging to the MPS or RES.

The specialized cells in these tissues may be responsible for systemic effects in organs or tissues secondary to the release of components or products from implants through various tissue–material interactions (e.g., corrosion products, wear debris, degradation products) or the presence of implants (e.g., microcapsule or nanoparticle drug-delivery systems).

Over the past decade, increasing numbers of studies have identified significant differences in macrophage phenotypic expression. This difference in macrophage function or activation, dictated by different environmental cues, has been

Table 2.4 Tissues and cells of MPS and RES

Tissues	Cells
Implant sites	Inflammatory macrophages
Liver	Kupffer cells
Lung	Alveolar macrophages
Connective tissue	Histiocytes
Bone marrow	Macrophages
Spleen and lymph nodes	Fixed and free macrophages
Serous cavities	Pleural and peritoneal macrophages
Nervous system	Microglial cells
Bone	Osteoclasts
Skin	Langerhans' cells
Lymphoid tissue	Dendritic cells

classified into various ways. Following on the T-cell literature, macrophages have been classified as M1 macrophages defined as classically activated or proinflammatory macrophages and M2 macrophages described as alternatively activated macrophages or anti-inflammatory/pro-wound-healing macrophages [14–16]. Others have attempted to identify three different macrophage classifications: classically activated macrophages, wound-healing macrophages, and regulatory macrophages [17]. In this classification, it is the regulatory macrophage that has anti-inflammatory activity whereas the wound-healing macrophage facilitates tissue repair. Attempts to classify macrophage activity are artificial and can be misleading given the wide variety of environmental cues that may activate macrophages and result in a wide variety of different forms of macrophage polarization, e.g., phenotypic expression. Mantovani best describes macrophage polarization, activity, or phenotypic expression as being a continuum ranging from M1 to M2 [18, 19].

The macrophage is probably the most important cell in chronic inflammation because of the great number of biologically active products it can produce. Important classes of products produced and secreted by macrophages include neutral proteases, chemotactic factors, arachidonic acid metabolites, reactive oxygen metabolites, complement components, coagulation factors, growth-promoting factors, cytokines, and acid. Phagolysosomes in macrophages can have acidity as low as pH of 4 and direct microelectrode studies of this acid environment have determined pH levels as low as 3.5. Moreover, only several hours are necessary to achieve these acid levels following adhesion of macrophages [20–23].

Growth factors such as platelet-derived growth factor (PDGF), fibroblast growth factor (FGF), transforming growth factor-β (TGF-β), TGF-α/epidermal growth factor (EGF), and interleukin-1 (IL-1) or tumor necrosis factor (TNF-α) are important to the growth of fibroblasts and blood vessels and the regeneration of epithelial cells. Effector molecules released by activated macrophages can initiate cell migration, differentiation, and tissue remodeling; and are involved in various stages of wound healing.

2.5 Granulation Tissue

Within 1 day following implantation of a biomaterial (e.g., injury), the healing response is initiated by the action of monocytes and macrophages. Fibroblasts and vascular endothelial cells in the implant site proliferate and begin to form granulation tissue, which is the specialized type of tissue that is the hallmark of healing inflammation. Granulation tissue derives its name from the pink, soft granular appearance on the surface of healing wounds, and its characteristic histologic features include the proliferation of new small blood vessels and fibroblasts. Depending on the extent of injury, granulation tissue may be seen as early as 3–5 days following implantation of a biomaterial.

The new small blood vessels are formed by budding or sprouting of preexisting vessels in a process known as neovascularization or angiogenesis [24, 25]. This process involves proliferation, maturation, and organization of endothelial cells into capillary vessels. Fibroblasts also proliferate in developing granulation tissue and are active in synthesizing collagen and proteoglycans. In the early stages of granulation tissue development, proteoglycans predominate but later collagen, especially type III collagen, predominates and forms the fibrous capsule seen with most biomaterials. Some fibroblasts in developing granulation tissue may have the features of smooth muscle cells, e.g., actin microfilaments. These cells are called myofibroblasts and are considered to be responsible for the wound contraction seen during the development of granulation tissue. In addition to contraction, myofibroblasts can invade and repair injured tissues by secreting an organizing extracellular matrix [26]. Recent studies indicate that myofibroblasts can originate from different precursor cells, the major contribution being from local recruitment of connective tissue fibroblasts; however, local mesenchymal stem cells, bone marrow-derived mesenchymal stem cells (fibrocytes), and cells derived from the epithelial-mesenchymal transition process may be an alternative source of myofibroblasts [27]. Macrophages are almost always present in granulation tissue. Other cells may also be present if chemotactic stimuli are generated.

The wound-healing response is generally dependent on the extent or degree of injury or defect created by the implantation procedure [28–31]. Wound healing by primary union or first intention is the healing of clean, surgical incisions in which the wound edges have been approximated by surgical sutures. This term does not apply in the context of host response to biomaterials. Healing under these conditions occurs without significant bacterial contamination and with a minimal loss of tissue. Wound healing by secondary union or second intention occurs when there is a large tissue defect that must be filled or there is extensive loss of cells and tissue. In wound healing by secondary intention, regeneration of parenchymal cells cannot completely reconstitute the original architecture and much larger amounts of granulation tissue are formed that result in larger areas of fibrosis or scar formation. Under these conditions, different regions of tissue may show different stages of the wound-healing process simultaneously. Wound healing by second intention is commonly seen with biomaterials and is related to the extent of provisional matrix formed between the implant and tissue.

Granulation tissue is distinctly different from granulomas, which are small collections of modified macrophages called epithelioid cells. Langhans' or foreign-body-type giant cells may surround nonphagocytosable particulate materials in granulomas. Foreign-body giant cells are formed by the fusion of monocytes and macrophages in an attempt to phagocytose the material.

2.6 Foreign-Body Reaction

The FBR to biomaterials is composed of foreign-body giant cells (FBGCs) and the components of granulation tissue (e.g., macrophages, fibroblasts, and capillaries in varying amounts), depending upon the form and topography of the implanted material [2, 32, 33]. Relatively flat and smooth surfaces such as that found on silicone breast prostheses have a foreign-body reaction that is composed of a layer of macrophages and foreign body giant cells one to two cells in thickness. Relatively rough surfaces such as those found on the outer surfaces of expanded poly tetrafluoroethylene (ePTFE) or Dacron vascular prostheses have a foreign-body reaction composed of macrophages and foreign-body giant cells at the surface. Fabric materials generally have a surface response composed of macrophages and foreign body giant cells, with varying degrees of granulation tissue subjacent to the surface response.

As previously discussed, the form and topography of the surface of the biomaterial determine the composition of the foreign-body reaction [34]. With biocompatible materials, the composition of the foreign-body reaction in the implant site may be controlled by the surface properties of the biomaterial, the form of the implant, and the relationship between the surface area of the biomaterial and the volume of the implant. For example, high-surface-to-volume implants such as fabrics, porous materials, particulate, or microspheres will have higher ratios of macrophages and foreign-body giant cells in the implant site than smooth-surface implants, which will have fibrosis as a significant component of the implant site [35, 36].

The FBR may persist at the tissue–implant interface for the lifetime of the implant. Generally, fibrosis (i.e., fibrous encapsulation) surrounds the biomaterial or implant with its interfacial foreign-body reaction, isolating the implant and foreign-body reaction from the local tissue environment. Early in the inflammatory and wound-healing response, the macrophages are activated upon adherence to the material surface [37].

Although it is generally considered that the chemical and physical properties of the biomaterial are responsible for macrophage activation, the subsequent events regarding the activity of macrophages at the surface are not clear. Tissue macrophages, derived from circulating blood monocytes, may coalesce to form multinucleated foreign-body giant cells. It is not uncommon to see very large foreign-body giant cells containing large numbers of nuclei on the surface of biomaterials. While these foreign-body giant cells may persist for the lifetime of the implant, it is not known if they remain activated, releasing their lysosomal constituents, or become quiescent [38].

Figure 2.2 demonstrates the progression from circulating blood monocyte to tissue macrophage to foreign-body giant cell development that is most commonly observed. Indicated in the figure are important biological responses that are considered to play

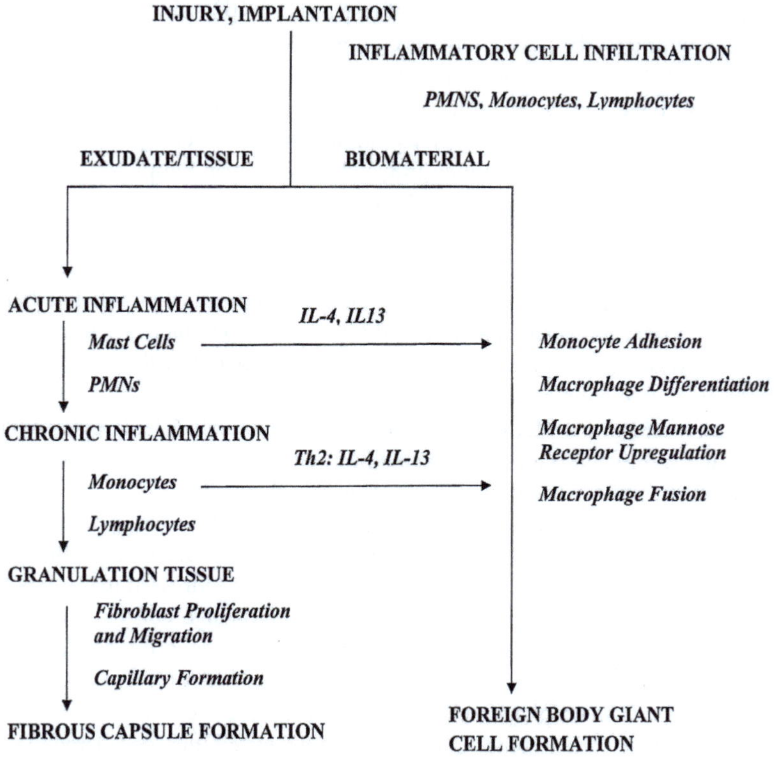

Fig. 2.3 Sequence of events involved in inflammatory and wound-healing responses leading to foreign body giant cell formation. This shows the potential importance of mast cells in the acute inflammatory phase and Th2 lymphocytes in the transient chronic inflammatory phase with the production of IL-4 and IL-13, which can induce monocytes/macrophage fusion to form foreign body giant cells

an important role in FBGC development. Material surface chemistry may control adherent macrophage apoptosis (i.e., programmed cell death) that renders potentially harmful macrophages nonfunctional, while the surrounding environment of the implant remains unaffected. The level of adherent macrophage apoptosis appears to be inversely related to the surface's ability to promote fusion of macrophages into FBGCs, suggesting a mechanism for macrophages to escape apoptosis. Figure 2.3 demonstrates the sequence of events involved in inflammation and wound healing when medical devices (i.e., biomaterials) are implanted. In general, the PMN predominant acute inflammatory response and the lymphocyte/monocyte predominant chronic inflammatory response resolve quickly (i.e., within 2 weeks) depending on the type and location of the implant. Studies using IL-4 or IL-13, respectively, demonstrate the role for Th2 helper lymphocytes and/or mast cells in the development of the foreign body reaction at the tissue/material interface [39, 40]. Integrin receptors of IL-4-induced FBGCs are characterized by the early constitutive expression of αVβ1 and the later induced expression of α5β1 and αXβ2, which indicate potential interactions with adsorbed complement C3, fibrin(ogen), fibronectin, Factor X, and vitronectin [39–47]. Interactions through indirect (paracrine) cytokine and chemokine signaling have shown a significant effect in enhancing adherent macrophage/FBGC activation at early times, whereas interactions via direct (juxtacrine) cell-cell mechanisms dominate at later times [48–51]. Th2 helper lymphocytes have been described as "anti-inflammatory" based on their cytokine profile, of which IL-4 is a significant component.

There is a common belief that mononuclear leukocytes, i.e., monocytes and macrophages, can proliferate in the healing response. However, this belief is based on rodent study observations and may not hold for human mononuclear leukocytes in the foreign body reaction to biomaterials [52–54]. No convincing evidence supporting the proliferation of mononuclear leukocytes and macrophages at implant sites in humans has been provided in the literature. This issue of mononuclear leukocyte replication addresses the significant issue of understanding species differences in considering the innate immune response to implanted biomaterials.

In vitro cell culture studies may play a role in determining the interactions of the various cell types involved in the innate immune response with candidate biomaterials for medical devices. However, several key issues must be addressed in the experimental design of these in vitro cell culture studies. Macrophage cell cultures are commonly used to determine the cytotoxicity as well as the capability for foreign body giant cell formation. Unfortunately, the vast majority of macrophage cell cultures utilize tumor-derived macrophage cell lines and rarely are the similarities and differences between these rodent tumor-derived macrophage cell lines compared to primary human blood-derived monocytes presented or discussed. Marked differences in phenotypic expression may be expected when rodent tumor-derived macrophage cell lines are used when compared to primary human monocyte/macrophage cell cultures. Table 2.5 presents a list of commonly utilized tumor-derived macrophage cell lines.

Table 2.5 Macrophage cell lines used in biomaterials studies

Cell Line	Source
HL-60	Human—Promyelocytic Leukemia
IC-21	Mouse—Transformed Peritoneal Macrophages
J774	Mouse—Histiocytic Lymphoma
J774A.1	Mouse—Histiocytic Lymphoma
P388D1	Mouse—Transformed Lymphoma
RAW	Mouse—Transformed Lymphoma
RAW 264.7	Mouse—Transformed Lymphoma
THP-1	Human—Acute Monocytic Lymphoma
U937	Human—Histiocytic Lymphoma

"Macrophages" mouse—peritoneal surface

2.7 Fibrosis/Fibrous Encapsulation

The end-stage healing response to biomaterials is generally fibrosis or fibrous encapsulation. However, there may be exceptions to this general statement (e.g., porous materials inoculated with parenchymal cells or porous materials implanted into tissue or bone). As previously stated, the tissue response to biomaterials is in part dependent upon the extent of injury or defect created in the implantation procedure and the amount of provisional matrix.

The ultimate goal of tissue engineering and regenerative medicine is replacement of injured tissue by cells that reconstitute normal tissue and organ structures. Numerous approaches, including stem cells, scaffolds, growth factors, etc., are currently being investigated. However, the relatively rapid responses of inflammation, wound healing, and the foreign body reaction, as well as other significant factors in tissue regeneration, present major challenges to the successful achievement of this goal. This is especially significant with the use of scaffold materials where migration and integration of the scaffold porosity is necessary in tissue engineering approaches.

Repair of biomaterial implant sites can involve two distinct processes: constructive remodeling, which is the replacement of injured tissue by parenchymal cells of the same type, or replacement by connective tissue that constitutes the fibrous capsule. These processes are generally controlled by either (1) the proliferative capacity of the cells in the tissue or organ receiving the implant and the extent of injury as it relates to the destruction, or (2) persistence of the tissue framework (i.e., extracellular matrix) of the implant site.

The regenerative capacity of cells allows them to be classified into three groups: labile, stable (or expanding), and permanent (or static) cells. Labile cells continue to proliferate throughout life; stable cells retain this capacity but do not continuously replicate; and permanent cells cannot reproduce themselves after birth. Perfect repair with restitution of normal structure can theoretically occur only in tissues consisting of stable and labile cells, whereas all injuries to tissues composed of permanent cells may give rise to fibrosis and fibrous capsule formation with very little restitution of the normal tissue or organ structure. Tissues composed of permanent cells (e.g., nerve cells and cardiac muscle cells) most commonly undergo an organization of the inflammatory exudate, leading to fibrosis. Tissues of stable cells (e.g., parenchymal cells of the liver, kidney, and pancreas); mesenchymal cells (e.g., fibroblasts, smooth muscle cells, osteoblasts, and chondroblasts); and vascular endothelial and labile cells (e.g., epithelial cells and lymphoid and hematopoietic cells) may also follow this pathway to fibrosis or may undergo resolution of the inflammatory exudate, leading to restitution of the normal tissue structure.

The condition of the underlying framework or supporting extracellular matrix (ECM) of the parenchymal cells following an injury plays an important role in the restoration of normal tissue structure. Retention of the framework ECM with injury may lead to restitution of the normal tissue structure, whereas destruction of the framework most commonly leads to fibrosis.

It is important to consider the species-dependent nature of the regenerative capacity of cells. For example, cells from the same organ or tissue but from different species may exhibit different regenerative capacities and/or connective tissue repair, as in endothelialization with vascular grafts and stents.

Local and systemic factors play a role in the wound-healing response to biomaterials or implants. Local factors include the anatomic site (tissue or organ) of implantation, the adequacy of blood supply, and the potential for infection. Systemic factors include nutrition, hematologic derangements, glucocorticoid administration, and preexisting diseases such as atherosclerosis, diabetes, and infection.

The end stage of wound healing/tissue repair with implanted medical devices is the fibrous capsule. Initially believed to be produced by infiltrating fibroblasts, it is now known that myofibroblasts and fibrocytes (resident and circulating mesenchymal progenitor cells) play a significant role in producing collagenous fibrosis, the main constituent of the fibrous capsule [55–59].

Finally, the implantation of biomaterials or medical devices may be best viewed from the perspective that the implant provides an impediment or hindrance to appropriate (normal) tissue or organ regeneration and healing. The fibrous capsule surrounding drug delivery devices also has been suggested to be a barrier to drug diffusion and inhibition of the function of drug delivery systems and biosensors (e.g., glucose sensors). However, that view may be short-sighted as recent studies with a wireless controlled drug delivery microchip for the delivery of an osteoporosis inhibitor of approximately 4000 molecular weight have been shown to produce clinically relevant blood levels for inhibition of osteoporosis [60]. Given our current limited ability to control the sequence of events following injury in the implantation procedure, restitution of normal tissue structures with function is rare.

The development of a fibrous capsule surrounding or encapsulating a biomaterial, i.e., scaffold, or medical device, i.e., pacemaker generator, has been commonly viewed as a rejection phenomenon in the end-stage healing of implanted biomaterials or medical devices. This perspective is medical device dependent and the formation of a fibrous capsule may be a positive or negative response. It is not necessarily a rejection phenomenon. Numerous biomaterials and medical devices require the development of an encapsulating fibrous capsule to stabilize the biomaterial or medical device within the tissue. Stabilization of devices within tissue markedly reduces motion of the implant relative to the tissue and has been shown to reduce the chronic inflammatory response produced by such motion when inadequate fibrous capsules have formed surrounding implanted biomaterials or medical devices. Inadequate fibrous capsule formation may lead to migration of a medical device and thus may be a failure mechanism. A classic example of migration as a failure mechanism has been observed with breast implants. Thus, the formation of a fibrous capsule is an important consideration in the design criteria of implanted biomaterials and medical devices.

Current studies directed toward developing a better understanding of the modification of the inflammatory response, stimuli providing for appropriate proliferation of permanent and stable cells, and the appropriate application of growth factors may provide keys to the control of inflammation, wound healing, and fibrous encapsulation of biomaterials.

2.8 Discussion and Perspectives

In spite of the significant advances that have been made in mechanistic understanding of the inflammatory, healing, and foreign body responses to biomaterials and medical devices over the past two decades, numerous challenges, lack of understanding which limit projection to clinical application still exist. The purpose of this discussion/perspectives section is to identify several of these problems that offer challenges/opportunities for the future.

Species differences in the various responses to the implantation of a biomaterial or medical device still exist. The translation from animals to humans is still poorly understood but significant considering that animal studies are a required precursor to clinical application. Current thought regarding the source of macrophages and their fused entity, i.e., foreign body giant cells, suggests that differentiated macrophages may be present due to self-renewal [52]. That is, resident macrophages are capable of proliferation. Studies that support this hypothesis have been conducted in nonhuman mammals and many major macrophage populations have been found to be derived from embryonic progenitors and are capable of renewal independent of hematopoietic stem cells. From a clinical applications perspective, this is a significant question as implant retrieval studies have identified macrophages and foreign body giant cells in the foreign body response to biomaterials and medical devices to be present at the tissue/material interface for approximately 30 years. As there is no compelling evidence that the macrophages in the foreign body reaction to biomaterials and medical devices implanted in humans are capable of self-renewal, i.e., proliferation, the turnover rate of these cells at the interface and the precursor cells that continue to populate the surface of the biomaterial or medical device with implants in humans [52, 61] remain unanswered questions. An example of significant species differences is the fact that human vascular grafts do not endothelialize their luminal surface, whereas higher vertebrates, including chimpanzees and baboons, do provide an endothelial lining in the healing response of vascular grafts. While putative evidence focuses on circulating stem cells in the blood to provide an endothelial lining, no evidence exists today to support this hypothesis.

The lack of a host response to an implanted biomaterial may be desirable in some applications; however, the holy grail of a biomaterial surface that does not adhere proteins or cells remains elusive. As noted earlier, almost immediately upon implantation, the humoral and cellular components of blood come in contact with implanted biomaterials or medical devices resulting in a provisional matrix. Recent studies have focused on inhibition of biomaterial-induced complement activation to reduce the protein adhesion phenomenon on the surface [62–64]. Inhibition of biomaterial-induced complement activation also would be expected to lead to a reduction in monocyte/macrophage adhesion to the biomaterial [40, 44]. However, the adhesion of monocytes/macrophages to biomaterial surfaces is far more complex as monocytes/macrophages express protein adhesion receptors (integrins) with at least three different types of beta chains (β1, β2, β3) that in turn can bind to a wide variety of proteins present in the provisional matrix. These blood-derived proteins include complement C3b fragments, fibrin, fibrinogen, fibronectin, factor X, and vitronectin.

Moreover, integrin expression by monocytes/macrophages is time-dependent and β1 integrins are not initially detected on adherent monocytes but begin to appear during macrophage development and are strongly expressed on fusing macrophages that form foreign body giant cells [42, 44]. Thus, monocyte/macrophage adhesion with subsequent macrophage fusion to form foreign body giant cells at the interface is far more complex given the relatively large number of adhesion proteins, their respective monocyte/macrophage receptors, and the time-dependent nature of receptor up-regulation on adherent macrophages and foreign body giant cells. Other mechanisms such as apoptosis or anoikis of adherent cells may be considered to reduce the adherent monocyte/macrophage/foreign body giant cell adhesion to biomaterial surfaces. Apoptosis is programmed cell death while anoikis is a term for apoptosis induced by cell detachment from its supportive matrix. Various biomaterial surface chemistries have identified apoptosis of adherent macrophages both in vitro and in vivo [49, 65–68]. These potential mechanisms for reducing cellular adhesion have been poorly studied and offer an opportunity for controlled and down-regulation of monocyte/macrophage/foreign body giant cell adhesion to biomaterial surfaces.

Regarding development of the fibrous capsule surrounding implants, fibrocytes, a subpopulation of circulating mesenchymal progenitor cells, have been identified as augmenting wound repair as well as producing different fibrosing disorders in humans [59]. Blood circulating fibrocytes can be recruited to sites of tissue or implant injury and differentiate into fibroblasts and myofibroblasts. Myofibroblasts are now considered to be a major contributor to fibrosis and may be responsible for the remodeling of granulation tissue collagen to fibrosis-dependent collagen, i.e., collagen type I. Recent studies have suggested that the mechanical properties of the biomaterial substrate can influence the contractile nature of myofibroblasts [26, 58, 69, 70].

A successful approach to the inhibition of inflammatory adhesion and activation has been the modification of biomaterial surfaces with CD47, a transmembrane molecular marker of "self." As inflammatory cells do not recognize these surfaces as being foreign, inflammatory cell adhesion is reduced with a down-regulation of expressed cytokines, an up-regulation of matrix metalloproteinases, and involvement of JAK/STAT signaling mechanisms [71, 72]. These findings suggest that both biomaterial degradation and fibrous capsule formation can be reduced with CD47 modification of biomaterial surfaces. Strict control of biomedical polymer morphology and porosity also has provided a means to down-regulate foreign body giant cell formation and fibrous capsule formation [34, 73]. These approaches can be expected to be useful in the development of scaffolds for clinical use.

2.9 Conclusions

Ultimately, the success or failure of medical devices, implants, and tissue-engineered constructs is modulated by the interaction between the characteristics of the biomaterial or medical device, patient conditions or factors, and surgical technique. These conditions may modulate the inflammatory, healing, and foreign body responses to biomaterials and medical devices resulting in their eventual failure. Infection remains a significant factor leading to implant failure. Recent studies suggest that individual patient genomic factors may predispose the patient to implant failure.

In the development of new biomaterials, tissue-engineered scaffolds with or without immunomodulary proteins or cells and new medical devices, the acute and chronic inflammatory response and foreign body reaction must be considered in the context of the materials, proteins, and cells that are utilized as well as the synthetic or native-derived materials that are utilized. These factors and their potential interactions must be considered in the experimental design and the goals and objectives of the study. Strengths and weaknesses of the experimental design should be identified prior to the initiation of any study designed to identify the immune response or the modulation of the immune response in a time-dependent nature. This is of special significance as the response to injury with the development of the acute and chronic inflammatory response usually occurs within the first two weeks following implantation and the foreign body reaction with adherent macrophages and foreign body giant cells, while developing within the first 2 weeks of implantation, are more persistent and in some cases have been identified to be present for the lifetime of the synthetic or biologically derived material. The investigation of tissue-engineered systems is of special significance as they have the potential to significantly modulate the immune response in a time-dependent fashion. It can be expected that each unique tissue-engineered system can lead to unique immune responses in a time-dependent fashion. While the objective of complex tissue-engineered systems is to provide regeneration of the tissue/organ under consideration, further investigation of the immune response and its modulation provides a future challenge in not only determining the relative regenerative capacity of the tissue-engineered system but also its biocompatibility and potential for success or failure.

References

1. Williams DF (2008) On the mechanisms of biocompatibility. Biomaterials 29(20):2941–2953
2. Anderson JM (2001) Biological responses to materials. Annu Rev Mater Res 31:81–110
3. Kumar V, Abbas AK, Fausto N et al (2005) Robbins and Cotran pathologic basis of disease, vol 15, 7th edn. Elsevier Saunders, Philadelphia, p 1525
4. Zdolsek J, Eaton JW, Tang L (2007) Histamine release and fibrinogen adsorption mediate acute inflammatory responses to biomaterial implants in humans. J Transl Med 5:31
5. Wiggins MJ, Wilkoff B, Anderson JM et al (2001) Biodegradation of polyether polyurethane inner insulation in bipolar pacemaker leads. J Biomed Mater Res 58(3):302–307
6. Christenson EM, Soofi W, Holm JL et al (2007) Biodegradable fumarate-based polyHIPEs as tissue engineering scaffolds. Biomacromolecules 8(12):3806–3814
7. Christenson EM, Dadsetan M, Wiggins M et al (2004) Poly(carbonate urethane) and poly(ether urethane) biodegradation: in vivo studies. J Biomed Mater Res A 69(3):407–416
8. Christenson EM, Anderson JM, Hiltner A (2004) Oxidative mechanisms of poly(carbonate urethane) and poly(ether urethane) biodegradation: in vivo and in vitro correlations. J Biomed Mater Res A 70(2):245–255

9. Henson PM (1971) The immunologic release of constituents from neutrophil leukocytes. II. Mechanisms of release during phagocytosis, and adherence to nonphagocytosable surfaces. J Immunol 107(6):1547–1557
10. Marchant RE, Anderson JM, Dillingham EO (1986) In vivo biocompatibility studies. VII. Inflammatory response to polyethylene and to a cytotoxic polyvinylchloride. J Biomed Mater Res 20(1):37–50
11. Yamaguchi K, Konishi H, Hara S et al (2001) Biocompatibility studies of titanium-based alloy pedicle screw and rod system: histological aspects. Spine J 1(4):260–268
12. MacEwan MR, Brodbeck WG, Matsuda T et al (2005) Monocyte/lymphocyte interactions and the foreign body response: in vitro effects of biomaterial surface chemistry. J Biomed Mater Res A 74(3):285–293
13. Brodbeck WG, Macewan M, Colton E et al (2005) Lymphocytes and the foreign body response: lymphocyte enhancement of macrophage adhesion and fusion. J Biomed Mater Res A 74(2):222–229
14. Mooney JE, Rolfe BE, Osborne GW et al (2010) Cellular plasticity of inflammatory myeloid cells in the peritoneal foreign body response. Am J Pathol 176(1):369–380
15. Gordon S, Pluddemann A (2013) Tissue macrophage heterogeneity: issues and prospects. Semin Immunopathol 35(5):533–540
16. Gordon S (2003) Alternative activation of macrophages. Nat Rev Immunol 3(1):23–35
17. Mosser DM, Edwards JP (2008) Exploring the full spectrum of macrophage activation. Nat Rev Immunol 8(12):958–969
18. Mantovani A, Sozzani S, Locati M et al (2002) Macrophage polarization: tumor-associated macrophages as a paradigm for polarized M2 mononuclear phagocytes. Trends Immunol 23(11):549–555
19. Mantovani A, Sica A, Sozzani S et al (2004) The chemokine system in diverse forms of macrophage activation and polarization. Trends Immunol 25(12):677–686
20. Silver IA, Murrills RJ, Etherington DJ (1988) Microelectrode studies on the acid microenvironment beneath adherent macrophages and osteoclasts. Exp Cell Res 175(2):266–276
21. Klebanoff SJ (2005) Myeloperoxidase: friend and foe. J Leukoc Biol 77(5):598–625
22. Jankowski A, Scott CC, Grinstein S (2002) Determinants of the phagosomal pH in neutrophils. J Biol Chem 277(8):6059–6066
23. Haas A (2007) The phagosome: compartment with a license to kill. Traffic 8(4):311–330
24. Nguyen LL, D'Amore PA (2001) Cellular interactions in vascular growth and differentiation. Int Rev Cytol 204:1–48
25. Browder T, Folkman J, Pirie-Shepherd S (2000) The hemostatic system as a regulator of angiogenesis. J Biol Chem 275(3):1521–1524
26. Hinz B, Phan SH, Thannickal VJ et al (2012) Recent developments in myofibroblast biology: paradigms for connective tissue remodeling. Am J Pathol 180(4):1340–1355
27. Micallef L, Vedrenne N, Billet F et al (2012) The myofibroblast, multiple origins for major roles in normal and pathological tissue repair. Fibrogenesis Tissue Repair 5(Suppl 1):S5
28. Pierce GF (2001) Inflammation in nonhealing diabetic wounds: the space-time continuum does matter. Am J Pathol 159(2):399–403
29. Mustoe TA, Pierce GF, Thomason A et al (1987) Accelerated healing of incisional wounds in rats induced by transforming growth factor-beta. Science 237(4820):1333–1336
30. Clark RAF (1996) The molecular and cellular biology of wound repair, vol 23, 2nd edn. Plenum Press, New York, p 611, 611 leaf of plates
31. Broughton G 2nd, Janis JE, Attinger CE (2006) The basic science of wound healing. Plast Reconstr Surg 117(7 Suppl):12S–34S
32. Anderson JM, Rodriguez A, Chang DT (2008) Foreign body reaction to biomaterials. Semin Immunol 20(2):86–100
33. Anderson JM, Cramer S (2015) Perspectives on the inflammatory, healing, and foreign body responses to biomaterials and medical devices. In: Badylak S (ed) Host response to biomaterials. The impact of host response on biomaterial selection. Elsevier, New York, pp 13–36
34. Bota PC, Collie AM, Puolakkainen P et al (2010) Biomaterial topography alters healing in vivo and monocyte/macrophage activation in vitro. J Biomed Mater Res A 95(2):649–657

35. Revell PA (2008) The combined role of wear particles, macrophages and lymphocytes in the loosening of total joint prostheses. J R Soc Interface 5(28):1263–1278
36. Charnley J (1970) The reaction of bone to self-curing acrylic cement. A long-term histological study in man. J Bone Joint Surg Br 52(2):340–353
37. Purdue PE (2008) Alternative macrophage activation in periprosthetic osteolysis. Autoimmunity 41(3):212–217
38. Brodbeck WG, Anderson JM (2009) Giant cell formation and function. Curr Opin Hematol 16(1):53–57
39. McNally AK, DeFife KM, Anderson JM (1996) Interleukin-4-induced macrophage fusion is prevented by inhibitors of mannose receptor activity. Am J Pathol 149(3):975–985
40. McNally AK, Anderson JM (1994) Complement C3 participation in monocyte adhesion to different surfaces. Proc Natl Acad Sci U S A 91(21):10119–10123
41. Nilsson B, Ekdahl KN, Mollnes TE et al (2007) The role of complement in biomaterial-induced inflammation. Mol Immunol 44(1–3):82–94
42. McNally AK, Macewan SR, Anderson JM (2007) alpha subunit partners to beta1 and beta2 integrins during IL-4-induced foreign body giant cell formation. J Biomed Mater Res A 82(3):568–574
43. McNally AK, Jones JA, Macewan SR et al (2008) Vitronectin is a critical protein adhesion substrate for IL-4-induced foreign body giant cell formation. J Biomed Mater Res A 86(2):535–543
44. McNally AK, Anderson JM (2002) Beta1 and beta2 integrins mediate adhesion during macrophage fusion and multinucleated foreign body giant cell formation. Am J Pathol 160(2):621–630
45. Jenney CR, Anderson JM (2000) Adsorbed serum proteins responsible for surface dependent human macrophage behavior. J Biomed Mater Res 49(4):435–447
46. Hynes RO, Zhao Q (2000) The evolution of cell adhesion. J Cell Biol 150(2):F89–F96
47. Hynes RO (2002) Integrins: bidirectional, allosteric signaling machines. Cell 110(6):673–687
48. Kyriakides TR (2015) Molecular events at tissue-biomaterial interface. In: Badylak SF (ed) Host response to biomaterials. The impact of host response on biomaterial selection. Elsevier, New York, pp 81–116
49. Jones JA, Chang DT, Meyerson H et al (2007) Proteomic analysis and quantification of cytokines and chemokines from biomaterial surface-adherent macrophages and foreign body giant cells. J Biomed Mater Res A 83(3):585–596
50. Chang DT, Colton E, Anderson JM (2009) Paracrine and juxtacrine lymphocyte enhancement of adherent macrophage and foreign body giant cell activation. J Biomed Mater Res A 89(2):490–498
51. Anderson JM, Jones JA (2007) Phenotypic dichotomies in the foreign body reaction. Biomaterials 28(34):5114–5120
52. Sieweke MH, Allen JE (2013) Beyond stem cells: self-renewal of differentiated macrophages. Science 342(6161):1242974
53. Robbins CS, Hilgendorf I, Weber GF et al (2013) Local proliferation dominates lesional macrophage accumulation in atherosclerosis. Nat Med 19(9):1166–1172
54. Badylak SF (2016) Tissue regeneration. A scaffold immune microenvironment. Science 352(6283):298
55. Wynn TA, Ramalingam TR (2012) Mechanisms of fibrosis: therapeutic translation for fibrotic disease. Nat Med 18(7):1028–1040
56. Wynn TA (2008) Cellular and molecular mechanisms of fibrosis. J Pathol 214(2):199–210
57. Hinz B, Mastrangelo D, Iselin CE et al (2001) Mechanical tension controls granulation tissue contractile activity and myofibroblast differentiation. Am J Pathol 159(3):1009–1020
58. Hinz B (2007) Formation and function of the myofibroblast during tissue repair. J Invest Dermatol 127(3):526–537
59. Bucala R (2012) Review series—inflammation & fibrosis. Fibrocytes and fibrosis. QJM 105(6):505–508
60. Farra R, Sheppard NF Jr, McCabe L et al (2012) First-in-human testing of a wirelessly controlled drug delivery microchip. Sci Transl Med 4(122):122ra121

61. Hashimoto D, Chow A, Noizat C et al (2013) Tissue-resident macrophages self-maintain locally throughout adult life with minimal contribution from circulating monocytes. Immunity 38(4):792–804
62. Morais JM, Papadimitrakopoulos F, Burgess DJ (2010) Biomaterials/tissue interactions: possible solutions to overcome foreign body response. AAPS J 12(2):188–196
63. Kourtzelis I, Rafail S, DeAngelis RA et al (2013) Inhibition of biomaterial-induced complement activation attenuates the inflammatory host response to implantation. FASEB J 27(7):2768–2776
64. Ekdahl KN, Lambris JD, Elwing H et al (2011) Innate immunity activation on biomaterial surfaces: a mechanistic model and coping strategies. Adv Drug Deliv Rev 63(12):1042–1050
65. Shive MS, Brodbeck WG, Anderson JM (2002) Activation of caspase 3 during shear stress-induced neutrophil apoptosis on biomaterials. J Biomed Mater Res 62(2):163–168
66. Brodbeck WG, Shive MS, Colton E et al (2001) Influence of biomaterial surface chemistry on the apoptosis of adherent cells. J Biomed Mater Res 55(4):661–668
67. Brodbeck WG, Patel J, Voskerician G et al (2002) Biomaterial adherent macrophage apoptosis is increased by hydrophilic and anionic substrates in vivo. Proc Natl Acad Sci U S A 99(16):10287–10292
68. Brodbeck WG, Colton E, Anderson JM (2003) Effects of adsorbed heat labile serum proteins and fibrinogen on adhesion and apoptosis of monocytes/macrophages on biomaterials. J Mater Sci Mater Med 14(8):671–675
69. Klingberg F, Hinz ES, White B (2013) The myofibroblast matrix: implications for tissue repair and fibrosis. J Pathol 229(2):298–309
70. Hinz B, Gabbiani G (2010) Fibrosis: recent advances in myofibroblast biology and new therapeutic perspectives. F1000 Biol Rep 2:78
71. Stachelek SJ, Finley MJ, Alferiev IS et al (2011) The effect of CD47 modified polymer surfaces on inflammatory cell attachment and activation. Biomaterials 32(19):4317–4326
72. Finley MJ, Clark KA, Alferiev IS et al (2013) Intracellular signaling mechanisms associated with CD47 modified surfaces. Biomaterials 34(34):8640–8649
73. Fukano Y, Usui ML, Underwood RA et al (2010) Epidermal and dermal integration into sphere-templated porous poly(2-hydroxyethyl methacrylate) implants in mice. J Biomed Mater Res A 94(4):1172–1186

Chapter 3
Macrophages: The Bad, the Ugly, and the Good in the Inflammatory Response to Biomaterials

Marta Scatena, Karen V. Eaton, Melissa F. Jackson, Susan A. Lund, and Cecilia M. Giachelli

Abstract Macrophages play a central role in guiding proper organ and tissue development, physiological healing, and in maintaining tissue homeostasis. Further, they are one of the major cell components of the inflammatory response. During healing, macrophages assume a temporal series of distinct phenotypes that guide tissue repair and restoration of tissue homeostasis. Macrophages then decline and the restored tissue is macrophage free. Dysfunction or imbalance in macrophage phenotypes results in compromised healing that is thought to be the root cause of inflammatory diseases. Implanted biomedical devices elicit a robust inflammatory response driven largely by dysfunctional macrophages, which show a significant shift in their physiological behavior. They do not progress through the temporal series of phenotypes and do not decline with time, rather remain with the biomedical device for the life of it (the bad). At the host-device interface macrophages fuse to create large cells, foreign body giant cells. These giant cells are believed to damage the biomedical device at a structural and functional level (the ugly). Significant effort has been put forward to understand the processes leading to the dysfunctional macrophage response to biomedical devices, as well as to design novel approaches to guide the macrophages through the temporal series of phenotypes of physiological healing (the good). In this chapter, the current understanding of the developmental origin of macrophages, their functions in physiological processes, and dysfunction in response to the foreign body will be presented and discussed, as well as approaches to guide them toward resolution of the foreign body-elicited inflammatory response.

Keywords Monocyte • Macrophage phenotype • Biomaterial • Pro-inflammatory • Pro-regeneration • Foreign body giant cell • Foreign body reaction

M. Scatena • K.V. Eaton • M.F. Jackson • S.A. Lund • C.M. Giachelli (✉)
Department of Bioengineering, University of Washington, Seattle, WA, USA
e-mail: ceci@uw.edu

B. Corradetti (ed.), *The Immune Response to Implanted Materials and Devices*,
DOI 10.1007/978-3-319-45433-7_3

3.1 Introduction

Monocytes, macrophages, and dendritic cells (DCs) are part of the mononuclear phagocyte system (MPS), a body-wide, specialized system of phagocytic cells. This system functions in the innate immune response, in support of the adaptive immune response and in the maintenance of tissue homeostasis. Monocytes are the precursors of macrophages responding to pathophysiological processes and are considered key cellular components of chronic inflammatory conditions like the host response to biomedical devices. DCs are specialized cells evolved mainly to stimulate and initiate the T and B lymphocyte-driven (adaptive) immune response. DC involvement in biomaterial healing is not well characterized and will not be discussed in this chapter.

Monocytes are identified by specific surface markers (CD115 and CD11b in mouse, and CD14 in human) and are found mainly in the bone marrow, blood, and spleen. In humans and mice at least two phenotypically and functionally distinct monocyte subsets are present, suggesting evolutionary conservation. They are categorized as "classical" and "patrolling" monocytes by differences in the expression of surface markers and functional features. "Classical" monocytes are thought to give rise to the macrophages populating the inflammatory response to the foreign material [1].

As described in detail in the following sections, monocyte-derived macrophages are the main inflammatory cell type found in the robust inflammatory response elicited by foreign materials. It is well documented that the wound-healing process ensuing material implantation does not follow the physiological path but rather a pathological course, whereby macrophages become dysregulated and are unable to guide the host tissue near and around the biomaterial to healing and restoration. Accordingly, the recruited macrophages rather than evolving from the initial phenotype that protects the host from potential infection and facilitates the removal of damaged tissue to the phenotype guiding tissue regeneration and homeostasis, remain locked in specific phenotypes and do not decline with time (the bad). This is generally thought to prolong and propagate inflammation. Further, the recruited macrophages fuse into foreign body giant cells (FBGCs). These cells are believed to develop in an attempt to destroy the foreign material; however, they just damage it and impede the biomedical device function (the ugly). Thus, new efforts to guide macrophages toward phenotypes that restore tissue hemostasis are needed (the good).

3.2 Developmental Origin

3.2.1 Monocytes and Macrophages Ontogeny and Lineage

The MPS has been proposed to arise from a temporal succession of macrophage progenitors. In mice, these start to develop first at embryonic day 8 from the primitive ectoderm of the yolk sac and give rise to macrophages without a monocytic progenitor, as identified by morphological characteristics, and by expression of macrophage markers. Later in development, haematopoiesis in fetal liver generates circulating monocytes during embryogenesis. Macrophages derived from the fetal liver resemble those that are present in adults. The fetal liver is initially seeded by haematopoietic progenitors from the yolk sac and subsequently from the hematogenic endothelium of the aorto-gonadal-mesonephros region of the embryo. In postnatal life, fetal liver haematopoiesis declines and is replaced by bone marrow haematopoiesis that becomes the source of circulating monocytes [2–4].

Studies have shown that monocytes, macrophages, and DCs are developmentally related and share a common bone marrow-derived precursor called the "monocyte-macrophage DC progenitor" (MDP). These cells have the ability to differentiate into phagocytic macrophages, antigen-presenting DCs, and osteoclasts that are specialized bone-resorbing cells [5–7]. In vitro, mouse cultures of MDPs progress toward macrophages in response to macrophage colony stimulating factor (M-CSF or Colony Stimulating Factor 1, CSF1). These cells show high expression of the macrophage markers CD11b and F4/80, and low expression of the dendritic cell marker CD11c. The addition of granulocyte macrophage colony stimulating factor (GM-CSF) and interleukin 4 (IL-4) skews the population toward DCs that, in turn, express high CD11c and MHC class II. In DC development, a common DC progenitor (CDP) derived from the MDP has been identified. It is able to develop into two DC subtypes, classical and plasmacytoid DCs, but not monocytes. In macrophage development occurring in the postnatal bone marrow, the common myeloid progenitor (cMoP) also derives from the MDP and gives rise to monocytes and their descendants but does not generate DCs [8, 9]. The developmental path leading to determination of the osteoclast lineage is less clear. Mouse and human bone marrow and spleen-derived cells expressing c-Kit and colony-stimulating factor 1 receptor (CSF1R) have been shown to be able to generate osteoclasts and compose 0.1–0.3 % of bone marrow cells [7, 10, 11]. Whether these cells derive from the cMoP or a separate osteoclast precursor branch derives from the MDP has yet to be established. We have recently found the early monocyte-macrophage marker CD68 is not expressed in osteoclast precursors suggesting a separation of the monocyte-macrophage and osteoclast lineage early on (Jackson, Scatena and Giachelli, in submitted).

A common myeloid progenitor has also been identified in murine spleen and peripheral blood. Peripheral monocyte populations are heterogeneous and have an inherent plasticity to generate different types of macrophages based on microenvironmental cues. Indeed, after release into the circulation, monocytes are found in two distinct phenotypes: "classical" and "patrolling." In the mouse system, it has been suggested that "patrolling," Ly6C$^-$ cells, originate from "classical," Ly6C$^+$ cells, monocytes, and that Ly6C$^-$ monocytes act to maintain vessel integrity and to detect pathogens, while Ly6C$^+$ monocytes are recruited predominantly to sites of infection and injury. This system is conserved in humans where peripheral-blood monocytes can also be divided into two subsets: CD14$^+$CD16$^-$ cells, or classical monocytes, and CD14$^+$CD16$^+$ cells or patrolling monocytes. Distinct chemokine-receptor expression profiles are also among the phenotypic differences that are recognized between these subsets: for example, CD14$^+$CD16$^+$ monocytes express CC-chemokine receptor 5 (CCR5), whereas CD14$^+$CD16$^-$ monocytes expressed CC-chemokine receptor 2(CCR2) [1].

Monocytes, which circulate in the bloodstream, have long been considered precursors of macrophages found in many uninjured and healthy tissues. However, recent studies show that tissue macrophages mostly originate from the yolk sac with few originating from the fetal liver. For example, lineage-tracing experiments have shown that brain macrophages (microglia) are primarily derived from yolk-sac progenitors, whereas Langerhans (skin associated macrophages) cells have a mixed origin as they derive from the yolk sac and the fetal liver. However, the major tissue-resident population of macrophages (defined as F4/80 bright in mice) in skin, spleen, pancreas, liver, brain, and lung arise from yolk sac progenitors. In a few tissues, such as kidney and lung, macrophages have a chimaeric origin being derived from the yolk sac and the bone marrow. These data indicate that there are at least three lineages of macrophages in the adult mouse, which arise at different stages of development (yolk sac, fetal liver, and bone marrow) and persist to adulthood. As tissue macrophages appear not to be derived from bone-marrow monocytes, by definition circulating patrolling monocytes act to maintain vessel integrity and to detect pathogens while classical monocytes are recruited predominantly to sites of infection or injury, or to tissues that have continuous cyclical recruitment of macrophages, such as the uterus [12, 13].

3.2.2 Monocytes and Macrophages Lineage Regulation

Regardless of their origin, the major lineage regulator of almost all macrophages is macrophage CSF1R. This class III transmembrane tyrosine kinase receptor is expressed on most, if not all, mononuclear phagocytic cells. Targeted ablation of the *Csf1r* gene causes severe depletion of macrophages in most tissues. Moreover, comparison of *Csf1r*-null mice with those homozygous for a spontaneous null mutation in its ligand, colony stimulating factor 1 (CSF1, *Csf1*$^{op/op}$ mice) demonstrated that all phenotypes in *Csf1r*-null mice were also found in the *Csf1*$^{op/op}$ mice, indicating that CSF1 has only a single receptor [14]. However, the phenotype of *Csf1r*-null mice is more severe than that of *Csf1*-null mice, which suggested the presence of another ligand that has been identified as IL-34. Targeted ablation of *Il34* results in loss of microglia and Langerhans cells, but had little impact on bone marrow, liver, or splenic macrophages. Other potential regulators of monocytes differentiation include GM-CSF and IL-3, which act as macrophage growth factors in tissue culture [15]. In addition, vascular endothelium growth factor A (VEGFA) is a regulator of monocytes because it can compensate for the loss of *Csf1* in osteoclast development in vivo [16].

At the transcriptional level, the most important factor regulating monocyte development is SFPI1 (also known as PU.1), a member of the ETS family whose loss following targeted mutation results in complete depletion of CD11b$^+$F4/80$^+$ macrophages in mice, including those derived from the yolk sac. Similarly, other members of the ETS family are also involved in macrophage differentiation, including *Ets2*, which positively regulates the *Csf1r* promoter. In adults, *Mafb* (also known as *v-Maf*) is required for the local proliferation that maintains resident macrophages, whereas *Gata2* is required for monocyte development but not for resident macrophage populations [17–19]. *Fos* and *Mitf* are required for the differentiation of osteoclasts [20].

3.2.3 Macrophage Function in Development

Macrophages regulate organ development and there is a general requirement for macrophages in tissue patterning and branching morphogenesis. In the mammary gland, macrophages are recruited to the growing ductal structure and their loss results in a slower rate of outgrowth and limited branching. Further, macrophages have also been implicated in maintaining the viability and function of mammary stem cells [21].

Macrophages also regulate angiogenesis. In the eye during the regression of the vasculature in the postnatal period, macrophages guide vascular endothelial cells to undergo apoptosis if these cells do not receive a counterbalancing survival signal. Macrophages appear to mediate vascular endothelial cell death directly via production of Wnt7B, and indirectly via the Wnt5A and Wnt11-dependent induction of the soluble VEGF receptor 1 (VEGFR1) that inhibits the endothelial survival factor VEGF, thereby reducing vascular morphogenesis so that the vascular system is appropriately patterned [22].

Macrophages also act on stem cells of different organs. In the gut they regulate the maintenance and regeneration of intestinal integrity via the regulation of gut stem cell, while in the bone marrow a subpopulation of macrophages in the haematopoietic niche regulates the dynamics of haematopoietic stem cell release and differentiation [13].

Brain development is also influenced by macrophages. Microglia are specialized brain macrophages. In the absence of CSF1R signaling there are no microglia, and the brains have substantial structural defects as they mature. Indeed, microglia express neuronal growth factor (NGF) that promotes neuron viability. Further, microglia modulate neuronal activity and prune synapses during development. The hypomorphic mutation in *CSF1R* in humans is responsible for hereditary diffuse leukoencephalopathy with spheroids that results from loss of myelin sheaves and axonal destruction [13].

Phagocytosis is clearly central to macrophage function even during development. During the development and the adult animal erythropoiesis, maturing erythroblasts are surrounded by macrophages that ingest the extruded erythrocyte nuclei. In the absence of macrophages erythropoiesis is blocked and survival is compromised. Macrophages also maintain the haematopoietic steady state through engulfment of neutrophils and erythrocytes in the spleen and liver [23].

3.3 Macrophages and Inflammation

3.3.1 The Wound-Healing Process

Macrophages are tissue-scavenging phagocytic cells found in all tissues that have a diverse set of functions. Besides their role during embryonic development, circulating monocytes and tissue macrophages play major roles in the adult. Their functions include tissue homeostasis, stimulating the immune system through production of pro-inflammatory cytokines, and termination of the inflammatory response.

One of the main functions of macrophages is phagocytosis. Indeed, macrophages perform phagocytic clearance of dying or dead cells during development and adult life. In primitive organisms, the phagocytic properties of macrophages are the host defense [24].

The physiological inflammatory response requires a highly orchestrated series of events characterized by four basic phases: reaction, regrowth, remodeling, and resolution. The acute and chronic inflammation response is triggered by certain pro-inflammatory molecules of invading microbes. However, sterile stimuli such as mechanical trauma, ischemia, chemicals, and antigens also trigger inflammation. In the event of infection or sterile inflammation, cytokines and chemokines are released, which increase permeability of capillaries and attract wandering neutrophils and monocytes. Neutrophils are the first cells to respond and arrive attracted by chemokines, such as extracellular adenosine triphosphate and other damage-associated molecular patterns [25, 26]. Neutrophils phagocytise debris, kill bacteria by releasing reactive oxygen species (ROS) and aid in the cleaning of the wound by secreting proteases that break down damaged tissue. Neutrophils are usually the most prevalent cell type during the first 48 h of inflammation. Once neutrophils finish cleaning up the infection, they undergo controlled apoptosis and infiltrating macrophages will engulf and degrade the neutrophils in a safe and controlled manner. Macrophage phagocytosis involves the ingestion of the dying cell, microbe, or cellular debris. The ingested particles become trapped in a phagosome, which fuses with a lysosome where enzymes and toxic peroxides digest the pathogen. During the next 2–5 days, macrophages are the most prevalent cell type at the site of inflammation [27].

Macrophages responding to injury and trauma mature from their circulating $CD14^{+}CD16^{+}$ in human and $Ly6C^{+}$ in mouse monocyte precursor cells at the site of inflammation and respond to key cytokine signals such as IL-4 and IFN-γ. When recruited to the site of injury, macrophages display two major phenotypes in response to the specific microenvironment that can be broadly defined as *M1 and M2*. M1 macrophages destroy foreign material at the damaged site and M2 macrophages promote growth, resolution, regeneration, and decline as tissue regenerates [28]. Macrophages of the M1 or pro-inflammatory subtype are the first to appear and aid in the natural debriding of the wound along with neutrophils. They are potent producers and inducers of inflammatory mediators, such as tumor necrosis factor alpha (TNFα). About 3 days postinfection, the M2 or pro-regeneration macrophage subtype becomes the prominent player at the site of injury where M2 macrophages secrete anti-inflammatory cytokines, such as IL-10, and perform efferocytosis of apoptotic pro-inflammatory macrophages and neutrophils. Pro-regeneration M2 macrophages also release factors, such as transforming growth factor beta (TGFβ), which recruit endothelial cells, epithelial cells, and fibroblasts; all key players in the next regrowth phase of healing. The regrowth phase can start as early as day 3 postinfection/injury and last until day 10 postinfection/injury. During this phase, new cells begin to lay down collagen and other extracellular matrix molecules. At the same time, new blood vessels are formed and begin to infiltrate into the wounded areas. The remodeling and resolution phases take place concurrently, in that the resolution phase decreases the number of fibroblasts and M2 pro-regeneration macrophages, as well as thins the dense capillary network. This occurs while the remodeling phase contracts the scar tissue and allows the scar to adjust to the tensions applied during everyday life [29].

Inflammatory resolution can take up to 21 days to complete; however, the remodeling phase may take up to 6–12 months to adjust for physical movements, depending on the tissue. In some cases, the resolution phase fails to resolve inflammation, which can ultimately lead to a chronic inflammatory state or constant fibrotic remodeling in the affected tissues. The dysregulation of the macrophage response has been implicated as a major player in these pathological processes [27]. Therefore, the ability to precisely regulate key players in the innate inflammatory response could help elucidate important mechanisms underlying dysregulated inflammation, which could possibly lead to remedies and an overall better understanding of inflammatory reactions and deadly inflammatory diseases.

3.3.2 Macrophages in Wound Healing

Pro-inflammatory M1 macrophages are the first to arise at the site of injury and propagate the initial response by releasing pro-inflammatory cytokines. They also produce ROS at the injury site in order to destroy foreign material. Thus, pro-inflammatory M1 macrophages are absolutely necessary in cleaning the injured and inflamed site before any healing can begin. On the other side of the spectrum, M2 macrophages promote growth and regeneration and are present following the pro-inflammatory macrophages decline. M2 macrophages are present toward the end of the inflammatory response and mainly function to end and resolve inflammation, stimulate healing, and restore tissue homeostasis characterized by proper vascularization and little to no fibrosis [30]. In chronic inflammatory diseases, pro-inflammatory M1 macrophages are the most prominent macrophage subtype in the tissue, with little pro-regeneration M2 macrophages present, and tissues show no signs of resolution and regeneration.

The pro-inflammatory/pro-regeneration (M1/M2) macrophage ratio can be helpful in determining the phase of the immune response. It has been hypothesized that a high M1/M2 macrophage ratio can lead to chronic inflammation due to the failure of M1 efferocytosis by deficient M2 cells [31]. If M1 cells are not efferocytized by M2 cells before they undergo necrosis/apoptosis pro-inflammatory M1 cytotoxic contents are released into the tissue and can cause further inflammation and recruitment of inflammatory cells. Left untreated, this condition could foster an infinite loop of necrosis and further inflammation. Conversely, it is hypothesized that fibrosis is the result of a low M1/M2 macrophage ratio in which there is an excess of TGFβ-producing M2 cells [32]. Due to high levels of TGFβ, there is a constant influx of collagen depositing fibroblasts and inflammation is never completely resolved. These notions strengthen the argument that it is necessary to have a balance of both pro-regeneration M2 and pro-inflammatory M1 cells and any skewing of this balance could potentially lead to chronic inflammation or fibrosis of the tissue.

3.3.3 Regulation and Functional Consequences of Distinct Macrophage Phenotypes

As mentioned in the previous section, macrophages mature and respond to microenvironmental cues with the activation of distinct functional phenotypes. Two broadly distinct states of polarization have been defined: a pro-inflammatory M1 and a pro-regeneration M2. These states mirror the Th1-Th2 polarization nomenclature of T-cells. Although, unlike Th1-Th2 polarization, macrophage polarization is not permanent but rather plastic and can be altered as stimuli change.

The phenotype of the pro-inflammatory M1 macrophages differs from the pro-regeneration M2 macrophage subsets by the expression of unique membrane receptors, and secretion of cytokines and chemokines. The concept of pro-inflammatory M1 cells was first introduced in the 1960s by Mackaness to describe the antigen-dependent microbicidal activity of macrophages toward *bacillus Calmette-Guerin* and *Listeria* upon secondary exposure to the pathogens. Thus, M1 macrophages constitute a potent arm of the immune system to fight infections [33].

The pro-inflammatory M1 phenotype is induced by IFN-γ, TNF-α, GM-CSF, bacterial products like toll receptor 4 (TLR4)-binding LPS, and other TLR ligands. IFN-γ is the main cytokine inducing the pro-inflammatory M1 phenotype and the main Th1 cell product. IFN-γ signals through the JAK1/STAT1 pathway. The pro-inflammatory M1 phenotype is characterized by the promotion of the Th1 response, the expression of pro-inflammatory cytokines, and the high production of reactive nitrogen and oxygen intermediates. The pro-inflammatory M1 phenotype is typically IL-12^{high} and IL-10^{low} and cytokine expression also includes IL-1, TNFα, and IL-6. Pro-inflammatory M1 polarization and subsequent NF-κB activation leads to transcription of inflammatory factors, such as various CXCL and CCL chemokine ligands. M1 macrophages also produce ROS, which originate primarily from the induced nitric oxide synthase (iNOS) enzyme. The production of NO by iNOS acts as an immune defense, as NO is a free radical with an unpaired electron that can react with superoxide and form peroxynitrate, which is a potent bactericidal agent [34]. At the transcriptional level, NF-κB, AP-1, STAT-1, and IRF5 have been identified as important transcription factors for the pro-inflammatory M1 phenotypic response [35].

Following pro-inflammatory M1 activation, the main inhibitory molecule for pro-inflammatory genes is IL-10. This interleukin activates STAT6 pathways, which sequester coactivator molecules required for the LPS/TLR4 pathway activation and prime the cell for a pro-regeneration M2 macrophage phenotype transition [36]. Ultimately, it is the balance of these cytokines, chemokines, and ROS that dictates the polarization state of macrophages. Unlike M1 cells, M2 cells are less-sensitive to pro-inflammatory stimuli and are actively involved in debris scavenging, angiogenesis, tissue remodeling, and secretion of anti-inflammatory cytokines. Further, they can assume multiple subphenotypes, which include M2a, M2b, and M2c. Hallmarks of pro-regeneration M2 macrophages are high production of IL-10, low expression of IL-12, and the production of Arginase, which deprives iNOS of its substrate. As far as the M2 subphenotypes, M2a macrophages are involved in the Th2 type immune response against parasites and are known to be pro-fibrotic. M2a macrophages are activated by IL-4, IL-10, and IL-13. These cells are characterized by their surface expression of IL-4R and FcεR, Dectin-1, CD163, CD206, and CD209. M2b macrophages are considered immunity-regulating and are induced by IL-1 and LPS. They produce IL-10, IL-1, IL-6, and TNF-α. M2c macrophages are induced by IL-10 and TGF-β. They are often referred to as anti-inflammatory and are known to be involved in tissue repair and remodeling. They produce large amounts of IL-10 and TGF-β and express receptors such as CD163, CD206, and RAGE [28, 37]. At the transcriptional level, factors, including STAT6, IRF4, and PPARγ, have been identified as important in the regulation of characteristic M2 macrophage target genes such as *Arg1*, *Mrc1*, and *Chi3l3*. Furthermore, it has been shown that PPARγ and STAT6 can cooperate to regulate many pro-regeneration M2 targets establishing the STAT6/PPARγ pathway as essential for the M2 phenotype [38].

Macrophage phenotype is in part regulated by the Kruppel-like factor (KLF) family of proteins. These are zinc finger transcription factors that regulate key cellular processes such as development, differentiation, proliferation, and programmed cell death. KLF2 has been identified as a tonic repressor of macrophage pro-inflammatory M1 phenotype, while KLF4 has been found to be essential for the macrophage pro-regeneration M2 phenotype in vitro and in vivo. On the other end, KLF6 has been shown to promote the pro-inflammatory M1 phenotype through cooperation with NF-κB and to inhibit the pro-regeneration M2 targets by suppressing PPAR-γ expression [39].

Increasingly, the M1 and M2 phenotypes are being perceived as two extremes of a continuum of functional states; however, the simplified M1/M2 macrophage nomenclature can be helpful in determining the phase of the immune response.

3.3.4 Macrophage Phenotype Plasticity

The phenotypes of pro-inflammatory M1 and pro-regeneration M2 macrophages in vivo and in vitro can be reversed to some extent. The pathology of the tissue has a large influence on the phenotypic state and influences whether the macrophages display a more inflammatory M1 phenotype or a more regeneration M2 phenotype

[40]. However, it still remains unclear whether the recruitment of circulating monocytes or the reprogramming of cells is the main mechanism for this phenotype switch. Nevertheless, pro-regeneration M2 macrophages can be skewed to a pro-inflammatory M1 macrophages phenotype by certain stimuli. For example, CCL16 chemokine delivery to the tumor environment combined with TLR9 ligand and anti-IL-10 antibody can induce macrophage redirection toward a tumor rejection pro-inflammatory M1-like phenotype [41]. Similarly, genetic deletion of p50 NF-κB allows for tumor-associated pro-regeneration M2 macrophages to express pro-inflammatory M1 markers in sufficient amounts to reduce tumor growth [42]. Lastly, delivery of IL-10 promotes a pro-regeneration M2 macrophages phenotype directly from pro-inflammatory M1 cells [43]. Thus, macrophages can be influenced to transition from a M2 to M1 phenotype, as well as be induced to transition from a M1 to M2 phenotype.

3.4 Macrophage Interaction with Biomaterial Implants

3.4.1 Macrophage Recruitment to the Biomaterial Surface

Implantation of biomaterials elicits a series of well-known molecular and cellular events that constitute the foreign body response (FBR) to biomaterials. A wide range of materials is used for implanted biomedical devices today. While these materials differ greatly in terms of chemical composition, mechanical properties, and surface topography, they all elicit a FBR when implanted in vivo. In wound healing, the resolution of the inflammatory response and subsequent tissue remodeling lead to the restoration of tissue homeostasis. In contrast, the FBR does not result in resolution of the inflammatory response and, instead it leads to encapsulation of the device with fibrotic tissue, which isolates it from the host and can negatively impact device structure and function [44, 45].

The FBR consists of blood-material interactions, provisional matrix deposition, acute inflammation, chronic inflammation, granulation tissue development, foreign body giant cell formation, and fibrosis. The initial phase of the FBR is characterized by blood-biomaterial interactions. This occurs immediately upon implantation and results in the nonspecific adsorption of proteins to the biomaterial surface. Adsorption of proteins to the surface creates cell adhesion sites and provides a way for immune cells, mainly neutrophils and monocytes-derived macrophages, to adhere to synthetic biomaterials that lack intrinsic ligands for cell receptors. Cell adhesion to the implant surface is dependent on the proteins that adsorb to the surface and much research has gone into developing surfaces that minimize or alter protein adsorption. The type, concentration, and surface orientation of adsorbed proteins are critical determinants of the early stages of the FBR. The presence of adsorbed proteins such as fibronectin, fibrinogen, vitronectin, complement C3, and albumin modulates inflammatory cell interactions with biomaterials and consequently is linked to downstream inflammatory responses [46–50].

Following blood-biomaterial interactions, immune cells are recruited to the surface of the biomaterial. As in the physiological response to injury, neutrophils are the first cells to arrive at the biomaterial surface, followed shortly after by macrophages. Neutrophils arrive first but are short lived and by 48–72 h post-implantation macrophages begin to predominate the inflammatory response [51]. Peripheral blood-derived and tissue resident macrophages are recruited to the implant site in response to cytokines and chemokines including TGF-β, PDGF, CXCL4, and IL-1β. Up to this step, the FBR correlates closely with the natural inflammatory reaction to invading microbes and sterile stimuli. Once they have reached the biomaterial surface, macrophages attach, undergo activation, and continue to propagate the FBR. In vitro studies have shown that when cultured on biomaterials macrophages assume a pro-inflammatory M1-like phenotype and produce cytokines including IL-1β, IL-6, TNF-α. The mechanism of macrophage M1 phenotype induction in response adhesion to biomaterials is unclear; however, some studies have indicated that biomaterial-derived microparticles taken up by macrophages activate a caspase-1-dependent release of IL-1β. This activation seems to be dependent on ROS generation, lysosomal destabilization, and K+ efflux. IL-1β activation of the NF-κB pathway would then propagate the pro-inflammatory M1 response [52].

Despite a microenvironment poised to propagate chronic inflammation, the population of macrophages surrounding a biomaterial implant is often made up of a mixture of cells possessing both M1 and M2 markers [52–54]. This intermediate macrophage phenotype has not yet been well characterized, but contrasts with the acute inflammatory response, which is characterized by distinct temporal phases of discrete M1 and M2 macrophage populations.

3.4.2 Macrophage-Derived Foreign Body Giant Cells

Foreign body giant cells (FBGCs) are multinucleated cells generated by fusion of macrophages and are a hallmark feature of the FBR. While these cells are not normally found in healthy tissues, they are abundant surrounding implanted biomaterials even years following implantation. As FBGCs typically do not succeed in phagocytosing large foreign materials, they remain at the biomaterial-tissue interface generating a sealed compartment between their surface and the underlying biomaterial. FBGCs secrete mediators such as ROS, degradative enzymes, and acid into this compartment, in a process referred to as "frustrated phagocytosis," which has been attributed to device failure. Indeed, some classes of polymers, including polyethylene and polypropylene, may undergo oxidative structural damage when exposed to reactive oxygen intermediates [44, 55, 56].

While much research has explored the formation of FBGCs, until recently little was known about their functional properties compared to their mononuclear macrophage counterparts. New studies have shed light onto the specialized functions of macrophage-derived multinucleated giant cells, which at least in vitro are indistinguishable from FBGCs. Macrophage-derived multinucleated giant cells can more efficiently phagocytose large particles and complement-coated particles compared to mono-nucleated macrophages. Interestingly, osteoclasts generated from monocytes by treatment with the receptor activator of nuclear factor kappa-B ligand (RANKL) and M-CSF treatment are also more efficient at taking up large particles, suggesting that this may be a general feature of multinucleated cells of the monocyte lineage. Macrophage-derived multinucleated giant cells have extensive membrane ruffles that contain preactivated complement CR3. These membrane ruffles provide abundant membrane for ingestion of large materials. Further, activated C3 fragments are found deposited on implanted biomaterials. This series of findings demonstrate that multinucleated giant cells are indeed more than the sum of their parts and may suggest that fusion at the biomaterial surface is indeed an attempt to ingest a very large particle [57]. It remains to be determined whether FBGCs possess other specialized functions. Additionally, given the remarkable plasticity of macrophages and the heterogeneity of macrophage phenotypes seen in vivo, it cannot be excluded that FBGCs may also display heterogeneity depending on the local tissue microenvironment.

Cell-cell fusion of macrophages to form multinucleated FBGCs requires a series of highly orchestrated steps. Macrophages must first become fusion competent, migrate toward each other, fuse, and then the resulting multinucleated cells must undergo intracellular cytoskeletal rearrangement [58, 59]. Macrophage fusion is dependent on the presence of fusion-inducing cytokines and a permissive surface. In vitro macrophages fuse to form FBGCs in the presence of the Th2 cytokines, IL-4 and IL-13, and in vivo IL-4 is necessary for FBGC formation by engaging the JAK/STAT partway [59, 60]. Many factors participate in macrophage fusion. These include monocyte chemotactic protein 1 (MCP-1), matrix metalloprotease 9 (MMP-9), epithelial cadherin (ECad), cell surface receptors and membrane proteins (DAP12, MR, CD44, CD47, DC-STAMP, tetraspanins), signal transducers (Rac1), and the matricellular protein Osteopontin [58, 61–64]. IL-4 upregulates DAP12, DC-STAMP, and ECad, while MCP-1 is thought to mediate rearrangement of the cytoskeleton through Rac-1. We have identified osteopontin, an extracellular matrix molecule initially isolated from bone but also produced by macrophages, as an inhibitor of FBGCs formation in vitro and in vivo. In vitro we found that osteopontin treatment can inhibit the IL-13-dependent formation of FBGCs derived from human peripheral blood monocytes. In vivo, we found the absence of osteopontin leads to more FBGCs at the biomaterial-host interface in a mouse model of biomaterial implantation [64] (Fig. 3.1).

Newer players in the field of FBGCs are microRNAs. MicroRNAs (miRs) are small noncoding RNAs that regulate gene expression at the post-transcriptional level. A single miR can influence post-transcriptional control of hundreds of target genes and regulate many cellular processes. It is, thus, not surprising that miRs play a role in macrophage differentiation from monocytes and homotypic cell fusion. miR-7a-1 targets DC-STAMP during FBGC formation and negatively regulates macrophage fusion [65]. Likewise, miR-223 is a negative regulator of IL-4-dependent macrophage fusion both in vitro and in vivo [66].

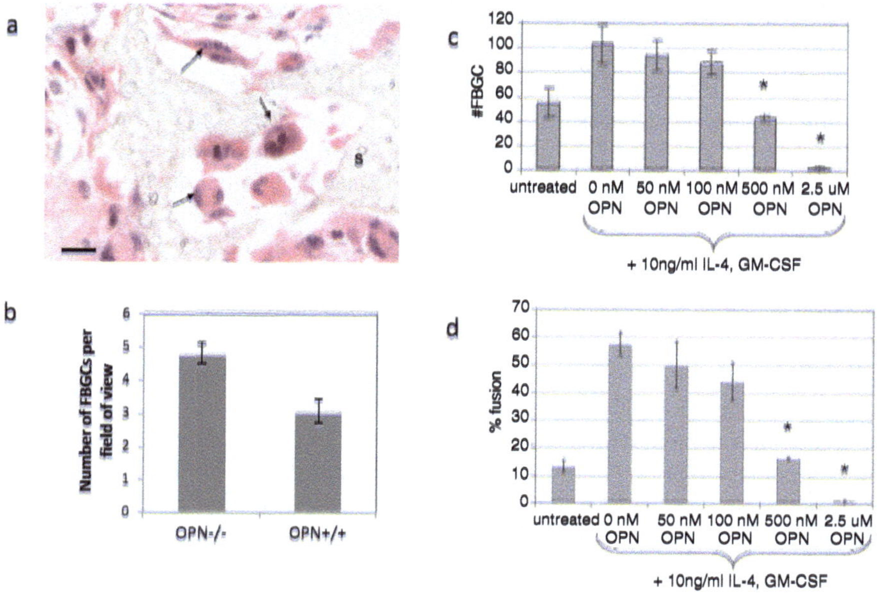

Fig. 3.1 Osteopontin inhibits FBCG formation. (**a**) FBGCs on PVA sponge implants. Implant section stained with H&E revealing FBGCs in association with the PVA sponge (S = sponge). Bar represents 20 μm. PVA sponges were implanted subcutaneously in $OPN^{+/+}$ and $OPN^{-/-}$ animals for 14 days. Two sections per implant were stained, 200 μm apart. In each section, we analyzed six random fields of view, for a total of 12 fields of view per mouse implant. FBGCs, defined as cells containing three or more nuclei, were counted manually. (**b**) Implants from $OPN^{-/-}$ mice had significantly more giant cells at their surfaces than those from $OPN^{+/+}$ mice (4.724 ± 0.333 vs. 2.991 ± 0.352, average number of FBGCs/field±SEM; $p<0.003$). $n=4$ mice per genotype. (**c, d**) Percentage of human FBGCs in vitro. Primary human peripheral monocytes were isolated from heparinized whole blood by density centrifugation, followed by adherence purification. Fusion was induced by treatment with IL-4 and GM-CSF in a 10-day assay. Nuclei in ten representative fields of view per well were counted and summed; each sample was performed in triplicate. Cells containing three or more nuclei were considered "giant." Results are shown for one donor, but are representative of a set of 3. (**c**) Number of FBGCs summed in ten fields of view. $^{*}p<0.0001$ ANOVA, $p<0.0001$ Bonferroni-Dunn post hoc comparison vs. positive control. (**b**) Percentage fusion defined as (number of nuclei contained within FBGC/total nuclei counted) × 100 %. $^{*}p<0.0001$ ANOVA, $p<0.0001$ Bonferroni-Dunn post hoc comparison vs. positive control. $n=3$ wells per sample. Standard deviations are shown. Reproduced from Tsai et al. Biomaterials *26*, 5835–5843

Fusion of macrophages to FBGCs is typically associated with a phenotype switch of the macrophages from an M1-like to a more M2-like phenotype. Indeed, they release pro-fibrotic factors such as TGF-β and platelet-derived growth factor (PDGF) that triggers the action of fibroblasts and endothelial cells. Activated fibroblasts synthesize and deposit collagen that often results in material encapsulation. However, FBGCs still produce pro-inflammatory RANTES and MCP-1, ROS and degradative enzymes, which are not associated with the M2 phenotype [67].

3.4.3 Macrophages and Fibrosis

Successful tissue repair requires resolution of inflammation through the release of anti-inflammatory mediators, cleavage of chemokines, down-regulation of inflammatory mediators and receptors, and apoptosis of immune cells. Although the mechanism is not well understood, it is assumed that continuous action of pro-inflammatory M1 macrophages and FBGCs results in prolonged fibroblast activation and excessive biomaterial-associated matrix deposition. Indeed, fibrosis is a common consequence of chronic inflammation as persistent inflammation can lead to fibrosis. After 2–4 weeks, implantable biomaterials become encapsulated in a nearly avascular, fibrous capsule termed the foreign body capsule (FBC). These capsules can reach thicknesses of 50–200 μm and completely surround implants in a network of dense, highly organized collagen fibers [68].

Macrophages regulate this process through the secretion of cytokines that promote fibroblast migration, proliferation, and activation (TGF-β, PDGF, IGF-1) and the production of matrix degrading enzymes, like MMPs, and their inhibitors, TIMPs that drive extracellular matrix deposition and remodeling. Macrophage-secreted cytokines such as IL-1β can contribute to a state of inflammation, while on the other hand macrophages can regulate fibrosis through the secretion factors that have been shown to suppress inflammation such as IL-10 and TGF-β. Thus, macrophage participation in fibrosis can contribute to either positive or negative outcomes depending on context specific polarization and the ability to shift polarized responses in a timely manner. In many cases an appropriate phenotypic switch in macrophage polarization is necessary for functional remodeling as macrophages are important regulators of the fibrotic response [69–71].

3.5 Strategies to Target Macrophage Function and Polarization in the FBR

There has been a growing interest in research regarding the effects of macrophage phenotype on the FBR to biomaterials [72–75]. As described above macrophages, unlike T-cells, appear to possess remarkable plasticity and are capable of phenotype switching in response to changes in local microenvironmental signals. This plasticity may represent a protective mechanism that allows the host to mount an appropriate response to foreign invaders, but also allows for the resolution of such a response without excessive inflammation.

The loss in function of medical devices over time is in large part due to the FBC that forms around the device. This capsule, which is a consequence of the dysregulated macrophage phenotype of the FBR, is characterized by extensive fibrosis and minimal vascularization of the surrounding tissue. To combat the deleterious effect of the FBR on implanted biomaterials, attempts have been made to mitigate some of the aspects of the FBR that are thought to cause the loss of function and/or structure of the biomedical device. The most commonly used metrics to determine the effectiveness of the various approaches are (1) the relative composition of pro-inflammatory M1 and pro-regenerative M2 macrophages; (2) the number of FBGCs; (3) the thickness and the density of the fibrotic tissue composing the capsule; and (4) the amount of vascularization. Even if these are independent measures and the cause and effect relationship is difficult to establish, they are nevertheless useful to assess the integration of biomaterial with the host and the regeneration of the host tissue with the understanding that the medical device function and structure would be preserved.

3.5.1 Macrophages and Biomaterial Biomimicry

To date, no material or coating has completely blocked the FBR or the formation of FBC surrounding implanted devices. One strategy to alter the surface of an implantable device is to coat the device with something that mimics the body. Ideally, these materials that simulate the cell membrane will be able to inhibit undesirable interactions, while allowing the selective and favorable binding of certain molecules. Most of these materials derive from hydrophobic phospholipids, which are an integral part of the cell membrane. The resulting phospholipid surface likely deceives the host, as it resembles a cell's own membrane and possibly appears less foreign to the host. Generally, phospholipid membranes have exhibited poor stability when grafted onto the surface of medical device implants. To combat this problem, phospholipids have been incorporated with polymers, which helps to increase stability. Furthermore, these materials seem to elicit a more favorable inflammatory response. For example, in vitro protein adsorption of PMB (poly(2-methacryloyloxyethyl phosphorylcho-line(MPC)-co-n-butylmethacrylate(BMA)s) phospholipid polymers, when compared to PET, PHEMA, or Tecoflex® 60, shows decreased protein adsorption, as well as decreased macrophage-like HL-60 cell adhesion which also expresses less IL-1β, a common pro-inflammatory M1 marker. By continuing to develop new and more stable phospholipid polymers that decrease the pro-inflammatory environment, it might be possible to disguise medical devices as the host's own tissue and elicit a decreased FBR response. However, the duration of this concealment will depend on the degradation properties of future phospholipid polymers, as the PMBs in this study only reduce the inflammatory response for less than 7 days [76].

Zwitterionic hydrogels are yet another approach in creating an ultralow biofouling surface. A specific zwitterionic hydrogel, known as poly(carboxybetaine methacrylate) (PCBMA), contains carboxybetaine. Since carboxybetaine is structurally similar to glycine betaine, which is in turn a naturally occurring osmotic regulator of living organisms, this zwitterionic hydrogel may be considered a biomimetic material. PCBMA hydrogels induce less collagen deposition in the FBC and increase blood vessel formation [77]. Additionally, even though both macrophage phenotypes are elicited by PCBMA hydrogels, this material contains fewer pro-inflammatory M1 macrophages and more pro-regeneration M2 macrophages following implantation. More importantly, this biomaterial inhibits the FBR for at least 3 months. The success of this biomaterial might have been due to the fact that the hydrogel was both biomimetic and zwitterionic. Development of future biomimetic hydrogels in combination with specific non-fouling techniques will likely have great potential in the field of biocompatible materials.

3.5.2 Biomaterial Modulation of Macrophage Phenotype

Understanding the control of pro-inflammatory M1 and pro-regeneration M2 macrophages through the modulation of biomaterial microenvironmental cues is a key step in the design of next generation informed biomaterials to enhance positive tissue remodeling, integration, and regeneration. Several studies indicate that macrophage phenotype is modulated by a wide variety of factors including substrate stiffness, surface patterning and roughness, and porosity. Some examples follow.

Studies using Arg-Gly-Asp (RGD)-functionalized polyethylene glycol hydrogels with varying stiffness (130–840 KPa) indicate a differential modulation of a subset of macrophage-derived cytokines. However, there is no definitive modulation of one macrophage phenotype vs. the other as both pro-inflammatory (IL-1β, IL-6, TNF-α) and anti-inflammatory (IL-10) factors correlate with the increased stiffness of the hydrogel. In vivo, increased stiffness appears to correlate with thicker FBC and recruitment of more macrophages perhaps because of mechanical mismatch between the material and the host tissue [78].

Cyclic strain of biomaterial may also be another way to modulate macrophage phenotype. When macrophages are plated on poly-ε-caprolactone bisurea strips induction of the pro-inflammatory M1 phenotype appears to be independent of strain. In contrast, pro-regeneration M2 cells increase with the degree of strain, suggesting the strain-induced elongation of the macrophages possibly skews toward the M2 phenotype as elongation correlates with M2 behavior (see below) [79].

Varying topographies are also active areas of investigation for the modulation of macrophage phenotype. It appears that surface topography can override surface chemistry, as macrophage elongation on the surface is inversely correlated with surface micropatterning. The wider the lines (500 nm to 2 μm) the more elongated the cells become. Further, elongation seems to promote the M2 macrophage phenotype as cells cultured on 20–50 μm lines or on PLLA and chitosan scaffolds with diagonal pore geometry facilitating cell elongation produce more M2-type cytokines. These effects seem to be dependent on a functioning cell contractile apparatus as inhibition of actin-myosin contraction dampens the M2 phenotype. In vivo studies show that micropatterned surfaces reduce cell adhesion and FBGCs when compared to planar substrates [80].

Macrophages prefer to adhere to rough rather than smooth surfaces. As far as macrophage phenotype, sandblasting- and acid etching-roughened titanium surfaces increase TNF-α secretion while decreasing production of the chemoattractants monocyte chemotactic protein 1 (MCP-1) and macrophage inflammatory protein-1a (MIP-1a) in RAW264.7 macrophages [81]. In contrast, nano-roughened titanium surfaces mitigate expression of TNFα, IL-1β, and NO in J744A.1 macrophages [82]. These contrasting findings may be the result of different cell lines used, time points, and variability in the surface roughness.

Electrospun polymer meshes are thought to closely simulate the complex extracellular matrix microenvironment of most native tissues. Both fiber diameter and alignment are found to affect macrophage adhesion and phenotype. Random alignment of a variety of electrospun fibers (PLLA, PCL and PDO) favors M1 cytokines production when compared to aligned fibers. At the same time, larger diameter favor M2 cytokine production. These findings may be also dependent on the relative porosity of the material as larger fibers produce a more porous structure. In vivo, electrospun nanofibers elicit a thinner FBC when compared to flat films, as do thinner fibers when compared to larger ones [83–86].

3.5.3 *Macrophage Spatiotemporal Distribution in the FBR*

While biomaterial-dependent modulation of macrophage phenotype is an important approach to gain control of the FBR healing process, knowledge of the macrophage phenotypes and of their dynamics in the natural progression of FBR may provide us with the necessary tools to guide the FBR toward resolution and regeneration as well as the full integration of the medical device with the host. It is becoming clear that more than one microenvironment can exist or is created within and around the biomaterial and that distinct macrophage phenotypes are associated with these distinct spaces. However, how the overall healing is affected still needs further investigation.

Extracellular matrix hydrogels have been commonly used as coatings to decrease biofouling, and more recently their effect on macrophage phenotype has been recognized. In a study that evaluated the spatial and temporal distribution of pro-inflammatory M1 and pro-regeneration M2 macrophage phenotypes in a rat model, it was shown that a portion of the macrophages present were of the M2 type; however, M1 macrophages were not observed around or within a subcutaneously implanted hexamethylenediiocyanate cross-linked dermal sheep collagen disk [87]. This study only identified macrophage phenotype with one marker, so it is possible that macrophages could have been expressing different markers that were representative of a mixed phenotype. More recently, proper balance between the pro-inflammatory or pro-regeneration macrophages is thought to result in the most favorable FBR outcome. However, the parameters of the balance are still not defined. Accordingly, several FBR studies have begun to look at the proportion of M1 and M2 macrophages in response to an implanted biomaterial.

Extracellular matrix hydrogel coatings have the ability to reduce the FBR to polypropylene meshes. Numerous studies have shown that both pro-inflammatory M1 and pro-regeneration M2 macrophages infiltrate the extracellular matrix hydrogel, yet there is still a higher proportion of M2 macrophages when compared to noncoated materials [72, 88]. Detailed spatiotemporal macrophage polarization was investigated in the response to a knitted heavy-weight polypropylene mesh [89]. The results reveal that hydrogel coatings composed of urinary bladder or dermal extracellular matrix are able to modulate the M1/M2 macrophage ratio by decreasing the overall amount of pro-inflammatory M1 macrophages at the surface of the implant, while at the same time the pro-regeneration M2 macrophages response remains similar compared to noncoated controls. The decrease in the M1/M2 ratio is immediately adjacent to the polypropylene mesh, where the M1 macrophages are usually most concentrated. This is also accompanied by a decrease in FBGCs at biomaterial host interface. This indicates that, indeed, it might be beneficial to have a specific balance of M1 and M2 macrophages during the FBR to promote a favorable outcome. In a study using different amounts of glutaraldehyde to crosslink gelatin hydrogels, the numbers of pro-inflammatory M1 macrophages were not affected by the degree of crosslinking. However, M2 macrophages increase with degree of crosslinking over time. Further, M2 macrophage number highly correlated with FBC thickness. Again, these studies suggest that different ratios of M1 and M2 macrophages may affect the FBR outcome [90]. Similarly, using sphere-templated pHEMA scaffolds of 34-μm pore size, a greater proportion of M1 macrophages are found located near the implant surface when compared to nontemplated scaffolds. The M2 macrophage phenotype is predominantly observed further from the implant surface and mostly detected in the FBC [54]. These studies illustrate the key concept that distinct zones of macrophage phenotypes can exist in microenvironments around an implant. Further, the sphere-templated scaffolds induce a more vascularized and less fibrotic FBC, by purportedly excluding the M2 macrophages from infiltrating the 34 μm pore network. In contrast, larger pores allow the penetration of M2 macrophages, which correlates with decreased angiogenesis and increased collagen deposition (fibrosis) within the pores. These findings suggest that biomaterial architecture can greatly influence the relative distribution of macrophage phenotypes and the density of new blood vessels. Additionally, the presence of M2 macrophages intermixed with M1 macrophages immediately adjacent to the implant surface and in the pores might be promoting the FBR further, when compared to a primarily M1 macrophage response. Future materials that elicit a decreased M1/M2 ratio environment and confine the M2 macrophage phenotype to zones not immediately adjacent to the implant surface might have promise in decreasing the FBR.

The examples above exemplify how biomaterial coating with extracellular matrix components and architecture can influence the relative ratio of M1/M2 macrophages as well as their spatial distribution. Another approach to control the macrophage phenotype composition is the sequential release of chemical signals. The design of scaffolds capable of sequential release of healing factors or molecules has been proposed and tested by several groups. Material can be engineered to achieve a biphasic pattern of cytokine release to activate monocytes and macrophages using monocyte regulatory factors. For example sequential release of MCP-1 and IL-4 from multidomain peptide hydrogel materials promotes the recruitment of monocytes/macrophages, and promotion of a pro-regeneration M2 environment [91]. However, the functional impact of this system on the FBR is not yet clear. Modified decellularized bone scaffolds releasing IFNγ at early times promote M1 macrophages, and a more sustained release of IL-4 promotes the M2 macrophages in vitro. M1 macrophages secrete VEGF and M2 macrophages secrete PDGF-BB that are two of the main factors promoting vessel sprouting and maturation respectively. In vivo this dual release scaffold shows increased vascularization; however, the relative ratio of M1/M2 macrophage phenotypes does not change regardless of the factor(s) released [92]. This may suggest that a balance of M1 and M2 phenotypes leads to healthy and robust vessel formation. The effects on the FBC are still to be determined.

3.5.4 Engineering Macrophages to Modulate Inflammation During the FBR

Engineering macrophages to control the inflammation response during the FBR is yet another approach. For instance, in vitro transduced macrophages with the IFNγ gene and delivered intratracheally to mice restore macrophage-dependent immune function in the lungs of immunodeficient mice [93]. However, this approach with constitutively active pro-inflammatory M1 macrophages probably has limited applications to the FBR as there is no "off switch" and persistent unabated M1 macrophages lead to chronic responses. Additionally, Oxford BioMedica has engineered human macrophages to express cytochrome P450. This cytochrome has the ability to convert a cancer prodrug into its active form in hypoxic conditions. The human-engineered macrophages tested in an avascular spheroid cancer model result in tumor cell death [94]. The success of this system is dependent on the hypoxia-driven expression of cytochrome P450 in macrophages and thus on a hypoxic microenvironment such as those found in tumors. Systematic or even local delivery of these hypoxia responsive engineered macrophages could possibly have large off-target effects, if any ischemic or inflammatory events exist nearby or if inflammation occurs during the cell therapy treatment. Thus, caution must be used.

In order to overcome off-target effects and persistence problems, we have engineered monocytes to polarize into pro-inflammatory M1 macrophages with the addition of a cell permeable drug [95]. These cells work by utilizing the chemical induced of dimerization (CID) system, in which the cytoplasmic portion of the TLR4 receptor is fused to a dimerization domain that has a binding site for the CID drug. When the CID drug is present, the dimerization of TLR4 and subsequent

TLR4-specific pathway activation occurs. The TLR4 pathway activation polarizes the monocyte into M1-like macrophages. These macrophages then express increased levels of pro-inflammatory markers, such as TNF-α, IL-6, and iNOS. Further, they prime endothelial cells for sprouting [95] (Fig. 3.2). The ability to tune these engineered macrophages to the M1 phenotype with either the addition or withdrawal of CID drug, as well as the ability to specifically activate only these engineered cells, independent of the local environment, provides a large added benefit. A potential FBR application could be to incorporate these engineered monocytes in an encompassing scaffold around the implanted device. Activation of engineered macrophages could then be achieved by the addition of the CID drug, which conditionally polarizes the engineered macrophages into a distinct functional phenotype.

One major difference between acute inflammation progression and the FBR process is the presence of both M1 and M2 macrophage markers at the same time [96, 97]. This observation suggests even further that the FBR is a dysregulated inflammation state. With this in mind, it might be beneficial to modulate the FBR to exhibit a distinct M1 macrophage phase followed by a distinct M2 macrophage phase, which would mirror the acute inflammatory response [92]. This could potentially decrease capsule formation by forcing a M1 macrophage environment with little to no M2 macrophages presence, and thus less production of TFGβ, a major product of M2 cells and driver of fibrosis. Withdrawal of the CID drug could then allow M1 macrophage deactivation, while M2 macrophage processes take over and guide resolution.

3.6 Conclusions

Lack of progression of macrophage polarization from a pro-inflammatory M1 to a pro-regeneration M2 phenotype is now believed to impair the healing process. At the same time, extreme M1 or M2 polarization may have deleterious consequences and lead to serious pathologies like rheumatoid arthritis (M1) and cancer (M2). It is also clear that biomaterial implantation alters the normal healing progression in macrophage phenotype (the bad), which results in the macrophage-derived formation of material damaging FBGCs (the ugly). The *bad* and the *ugly* have led to great efforts in understanding the biomaterial-host interaction at the cellular and molecular levels. These new insights, together with critical new knowledge in the field of macrophage biology, are allowing the development of new strategies aimed to promote seamless biomaterial integration as well as proper healing and regeneration of the surrounding tissues (the good). Indeed, an increasing number of studies in the field of biomaterial and regenerative medicine have shown that macrophage phenotype can be modulated by biomaterials or with cell therapy approaches resulting in better biomaterial-host interaction, tissue regeneration, and improved long-term functional outcomes.

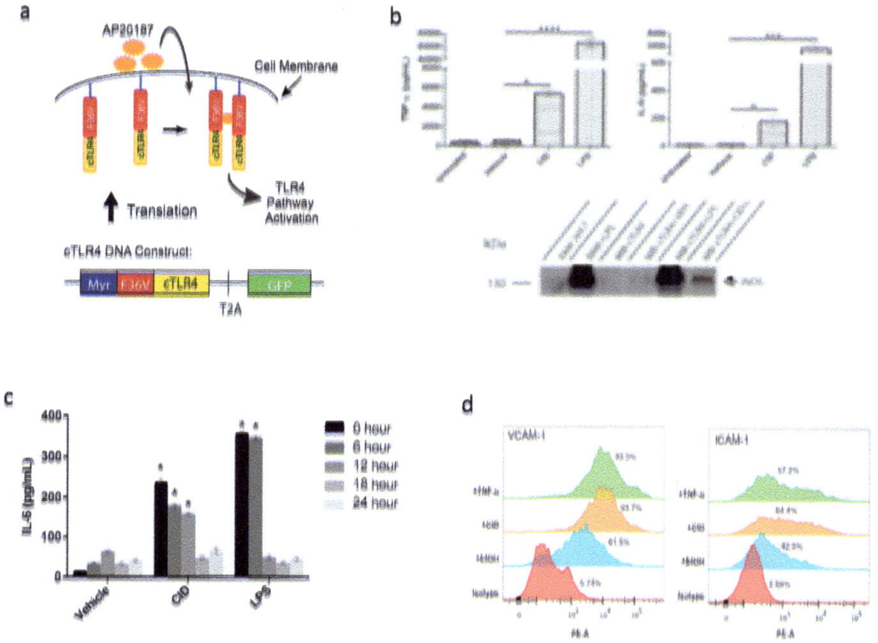

Fig. 3.2 Engineered monocytes polarize to pro-inflammatory M1 macrophages with addition of a cell permeable drug. (**a**) Diagram of the cytoplasmic portion of the TLR4 receptor is fused to a dimerization domain (cTLR4) construct. The cTLR4 construct contains: a myristolation domain (Myr), an engineered dimerization domain (F36V), the cytoplasmic portion of the TLR4 domain (cTLR4), a T2A ribosome skipping sequence, and a GFP tag. Binding of the permeable drug (CID) results in dimerization of the engineered construct and activation of the TLR4 pathway. (**b**) CID-treated cTLR4 engineered monocytes exhibit increased expression of TNFα, IL-6, and iNOS. Bar graphs show the levels of TNFα and IL-6 of CID-treated (50 nM) cTLR4 engineered monocytes compared to untreated, vehicle (100 % EtOH), and LPS-treated cells (100 ng/mL). Cell media was collected following 24 h treatment. Western blot shows intensity of iNOS expression (130 kDa) for CID and LPS-treated cTLR4 engineered monocytes when compared to controls. Cells were lysed following 24 h treatment. (**c**) cTLR4 engineered monocytes return to baseline levels 18 h following CID drug withdrawal. CID drug withdrawal experiment determined by IL-6 expression. Cells were treated with CID drug (50 nM), LPS (100 ng/mL), or vehicle for 24 h and then left untreated for up to 24 h. Media was collected at the indicated timepoints after complete CID drug withdrawal and IL-6 levels were measured at each timepoint to determine activation intensity. (**d**) Medium from CID-treated cTLR4 engineered monocytes upregulates VCAM-1 and ICAM-1 on endothelial cells. Endothelial activation determined by flow cytometry. Conditioned medium from cTLR4 engineered monocytes treated for 6 h with TNFα, CID, or vehicle was transferred to plated bEnd.3 endothelial cells and left to incubate for 12 h. Following the 12 h incubation, bEnd.3 cells were then trypsinized and stained for both VCAM-1 and ICAM-1. Histograms showing intensity of VCAM-1 and ICAM-1 expression on bEnd.3 cells incubated with cTLR4 engineered monocyte conditioned medium treated with TNFα (20 ng/mL), CID (50 nM), or vehicle. Reproduced from Eaton et al. Exp Cell Res *339*, 300–309

References

1. Thomas G, Tacke R, Hedrick CC et al (2015) Nonclassical patrolling monocyte function in the vasculature. Arterioscler Thromb Vasc Biol 35(6):1306–1316
2. McGrath KE, Palis J (2005) Hematopoiesis in the yolk sac: more than meets the eye. Exp Hematol 33(9):1021–1028
3. McGrath KE, Frame JM, Fegan KH et al (2015) Distinct sources of hematopoietic progenitors emerge before HSCs and provide functional blood cells in the mammalian embryo. Cell Rep 11(12):1892–1904
4. McGrath KE, Frame JM, Palis J (2016) Early hematopoiesis and macrophage development. Semin Immunol 27(6):379–387
5. Miyamoto T (2011) Regulators of osteoclast differentiation and cell-cell fusion. Keio J Med 60(4):101–105
6. Watowich SS, Liu YJ (2010) Mechanisms regulating dendritic cell specification and development. Immunol Rev 238(1):76–92
7. Jacome-Galarza CE, Lee SK, Lorenzo JA et al (2013) Identification, characterization, and isolation of a common progenitor for osteoclasts, macrophages, and dendritic cells from murine bone marrow and periphery. J Bone Miner Res 28(5):1203–1213
8. Ginhoux F, Guilliams M (2016) Tissue-resident macrophage ontogeny and homeostasis. Immunity 44(3):439–449
9. Fogg DK, Sibon C, Miled C et al (2006) A clonogenic bone marrow progenitor specific for macrophages and dendritic cells. Science 311(5757):83–87
10. Lee J, Breton G, Oliveira TY et al (2015) Restricted dendritic cell and monocyte progenitors in human cord blood and bone marrow. J Exp Med 212(3):385–399
11. Xiao Y, Zijl S, Wang L et al (2015) Identification of the common origins of osteoclasts, macrophages, and dendritic cells in human hematopoiesis. Stem Cell Reports 4(6):984–994
12. Hoeffel G, Ginhoux F (2015) Ontogeny of tissue-resident macrophages. Front Immunol 6:486
13. Lavin Y, Mortha A, Rahman A et al (2015) Regulation of macrophage development and function in peripheral tissues. Nat Rev Immunol 15(12):731–744
14. Cecchini MG, Hofstetter W, Halasy J et al (1997) Role of CSF-1 in bone and bone marrow development. Mol Reprod Dev 46(1):75–83, discussion 83-84
15. Nakamichi Y, Udagawa N, Takahashi N (2013) IL-34 and CSF-1: similarities and differences. J Bone Miner Metab 31(5):486–495
16. Niida S, Kondo T, Hiratsuka S et al (2005) VEGF receptor 1 signaling is essential for osteoclast development and bone marrow formation in colony-stimulating factor 1-deficient mice. Proc Natl Acad Sci U S A 102(39):14016–14021
17. Dahl R, Walsh JC, Lancki D et al (2003) Regulation of macrophage and neutrophil cell fates by the PU.1:C/EBPalpha ratio and granulocyte colony-stimulating factor. Nat Immunol 4(10):1029–1036
18. Dahl R, Simon MC (2003) The importance of PU.1 concentration in hematopoietic lineage commitment and maturation. Blood Cells Mol Dis 31(2):229–233
19. Burda P, Laslo P, Stopka T (2010) The role of PU.1 and GATA-1 transcription factors during normal and leukemogenic hematopoiesis. Leukemia 24(7):1249–1257
20. Nomura S, Sakuma T, Higashibata Y et al (2001) Molecular cause of the severe functional deficiency in osteoclasts by an arginine deletion in the basic domain of Mi transcription factor. J Bone Miner Metab 19(3):183–187
21. Epelman S, Lavine KJ, Randolph GJ (2014) Origin and functions of tissue macrophages. Immunity 41(1):21–35
22. Corliss BA, Azimi MS, Munson JM et al (2016) Macrophages: an inflammatory link between angiogenesis and lymphangiogenesis. Microcirculation 23(2):95–121
23. Korolnek T, Hamza I (2015) Macrophages and iron trafficking at the birth and death of red cells. Blood 125(19):2893–2897

24. Mehta NG (1976) Recognition of self and nonself, the crucial role of phagocytosis and lysosomal destruction of antigen. Med Hypotheses 2(4):141–146
25. Billadeau DD (2008) PTEN gives neutrophils direction. Nat Immunol 9(7):716–718
26. McDonald B, Pittman K, Menezes GB et al (2010) Intravascular danger signals guide neutrophils to sites of sterile inflammation. Science 330(6002):362–366
27. Lawrence T, Gilroy DW (2007) Chronic inflammation: a failure of resolution? Int J Exp Pathol 88(2):85–94
28. Mantovani A, Sica A, Locati M (2005) Macrophage polarization comes of age. Immunity 23(4):344–346
29. MacLeod AS, Mansbridge JN (2016) The innate immune system in acute and chronic wounds. Adv Wound Care (New Rochelle) 5(2):65–78
30. Gordon S, Martinez FO (2010) Alternative activation of macrophages: mechanism and functions. Immunity 32(5):593–604
31. Ward C, Dransfield I, Chilvers ER et al (1999) Pharmacological manipulation of granulocyte apoptosis: potential therapeutic targets. Trends Pharmacol Sci 20(12):503–509
32. Murray LA, Kramer MS, Hesson DP et al (2010) Serum amyloid P ameliorates radiation-induced oral mucositis and fibrosis. Fibrogenesis Tissue Repair 3:11
33. Mackaness GB (2014) Pillars article: the immunological basis of acquired cellular resistance. J. Exp. Med. 1964. 120: 105–120. J Immunol 193:3222–3237
34. Knight JA (2000) Review: free radicals, antioxidants, and the immune system. Ann Clin Lab Sci 30(2):145–158
35. Olefsky JM, Glass CK (2010) Macrophages, inflammation, and insulin resistance. Annu Rev Physiol 72:219–246
36. Bouhlel MA, Derudas B, Rigamonti E et al (2007) PPARgamma activation primes human monocytes into alternative M2 macrophages with anti-inflammatory properties. Cell Metab 6(2):137–143
37. Mantovani A, Biswas SK, Galdiero MR et al (2013) Macrophage plasticity and polarization in tissue repair and remodelling. J Pathol 229(2):176–185
38. Tugal D, Liao X, Jain MK (2013) Transcriptional control of macrophage polarization. Arterioscler Thromb Vasc Biol 33(6):1135–1144
39. Date D, Das R, Narla G et al (2014) Kruppel-like transcription factor 6 regulates inflammatory macrophage polarization. J Biol Chem 289(15):10318–10329
40. Sica A, Mantovani A (2012) Macrophage plasticity and polarization: in vivo veritas. J Clin Invest 122(3):787–795
41. Guiducci C, Vicari AP, Sangaletti S et al (2005) Redirecting in vivo elicited tumor infiltrating macrophages and dendritic cells towards tumor rejection. Cancer Res 65(8):3437–3446
42. Saccani A, Schioppa T, Porta C et al (2006) p50 nuclear factor-kappaB overexpression in tumor-associated macrophages inhibits M1 inflammatory responses and antitumor resistance. Cancer Res 66(23):11432–11440
43. Boehler RM, Kuo R, Shin S et al (2014) Lentivirus delivery of IL-10 to promote and sustain macrophage polarization towards an anti-inflammatory phenotype. Biotechnol Bioeng 111(6): 1210–1221
44. Anderson JM, Rodriguez A, Chang DT (2008) Foreign body reaction to biomaterials. Semin Immunol 20(2):86–100
45. Luttikhuizen DT, Harmsen MC, Van Luyn MJ (2006) Cellular and molecular dynamics in the foreign body reaction. Tissue Eng 12(7):1955–1970
46. Shen M, Garcia I, Maier RV et al (2004) Effects of adsorbed proteins and surface chemistry on foreign body giant cell formation, tumor necrosis factor alpha release and procoagulant activity of monocytes. J Biomed Mater Res A 70(4):533–541
47. Kourtzelis I, Rafail S, DeAngelis RA et al (2013) Inhibition of biomaterial-induced complement activation attenuates the inflammatory host response to implantation. FASEB J 27(7):2768–2776
48. Tang L (1998) Mechanisms of fibrinogen domains: biomaterial interactions. J Biomater Sci Polym Ed 9(12):1257–1266

49. Jenney CR, Anderson JM (2000) Adsorbed serum proteins responsible for surface dependent human macrophage behavior. J Biomed Mater Res 49(4):435–447
50. Brodbeck WG, Nakayama Y, Matsuda T et al (2002) Biomaterial surface chemistry dictates adherent monocyte/macrophage cytokine expression in vitro. Cytokine 18(6):311–319
51. Patel JD, Krupka T, Anderson JM (2007) iNOS-mediated generation of reactive oxygen and nitrogen species by biomaterial-adherent neutrophils. J Biomed Mater Res A 80(2):381–390
52. Moore LB, Sawyer AJ, Charokopos A et al (2015) Loss of monocyte chemoattractant protein-1 alters macrophage polarization and reduces NFkB activation in the foreign body response. Acta Biomater 11:37–47
53. Moore LB, Kyriakides TR (2015) Molecular characterization of macrophage-biomaterial interactions. Adv Exp Med Biol 865:109–122
54. Sussman EM, Halpin MC, Muster J et al (2014) Porous implants modulate healing and induce shifts in local macrophage polarization in the foreign body reaction. Ann Biomed Eng 42(7):1508–1516
55. McNally AK, Anderson JM (2015) Phenotypic expression in human monocyte-derived interleukin-4-induced foreign body giant cells and macrophages in vitro: dependence on material surface properties. J Biomed Mater Res A 103(4):1380–1390
56. Brodbeck WG, Anderson JM (2009) Giant cell formation and function. Curr Opin Hematol 16(1):53–57
57. Milde R, Ritter J, Tennent GA et al (2015) Multinucleated giant cells are specialized for complement-mediated phagocytosis and large target destruction. Cell Rep 13(9):1937–1948
58. Helming L, Gordon S (2009) Molecular mediators of macrophage fusion. Trends Cell Biol 19(10):514–522
59. McInnes A, Rennick DM (1988) Interleukin 4 induces cultured monocytes/macrophages to form giant multinucleated cells. J Exp Med 167(2):598–611
60. DeFife KM, Jenney CR, McNally AK et al (1997) Interleukin-13 induces human monocyte/macrophage fusion and macrophage mannose receptor expression. J Immunol 158(7):3385–3390
61. Helming L, Winter J, Gordon S (2009) The scavenger receptor CD36 plays a role in cytokine-induced macrophage fusion. J Cell Sci 122(Pt 4):453–459
62. Yagi M, Miyamoto T, Sawatani Y et al (2005) DC-STAMP is essential for cell-cell fusion in osteoclasts and foreign body giant cells. J Exp Med 202(3):345–351
63. Yagi M, Miyamoto T, Toyama Y et al (2006) Role of DC-STAMP in cellular fusion of osteoclasts and macrophage giant cells. J Bone Miner Metab 24(5):355–358
64. Tsai AT, Rice J, Scatena M et al (2005) The role of osteopontin in foreign body giant cell formation. Biomaterials 26(29):5835–5843
65. Sissons JR, Peschon JJ, Schmitz F et al (2012) Cutting edge: microRNA regulation of macrophage fusion into multinucleated giant cells. J Immunol 189(1):23–27
66. Moore LB, Sawyer AJ, Saucier-Sawyer J et al (2016) Nanoparticle delivery of miR-223 to attenuate macrophage fusion. Biomaterials 89:127–135
67. Kyriakides TR, Foster MJ, Keeney GE et al (2004) The CC chemokine ligand, CCL2/MCP1, participates in macrophage fusion and foreign body giant cell formation. Am J Pathol 165(6):2157–2166
68. Ratner BD (2016) A pore way to heal and regenerate: 21st century thinking on biocompatibility. Regen Biomater 3(2):107–110
69. Wynn TA (2004) Fibrotic disease and the T(H)1/T(H)2 paradigm. Nat Rev Immunol 4(8):583–594
70. Wynn TA, Barron L (2010) Macrophages: master regulators of inflammation and fibrosis. Semin Liver Dis 30(3):245–257
71. Wynn TA, Vannella KM (2016) Macrophages in tissue repair, regeneration, and fibrosis. Immunity 44(3):450–462
72. Brown BN, Ratner BD, Goodman SB et al (2012) Macrophage polarization: an opportunity for improved outcomes in biomaterials and regenerative medicine. Biomaterials 33(15):3792–3802

73. Brown BN, Badylak SF (2013) Expanded applications, shifting paradigms and an improved understanding of host-biomaterial interactions. Acta Biomater 9(2):4948–4955
74. Alvarez MM, Liu JC, Trujillo-de Santiago G et al (2016) Delivery strategies to control inflammatory response: modulating M1-M2 polarization in tissue engineering applications. J Control Release
75. Boersema GS, Grotenhuis N, Bayon Y et al (2016) The effect of biomaterials used for tissue regeneration purposes on polarization of macrophages. Biores Open Access 5(1):6–14
76. Sawada S, Sakaki S, Iwasaki Y et al (2003) Suppression of the inflammatory response from adherent cells on phospholipid polymers. J Biomed Mater Res A 64(3):411–416
77. Zhang L, Cao Z, Bai T et al (2013) Zwitterionic hydrogels implanted in mice resist the foreign-body reaction. Nat Biotechnol 31(6):553–556
78. Blakney AK, Swartzlander MD, Bryant SJ (2012) The effects of substrate stiffness on the in vitro activation of macrophages and in vivo host response to poly(ethylene glycol)-based hydrogels. J Biomed Mater Res A 100(6):1375–1386
79. Ballotta V, Driessen-Mol A, Bouten CV et al (2014) Strain-dependent modulation of macrophage polarization within scaffolds. Biomaterials 35(18):4919–4928
80. Almeida CR, Serra T, Oliveira MI et al (2014) Impact of 3-D printed PLA- and chitosan-based scaffolds on human monocyte/macrophage responses: unraveling the effect of 3-D structures on inflammation. Acta Biomater 10(2):613–622
81. Refai AK, Textor M, Brunette DM et al (2004) Effect of titanium surface topography on macrophage activation and secretion of proinflammatory cytokines and chemokines. J Biomed Mater Res A 70(2):194–205
82. Lee CH, Kim YJ, Jang JH et al (2016) Modulating macrophage polarization with divalent cations in nanostructured titanium implant surfaces. Nanotechnology 27(8):085101
83. Sanders JE, Stiles CE, Hayes CL (2000) Tissue response to single-polymer fibers of varying diameters: evaluation of fibrous encapsulation and macrophage density. J Biomed Mater Res 52(1):231–237
84. Garg K, Pullen NA, Oskeritzian CA et al (2013) Macrophage functional polarization (M1/M2) in response to varying fiber and pore dimensions of electrospun scaffolds. Biomaterials 34(18):4439–4451
85. Garg T, Goyal AK (2014) Biomaterial-based scaffolds—current status and future directions. Expert Opin Drug Deliv 11(5):767–789
86. Cao D, Wu YP, Fu ZF et al (2011) Cell adhesive and growth behavior on electrospun nanofibrous scaffolds by designed multifunctional composites. Colloids Surf B Biointerfaces 84(1):26–34
87. van Putten SM, Ploeger DT, Popa ER et al (2013) Macrophage phenotypes in the collagen-induced foreign body reaction in rats. Acta Biomater 9(5):6502–6510
88. Valentin JE, Stewart-Akers AM, Gilbert TW et al (2009) Macrophage participation in the degradation and remodeling of extracellular matrix scaffolds. Tissue Eng Part A 15(7):1687–1694
89. Wolf MT, Dearth CL, Ranallo CA et al (2014) Macrophage polarization in response to ECM coated polypropylene mesh. Biomaterials 35(25):6838–6849
90. Yu T, Wang W, Nassiri S et al (2016) Temporal and spatial distribution of macrophage phenotype markers in the foreign body response to glutaraldehyde-crosslinked gelatin hydrogels. J Biomater Sci Polym Ed 27(8):721–742
91. Kumar D, Mutreja I, Keshvan PC et al (2015) Organically modified silica nanoparticles interaction with macrophage cells: assessment of cell viability on the basis of physicochemical properties. J Pharm Sci 104(11):3943–3951
92. Spiller KL, Nassiri S, Witherel CE et al (2015) Sequential delivery of immunomodulatory cytokines to facilitate the M1-to-M2 transition of macrophages and enhance vascularization of bone scaffolds. Biomaterials 37:194–207
93. Wu M, Hussain S, He YH et al (2001) Genetically engineered macrophages expressing IFN-gamma restore alveolar immune function in scid mice. Proc Natl Acad Sci U S A 98(25):14589–14594

94. Griffiths L, Binley K, Iqball S et al (2000) The macrophage—a novel system to deliver gene therapy to pathological hypoxia. Gene Ther 7(3):255–262
95. Eaton KV, Yang HL, Giachelli CM et al (2015) Engineering macrophages to control the inflammatory response and angiogenesis. Exp Cell Res 339(2):300–309
96. Madden LR, Mortisen DJ, Sussman EM et al (2010) Proangiogenic scaffolds as functional templates for cardiac tissue engineering. Proc Natl Acad Sci U S A 107(34):15211–15216
97. Badylak SF, Valentin JE, Ravindra AK et al (2008) Macrophage phenotype as a determinant of biologic scaffold remodeling. Tissue Eng Part A 14(11):1835–1842

Chapter 4
Understanding Nanoparticle Immunotoxicity to Develop Safe Medical Devices

Marina A. Dobrovolskaia

Abstract Nanotechnology has a potential to transform healthcare by improving the quality of drugs and medical devices. The benefits of using nanomaterials in medical devices include improving device durability, decreasing bacterial adhesion and biofilm formation, and providing slow and controlled release of device-associated drugs. Using nanoparticles as implantable materials and components of medical devices also poses some safety concerns and regulatory challenges. This chapter reviews hematological and immunological toxicities relevant to the nanomaterials used in medical devices. Regulatory challenges, translational considerations, and literature case studies pertinent to the immunological safety of nanotechnology-based devices are also discussed.

Keywords Immunotoxicity • Nanomaterials • Healthcare • Environmental health and safety issues

4.1 Introduction

A wide range of nanomaterials is considered for use in medical devices. Examples of nanoparticles used in medical implants and devices include dendrimers, polymers, liposomes, micelles, nanocapsules, metallic nanoparticles, nanocomposite materials, and nanoemulsions. According to a recent study by Etheridge et al., there were more therapeutics than devices among nanotechnology-based products which received regulatory approval before the year of 2000 [1]. A steep growth in the development of nanotechnology-based devices was noted in the first decade of the twenty-first century while the growth of nanotherapeutics remained steady [1]. Popular areas of research included enhancement of in vivo imaging, in vitro diagnostics, bone substitutes, dental composites, medical dressings and textiles, thermal

M.A. Dobrovolskaia (✉)
Nanotechnology Characterization Laboratory, Cancer Research Technology Program, Leidos Biomedical Research Inc., Frederick National Laboratory for Cancer Research, National Cancer Institute at Frederick, P.O.Box B, Frederick, MD 21702, USA
e-mail: marina@mail.nih.gov

B. Corradetti (ed.), *The Immune Response to Implanted Materials and Devices*,
DOI 10.1007/978-3-319-45433-7_4

ablation mediated cancer treatment, drug delivery, tissue engineering, coatings for implanted and surgical devices [1]. The proportion of nanoparticle-based medicines and medical devices is also different between investigational and approved products. Nanotechnology-based devices dominate investigational products while drugs lead among commercial nanomaterials (Fig. 4.1a). Among investigational nanomaterials liposomes, polymeric nanoparticles, emulsions, and solid nanoparticles prevail other types of nanomaterials, while solid nanoparticles and nanocomposites dominate clinically used nanotechnology-based products (Fig. 4.1b).

Translation of nanotechnology-based devices and implantable materials requires a thorough investigation of their safety profiles. Among other critical parameters are the potential nanoparticle toxicity to the immune system and (in)compatibility with blood. Nanoparticle effects on the immune system may produce different outcomes. Several outcomes including immunostimulatory, immunosuppressive, and immunomodulatory reactions have been described. The consequences of these effects on the immune system to the host can be good or bad, and depend on the intended use of the engineered nanomaterials. For example, desirable immunostimulation can be used in applications requiring the immune activation (e.g., to improve the immunogenicity

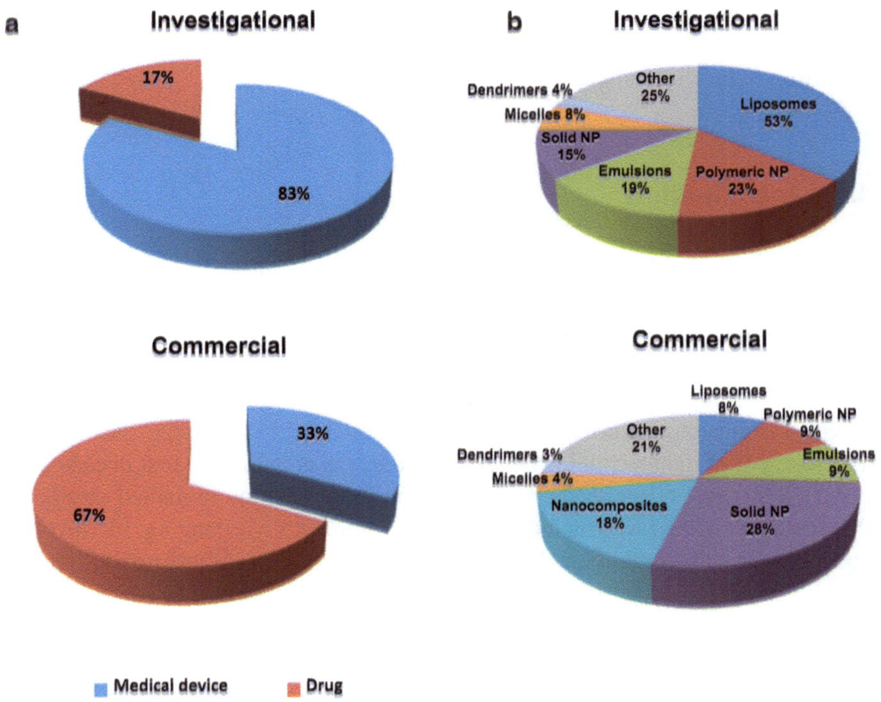

Fig. 4.1 A variety of nanotechnology-based devices and drugs. (**a**) The proportion of nanoparticle-based medicines and medical devices in investigational and commercial stages. (**b**) Types of nanomaterials in nanotechnology-enabled medicines and medical devices. The graphs are built based on the information reviewed in the reference [1]

of vaccines). In contrast, unintended immunostimulation may lead to side effects that include common dose limiting toxicities such as hypersensitivity reactions, anaphylaxis, and thrombosis [2]. Similarly, intended immunosuppression provides therapeutic benefits to the treatment of inflammatory disorders and autoimmune diseases, improves transplant acceptance, and prevents allergic reactions; the unwanted immunosuppression may lower host's response to infected and malignant cells, and affect the normal function of the bone marrow and thymus [2]. Nanoparticle physicochemical properties (size, charge, hydrophobicity, and presence of targeting moieties) determine particle interaction with the components of the immune system and influence the immunotoxicity [3]. Despite the fact that nanoparticles can be immunotoxic, the toxicities described so far are not unique to their nano size. Therefore, immunotoxicity of engineered nanomaterials is assessed using a current portfolio of methods established for low molecular weight drugs, therapeutic nucleic acids, and macromolecules [4]. Review of available data suggests that engineered nanomaterials are intrinsically not more immunotoxic than traditional drugs currently in use. Moreover, there are an increasing number of examples demonstrating a decrease in the immunotoxicity of traditional drugs after reformulation using nanotechnology platforms. For example, fever and hypotension were the adverse reactions limiting clinical use of tumor necrosis factor alpha (TNF-α) [5], while none such toxicities were observed with TNF-α formulated using PEGylated colloidal gold [6]. Similarly, formulation of the anticancer drug paclitaxel (Taxol®) caused severe hypersensitivity reactions in sensitive individuals, while nanoparticle formulation of the same drug (Abraxane®) did not [7]. Despite some success stories, a common translational challenge encountered with nanotechnology-formulated drugs is an alteration in the drug toxicity profile caused by the change in biodistribution due to the nanocarrier. In such a scenario, a nephrotoxicity may be solved in exchange for hepatotoxicity. For example, common toxicities limiting clinical use of doxorubicin are myelosuppression and cardiotoxicity [8]. Reformulation of this drug using PEGylated liposome shifted biodistribution away from bone marrow, and heart muscle thus alleviating toxicity to these organs but resulted in accumulation in the skin, which consequently lead to palmar-plantar erythrodysesthesia [9]. In contrast, reformulation of doxorubicin using cyanoacrylate nanocarrier led to drug accumulation in the kidney with subsequent renal toxicity [10].

When nanoparticles are used to prepare or coat medical devices, a relevant safety concern is a potential particle release from the device and distribution to other organs, distant from the implanted material or device, with subsequent off-target toxicity. For example, nanosized particles released into the air from dental composite materials were suggested to result in pulmonary toxicity due to the inhalation by both patients and dentists [11]. Oxidative stress often mediates pulmonary toxicity of nanomaterials due to the particle uptake by alveolar macrophages [12–14]. Antimicrobial properties, of silver nanoparticles, lead to the growing use of these nanomaterials in the coatings of surgical instruments, in dental composites, wound sutures, bone prostheses, and endotracheal tubes (reviewed in references [15, 16]). One of the safety concerns associated with the use of silver nanoparticles in dental implants is due to a concern that patients may ingest silver nanoparticles released from dental implants or composites and that such ingestion may result in an alteration in the gut microbiota. It has also been suggested that the release of particles from devices and implantable materials contained within the body (e.g., in joints and bones, or on the skin) may lead to toxicity. Furthermore, such toxicity may occur through different mechanisms involving, among others, an ion release from metallic nanoparticles and particle agglomeration (for the detailed review, please see [17]).

Below I will summarize literature data demonstrating the link between nanoparticle physicochemical properties and immunotoxicity, and outline key considerations for preclinical studies aimed at understanding immunotoxicity of engineered nanomaterials. Although many examples used in this book chapter are derived from studies using nanocarriers and nanotechnology-formulated drugs, the knowledge gained from these studies can also be used to understand both the parameters responsible for the nanoparticle immunotoxicity and the framework needed to establish the safety of nanotechnology-based medical devices.

4.2 Hematological Compatibility

If a nanoparticle is intended for clinical applications, the first two principal questions are (a) what is the intended use, and (b) what is the expected route of administration. Understanding of nanoparticle interaction with blood components is an important element in the initial safety assessment when a formulation is either directly injected into the blood or otherwise distributes into systemic circulation after administration through other routes. Hemolysis (damage of red blood cells), thrombogenic, and anticoagulant properties as well as effects on the complement system are commonly used to evaluate the hematological toxicity of nanoparticles.

4.2.1 Hemolysis

This term is referring to a nanoparticle potential to damage erythrocytes. Loss of red blood cells following the hemolysis may lead to anemia, and the release of iron-containing protein hemoglobin into circulation may result in nephrotoxicity [18]. Therefore, optimizing nanoparticle formulation to avoid hemolysis is essential to prevent such complications. Nanoparticle interaction with red blood cells depends on the particle physicochemical properties, which may differ between different classes on engineered nanomaterials. For example, hemolytic activity of silver colloids with identical surface charge was shown to depend on particle surface area, in that particles with smaller size and larger surface to volume ratio were more toxic in hemolysis assay than their larger counterparts [19]. The mechanism of toxicity of anionic silver nanoparticles was attributed to the greater release of silver ions from particles with greater surface area [19]. Hemolysis caused by silica nanoparticles is attributed to the particle surface charge, porosity, and shape. For example, cationic particles were shown to be more damaging to erythrocytes; particles with large aspect ratio were less hemolytic than their spherical counterparts and particles with low aspect ratio [20]. The shape was also named a primary parameter determining hemocompatibility of gold colloids; however, in this case, the spherical shape was found to be more compatible with blood [21]. Nanotechnology was shown to benefit formulation of hydrophobic macromolecules with reduced hemolytic activity by improving solubility and reducing zeta potential through the preparation of polymer-based nanosuspensions [22, 23]. There are a lot of experimental nuances in each study, which may make a comparison between them inaccurate. Use of surfactants during particle synthesis, variations in material purity, stability under physiological conditions, interaction with plasma proteins, use of different methods to study hemolysis are a limited set of variabilities, which may influence the interpretation of the test results. While it is clear that nanoparticle size, surface area, charge, shape, and porosity may influence nanoparticle compatibility with erythrocytes, naming one key factor that would apply to all nanocarriers is impossible without considering other relevant details. The use of both in vitro and in vivo methods to test nanoparticle interaction with erythrocyte is proven to be informative in selecting biocompatible therapeutic nanocarriers.

4.2.2 Thrombogenicity

This toxicity refers to the particle ability to promote blood clotting through interaction with various elements of the blood coagulation system, which includes platelets (also known as thrombocytes), and plasma coagulation factors. Endothelial cells, leukocytes, and certain abnormal cells (e.g., cancer cells) contribute to blood clotting through their ability to express procoagulant activity (PCA) complex on the cell surface which promotes associates of plasma coagulation factors thus promoting blood clotting. Similar to hemolysis, nanoparticle toxicity to platelets, plasma coagulation cascade, and cells contributing to the blood clotting is determined by the nanoparticle physicochemical properties. More detailed overview of nanoparticle pro-coagulant activities is provided elsewhere [24, 25]. Here I review only a few examples of polymeric nanomaterials to demonstrate the link between nanoparticle structure and compatibility with the blood coagulation system. Mechanisms of platelet activation by polystyrene latex nanoparticles were shown to depend on the nanoparticle surface charge in that perturbation of the cellular membrane and up-regulation of adhesion receptors were observed with cationic and anionic latex nanoparticles, respectively [26]. Similarly, amine-terminated nanoparticles bound coagulation factors VII and IX, which resulted in a decrease in thrombin generation. This effect was also size-dependent in that smaller particles were found to be more prominent than for their larger counterparts. Nanoparticles with negative surface charge induced contact activation of the intrinsic coagulation pathway [27]. This property was also dependent on particle size in that it was observed only with large particles. Surface curvature was suggested to contribute to the difference observed between small and large particles. Oslakovic et al. hypothesized that steric hindrance prevented the assembly of the multicomponent protein complex initiating the intrinsic pathway on the surface of small nanoparticles [27]. In contrast to their anionic and neutral counterparts, cationic dendrimers were shown to be cytotoxic in vitro [28]. Cationic dendrimers resulted in time-dependent changes in permeability of the cellular membrane of endothelial cells [29]. Both systemic and oral administration of amine-terminated polyamidoamine (PAMAM) dendrimers led to disseminated intravascular coagulation in mice [30]. This toxicity was dose-dependent and observed when dendrimers were injected at dose levels exceeding their maximum tolerated dose [30]. The mechanism of this toxicity is very complex and besides endothelial cells, involves coagulation factors, leukocytes, and platelets. Induction of both platelet aggregation and leukocyte procoagulant activity by PAMAM dendrimers in vitro is determined by the particle size and zeta potential [31–33]. Only cationic PAMAM dendrimers, but not their anionic and neutral counterparts, induced platelet aggregation [32, 33] and leukocyte procoagulant activity [31] in vitro. Cationic PAMAM dendrimers of higher generations (e.g., larger size) were more toxic to platelets and leukocytes than particles of lower generations (e.g., smaller size) [31–33]. The mechanism of platelet aggregation induced by cationic PAMAM dendrimers also involved disturbance of membrane integrity [32].

Thrombogenic properties of engineered nanomaterials can be desirable and benefit therapy of certain pathologies in which normal coagulation is altered, due to for example blood loss, induced or genetically determined low platelet count, or missing coagulation factors. For example, 170 nm poly(lactic-co-glycolic acid)-poly-L-lysine (PLGA-PLL) nanospheres were engineered for the use as artificial platelets to induce platelet aggregation at the sites of tissue injury. The RGD (Arginine-Glycine-Aspartic Acid) peptide, which is known for its interaction with activated platelets but not with inactive platelets, was used as a targeting moiety on the surface of the particles to effectively induce injury-associated clotting [34]. These nanoparticles had longer shelf-life and stopped bleeding more efficiently than rFVIIa and allogenic platelets [34]. PLGA-nanosheets conjugated with dodecapeptide H12(HHLGGAKQAGDV), a fibrinogen sequence located in γ-chain of fibrinogen, is another example of nanomaterials engineered to promote blood coagulation [35]. Nanoparticle targeting to the injured blood vessel was also achieved by conjugating GPIb targeting P-selectin and von Willebrand factor [36]. Furthermore, nanoparticles can be used to formulate recombinant coagulation factors to improve their stability and decrease the immunogenicity. For example, rFVII was conjugated to human serum albumin (HSA) attached to the surface of maghemite particles without a loss in the protein activity [37]. Factors VIIa and VIII were also formulated using PEGylated liposomes (97:3 ratio of 1-palmitoyl-2-oleoyl phosphatidylcholine (POPC) to 1,2 distearoyl-sn-glycero-3-phosphatidylethanol-amine-N-[methoxy(polyethylene glycol)-2000]). This reformulation into a nanoliposomes helped to achieve longer circulation time and reduce the number of injections needed to maintain therapeutic efficacy of these factors [38–40].

4.2.3 Anticoagulant Activity

In health, the blood clotting and anticoagulation are balanced to prevent blood loss and thrombosis, respectively. Healthy anticoagulation is needed to dissolve the blood clot to avoid thrombosis. However, unbalanced anticoagulation can lead to hemorrhage. The latter is as bad as thrombosis; therefore, understanding anticoagulant properties of nanotechnology carriers is as important as the understanding of their thrombogenicity. Below I will review few examples demonstrating the use of nanoparticles as carriers to anticoagulant drugs. Liposomes, carbon, and polymer nanoparticles were shown to be engineered to benefit therapeutic intervention of pathological conditions in which anticoagulant activity is desirable. Nanoparticles in this type of applications are used to reduce side effects and increase the efficacy of thrombolytic agents (e.g., urokinase, streptokinase, tissue-type plasminogen activator (t-PA)) [41–44]. For example, liposomes can increase metabolic stability of thrombolytic agents in blood [42, 45–47]. Streptokinase formulated into liposomes was more efficient in dissolving blood clots than unformulated enzyme in a rat model of thrombosis [48]. Better efficacy of the enzyme, in this case, was attributed to the extended circulation time as well as to the higher accumulation in the clot area due to the higher ability of the liposome-formulated drug to penetrate clots [48]. Other examples of the beneficial application of nanomaterials to prevent blood clotting are reviewed elsewhere [49].

4.2.4 Complement Activation

Activation of the complement system, represented by a group of approximately 30 plasma proteins, results in a release of the potent immunostimulants called anaphylatoxins (C3a, C4a, and C5a), which are responsible for the life-threatening anaphylaxis reactions. These responses include changes in pulmonary and blood pressure, affect heart rate, result in edema and other symptoms similar to the immediate type allergies but occur independently of IgE. To distinguish this complement-mediated toxicity from allergy a term Complement Activation Related PseudoAllergy (CARPA) was proposed by Dr. Janos Szebeni [50, 51]. CARPA complicates clinical translation and use of particular types of nanomaterials including but not limited to PEGylated liposomes and polymeric nanoparticles. Szebeni et al. reported that human CARPA could be reliably modeled in vivo using pigs and dogs; the studies by his groups also suggested that CARPA reactions in rats can be reproduced only with higher doses of engineered nanomaterials, therefore, advising not to use rats for in vivo assessment of this toxicity [51]. The use of the human serum and plasma as a matrix for the complement activation assays are commonly recognized models for in vitro evaluation of the complement activation by nanoparticles. For example, complement activation by poly(ethylenimine) and poly(ethyleneimine)-graft-poly(ethylene glycol) block copolymers in human serum in vitro was found to correlate with the findings of the in vivo study conducted using Yorkshire pigs [50].

A group led by Dr. Lanza demonstrated that complement activation by perfluorocarbon-based nanoemulsions can also be reliably detected using human serum [52]. In vitro assays utilizing human serum were successfully used to detect complement activation by iron oxide nanoparticles [53], lipid nanocapsules [54], and polymeric nanoparticles [55, 56]. Activation of the complement system by nanomaterials is determined by the particle size and surface properties. Most interestingly, the density, length, terminal groups, and conformation of the nanoparticle-surface bound polymers (e.g., PEG) were reported as the key determinants of the complement activation by the engineered nanomaterials. Case studies and literature examples supporting this statement have recently been reviewed in excessive detail by Drs. Caroline Salvador-Morales and Robert Sim (please refer to the following reference [57]).

4.2.5 Protein "Corona"

This terminology refers to the proteins bound to the particle surface after the particle is exposed to biological matrices such as blood. The consequences of nanoparticle-protein interaction affect both the particle and the protein. For example, protein corona can affect nanoparticle size and charge, and may also promote particle aggregation [58, 59]. Likewise, nanoparticle size, surface properties, and curvature determine the formation of the protein corona and can alter protein conformation, stability, and function [60–62]. Certain plasma proteins (e.g., lipoproteins Apo B and ApoE, C3, C4, C5 components of the complement system, immunoglobulins, and coagulation protein fibrinogen) are commonly found in the protein corona of various nanomaterials [62]. Protein binding to nanoparticles is similar to that described for biomaterials and surfaces. It is also generally accepted that interaction with a surface may change protein conformation and result in activation or inhibition of the protein activity, the appearance of otherwise masked epitopes and other structural and functional changes [59, 63, 64]. Protein corona affects particle circulation time, uptake into cells, and clearance rates [64–70]. However, the fact of finding a protein in association with a nanoparticle surface per se cannot be used to predict the toxicity of nanoparticle [71]. Therefore, estimation of a total protein binding provides useful information about particle biodistribution, while specialized hematotoxicity tests are more reliable to estimate nanoparticle hematotoxicity.

4.3 Effects on the Immune Cell Function

Unintended inhibition of the immune cell function may decrease the host resistance to infections and cancer, lead to thymic suppression and myelosuppression. Therefore, identification of undesirable immunosuppressive properties of engineered nanomaterials is a major component of establishing their safety profile. It is recognized that nanoparticle physicochemical properties determine their interactions with the immune system. Unintended immunosuppression may weaken host resistance to infections and cancer. For example, imaging Resovist® composed of iron oxide nanoparticles, when administered to Balb/c mice in a single i.p. dose 1 h before challenge with model antigen (OVA), lowered the titer of OVA-specific antibodies. Splenocytes from the animal treated with Resovist demonstrated significantly lower production of IFN-γ and IL-4 than splenocytes from control animals [72]. Inhalation of Multi-Walled Carbon Nanotubes (MWCNT) resulted in suppression of antibody production and T-lymphocyte proliferation in response to the SRBC challenge [73]. Since inhaled MWCNT do not enter systemic circulation, it was hypothesized that the observed immunosuppressive effect did not result from the direct interaction between the carbon nanotubes and spleen cells. Through the series of experiments Mitchell LA et al. demonstrated that inhaled MWCNT induced production of TGFβ in alveolar macrophages. Systemically distributed TGFβ induced production of splenic IL-10 and activation of the cyclooxygenase pathway, which altogether resulted in suppression of antibody production [74].

Some nanoparticles may affect the antigen presenting capacity of dendritic cells. For example, poly(vinyl alcohol)-coated superparamagnetic iron oxide nanoparticles (PVA-SPIONs) affected antigen processing without inhibiting its uptake by the dendritic cells [75]. Phagocytic cells are sensitive to nanoparticle toxicity due to their greater ability to engulf the particles. For example, quantum dots (QD) at noncytotoxic concentration were not taken up by hepatocytes but accumulated in macrophages in vitro. Consequently, the functional activity of macrophages was reduced; the mechanism of the reduction was attributed to the alteration of normal cytoskeleton function in macrophages [76]. Toxicity of nanoparticle to hematopoietic progenitors was reported for antimony oxide (Sb2O3) and cobalt nanoparticles [77].

Myelosuppression is a common dose-limiting toxicity of cytotoxic oncology drugs. The main intention of using nanoparticles delivery of cytotoxic drugs is to decrease the toxicity of the latter due to precise targeting, slow release and decrease in dose. However, not all nanoparticles can achieve this goal. One has to keep in mind that nanoparticles per se may be harmless to cells, but may enhance myelosuppressive effects of drugs they carry. For example, doxorubicin conjugated to polyisobutyl (PIBCA) and polyisohexylcyanoacrylate (PIHCA) nanoparticles was significantly more myelosuppressive than the free drug [78]. Moreover, the severity of the myelosuppression was carrier dependent in that it was greater in PIHCA than in PIBCA nanoparticles. This unfortunate effect was due to accumulation/targeting and retention of the conjugated doxorubicin in bone marrow and spleen. Passivation of nanoparticle surface with hydrophilic polymers such as PEG is recognized as a reliable way of increasing "stealthiness" of the particles leading to lower accumulation in mononuclear phagocytic cells. Intuitively, nanoparticles increasing myelosuppression of cytotoxic drugs due to increased phagocytic uptake can be engineered to avoid this type of toxicity.

Primary immune cells may exert different sensitivity to the same type of nanoparticle. Some nanoparticles are toxic to monocytes but do not affect the viability of lymphocytes. For example, the sensitivity of cells to the toxicity of zinc oxide nanoparticles is greater in monocytes than in Natural Killer (NK) cells, while NK cells, in turn, are more sensitive to zinc oxide particles than T and B lymphocytes [79]. This finding is similar to another study, which demonstrated that zinc oxide nanoparticles at the concentrations leading to low toxicity in human peripheral blood mononuclear cells were extremely cytotoxic to monocyte-derived dendritic cells [80]. An elevation of zinc ion concentration inside cells and subsequent mitochondrial dysfunction triggering apoptosis was proposed as the mechanism of cytotoxicity of the zinc oxide nanoparticles [81]. Another metal oxide nanoparticle (e.g., TiO2) was found to be not cytotoxic in vitro, but inhibited function of various immune cells including macrophages, T-, B-, and NK cells in vivo, and resulted in greater susceptibility of animals to melanoma challenge [80, 82].

4.4 Preclinical Characterization

The transition of nanotechnology-based concepts from bench to bedside requires thorough preclinical characterization that includes assessment of nanoparticle physicochemical properties, efficacy, and safety profiles. Common challenges in the preclinical development of nanoparticle-based products include but not limited to the low quantity of the nanomaterials, batch-to-batch inconsistency, issues with sterility, endotoxin contamination, sterilization, and depyrogenation. To assess nanoparticle safety, and efficacy, both in vitro and in vivo methods are used. While in vivo tests are preferred during IND- and IDE-enabling studies, some in vitro methods were found to be useful in supplementing the in vivo data during later stages of the development as well as in providing quicker and relatively cheaper selection of the lead candidate with desirable safety profile during the initial safety assessment. The selection of appropriate methods, as well as the correlation between in vitro and in vivo immunotoxicity tests, has been reviewed in more detail earlier (please consider references [83–86]). Common pitfalls in preclinical development of nanotechnology-based products have also been recently described [87]. Below I will review few of these challenges that are directly relevant to the preclinical evaluation of nanoparticle immunotoxicity.

4.4.1 Sterility and Endotoxin

One-third of nanotechnology-enabled products fail in preclinical stage due to the endotoxin contamination [87]. Endotoxin, which is a component of the cell wall of gram-negative bacteria, is a potent immunostimulant. When present in nanoformulations, it may confound the results of efficacy and toxicity studies, lead to erroneous data and wrong conclusions, and overall complicate the interpretation of preclinical studies. Moreover, some nanoparticles are not immunotoxic but may exaggerate the immunotoxicity of endotoxin. For example, my group has demonstrated that cationic PAMAM dendrimers do not induce PCA in normal leukocytes, while enhance the PCA induced in these cells by endotoxin [31]. We also found that the mechanism of the exaggeration of endotoxin-induced PCA by dendrimers involves inhibition of the negative regulation and among other potential factors includes inhibition of the PI3K [88]. Exaggeration of other endotoxin-mediated inflammatory reactions is not unique to PAMAM dendrimers as other research groups reported these effects with other types of nanomaterials [89–91]. Since endotoxin contamination of nanomaterials is a very common translational problem and cationic moieties on nanoparticle surface are commonly used to perform conjugation of drugs and targeting ligands, exposure of unreacted amines on the particle surface has to be monitored along with the due diligence to keep such nanoparticle-based products essentially endotoxin-free. This care is necessary to avoid undesirable inflammation-mediated toxicities resulting from endotoxin-triggered reactions exaggerated by the presence of cationic nanomaterials. For more details about challenges with endotoxin detection in nanomaterials the reader is referred to several recent publications [86, 87, 92–96]

4.4.2 Sterilization

Sterility is required for all nanotechnology-based drugs and medical devices to avoid complications and health risks due to the microbial contamination. The sterility is achieved through either establishing sterile synthesis method or a terminal sterilization. Various sterilization methods are available and, among others, include the following most common procedures: gamma irradiation, filtration, heat/steam-, ethylene oxide-, and UV-based sterilization. The complex nature of the nanotechnology-based drugs and devices creates an additional challenge in the preclinical development of these products. High temperature associated with autoclaving is not damaging to carbon nanotubes and fullerenes. However, liposomes, emulsions, and other nanoparticles consisting of materials with low melting points and/or containing biologics (recombinant proteins or antibodies) as targeting moieties or active pharmaceutical ingredients are destroyed by autoclaving [97]. Ethylene oxide is damaging to polymeric nanomaterials, while gamma irradiation may affect a particular type of metal colloids [97]. For example, citrate-stabilized colloidal gold nanoparticles can be sterilized by gamma irradiation without affecting physicochemical properties of these particles. In contrast, citrate-stabilized silver colloids lose their spherical morphology and become toxic to platelets after sterilization using gamma irradiation (Fig. 4.2 and reference [98]). Franca et al. conducted a study to evaluate the effect of different sterilization techniques (UV light, autoclaving, ethylene oxide, formaldehyde, and hydrogen peroxide gas plasma treatments) on stability and biocompatibility of gold nanoparticles with different surface functionalities [99]. The main conclusion from this study was that generalization regarding the applicability of sterilization methods to nanomaterials with identical core material is impossible due to the differences in composition and properties of surface

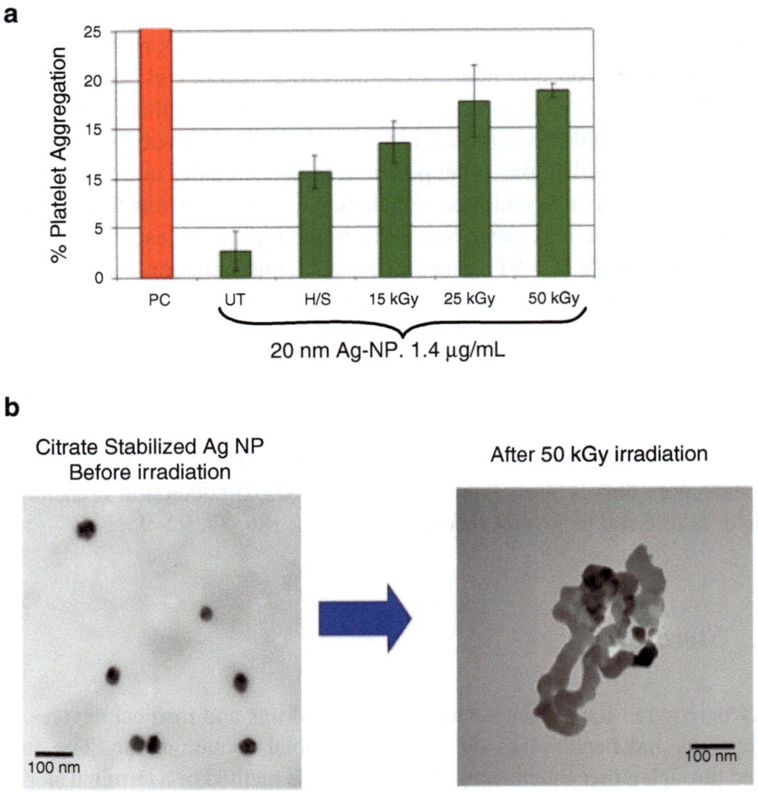

Fig. 4.2 Sterilization stability of nanoparticles. Citrate stabilized silver nanoparticles with a nominal size of 20 nm were untreated or subjected to sterilization by gamma irradiation or autoclave. Particle hemocompatibility (**a**) and size (**b**) were assessed before and after sterilization. The results demonstrated that gamma-irradiation affects particle integrity and biocompatibility

moieties [99]. This study also emphasized that each nanotechnology-based product should be considered on a case-by-case basis. More examples and lessons learned from sterilization of nanomaterials are available elsewhere [97].

4.5 Regulatory Considerations

Several devices containing nanotechnology components have already been approved by the US FDA for medical use. They include silver nanoparticle-containing devices ON-Q SilverSoaker catheter, Actisorb Silver 220 antibacterial binding wound dressing, Modern Medical Antibacterial wound dressing, and ActiCoat Moisture Control Dressing; products Condyloform II NFC and Universal NanoCeramic dental restorative agent that are based on silica nanoparticles; DePuy ASR modular acetabular cup system containing hydroxyapatite, NanoTite dental implant containing titania nanoparticles, and NanoCheck DAT 5M containing gold nanoparticles [100]. US FDA does not treat nanotechnology products differently from other drugs and medical devices [100]. By the 21CFR 314.50 (d), US FDA requires a full description of physical and chemical characteristics and stability for any drug product and device. Particle size, crystalline form, surface area, volume, coatings, identity, strength, quality, purity, potency, bioavailability, manufacturing process and controls, analytical procedures, efficacy, and safety, including immunological toxicity, are among many parameters required to be included in IND and IDE applications of products containing nanotechnology [100]. The detailed list of the FDA regulations can be found here http://www.fda.gov/regulatoryinformation/guidances/default.htm. Below I briefly discuss the regulations pertinent to the immunotoxicity assessment. The requirements and considerations for the planning and designing the immunotoxicity studies are described in the ICH S8 guidance for small molecular drugs and the ICH S6 for biological therapeutics. Nanotechnology-based drugs and devices are considered complex drugs and combination products, respectively. As such, the immunotoxicity guidances applicable to each element of the combination product or complex drug product are also applied to the nanotechnology-based products [101]. ICH S8 recommends hematological and clinical chemistry tests, as well as assessment of immune organs cellularity and weight. Functional tests to assess potential immunosuppression are also recommended to supplement traditional toxicity evaluation. Important information derived from these tests includes dose at which nanoparticle immunotoxicity occurs, the severity of the effect, stability, duration and reversibility of the effect. The ICH S6 guidance supplements considerations described in the ICH S8 by describing the importance and recommendations for conducting immunogenicity studies. For a more detailed view of the FDA recommendations for the evaluation of nanoparticle immunotoxicity, please refer to the recent publication by Dr. James Weaver et al. [101].

4.6 Conclusions

The body of knowledge of nanoparticle properties determining particle toxicity to the immune system has grown exponentially over the past decade. Structure-activity relationship and mechanistic studies uncovered critical parameters underlying nanoparticle reactivity with the immune cells and influencing immune function. The majority of such studies concentrated on nanoparticle-based platforms and nanotechnology-formulated drugs. Nevertheless, this knowledge applies to nanotechnology-based devices as nanoparticles used for device coating may dissociate from the device and distribute to blood and other organs where their toxicity will follow the same trends described for individual types of nanomaterials. Future work will concentrate on broadening our understanding of immunocompatibility of nanotechnology-based devices and among other parameters address the effects on the immune system function, long-term immunotoxicity, and understanding of its mechanisms.

References

1. Etheridge ML, Campbell SA, Erdman AG et al (2013) The big picture on nanomedicine: the state of investigational and approved nanomedicine products. Nanomedicine 9(1):1–14
2. Zolnik BS, Gonzalez-Fernandez A, Sadrieh N et al (2010) Nanoparticles and the immune system. Endocrinology 151(2):458–465
3. Di Gioacchino M, Petrarca C, Lazzarin F et al (2011) Immunotoxicity of nanoparticles. Int J Immunopathol Pharmacol 24(1 Suppl):65S–71S
4. Dobrovolskaia MA, Germolec DR, Weaver JL (2009) Evaluation of nanoparticle immunotoxicity. Nat Nanotechnol 4(7):411–414
5. Steinmetz T, Schaadt M, Gahl R et al (1988) Phase I study of 24-hour continuous intravenous infusion of recombinant human tumor necrosis factor. J Biol Response Mod 7(5):417–423
6. Libutti SK, Paciotti GF, Byrnes AA et al (2010) Phase I and pharmacokinetic studies of CYT-6091, a novel PEGylated colloidal gold-rhTNF nanomedicine. Clin Cancer Res 16(24):6139–6149
7. Gradishar WJ, Tjulandin S, Davidson N et al (2005) Phase III trial of nanoparticle albumin-bound paclitaxel compared with polyethylated castor oil-based paclitaxel in women with breast cancer. J Clin Oncol 23(31):7794–7803
8. Chatterjee K, Zhang J, Honbo N et al (2010) Doxorubicin cardiomyopathy. Cardiology 115(2):155–162
9. Lorusso D, Di Stefano A, Carone V et al (2007) Pegylated liposomal doxorubicin-related palmar-plantar erythrodysesthesia ('hand-foot' syndrome). Ann Oncol 18(7):1159–1164
10. Manil L, Couvreur P, Mahieu P (1995) Acute renal toxicity of doxorubicin (adriamycin)-loaded cyanoacrylate nanoparticles. Pharm Res 12(1):85–87
11. Van Landuyt KL, Hellack B, Van Meerbeek B et al (2014) Nanoparticle release from dental composites. Acta Biomater 10(1):365–374
12. Inoue K, Takano H, Yanagisawa R et al (2006) Effects of airway exposure to nanoparticles on lung inflammation induced by bacterial endotoxin in mice. Environ Health Perspect 114(9):1325–1330
13. Kayat J, Gajbhiye V, Tekade RK et al (2011) Pulmonary toxicity of carbon nanotubes: a systematic report. Nanomedicine 7(1):40–49
14. Li JJ, Muralikrishnan S, Ng CT et al (2010) Nanoparticle-induced pulmonary toxicity. Exp Biol Med (Maywood) 235(9):1025–1033
15. Cheng L, Zhang K, Weir MD et al (2015) Nanotechnology strategies for antibacterial and remineralizing composites and adhesives to tackle dental caries. Nanomedicine (Lond) 10(4):627–641
16. Correa JM, Mori M, Sanches HL et al (2015) Silver nanoparticles in dental biomaterials. Int J Biomater 2015:485275
17. Nunez-Anita RE, Acosta-Torres LS, Vilar-Pineda J et al (2014) Toxicology of antimicrobial nanoparticles for prosthetic devices. Int J Nanomedicine 9:3999–4006
18. Cohn SM (2004) Alternatives to blood in the 21st century. Crit Care 8(Suppl 2):S15–S17
19. Choi J, Reipa V, Hitchins VM et al (2011) Physicochemical characterization and in vitro hemolysis evaluation of silver nanoparticles. Toxicol Sci 123(1):133–143
20. Yu T, Malugin H, Ghandehari A (2011) Impact of silica nanoparticle design on cellular toxicity and hemolytic activity. ACS Nano 5(7):5717–5728
21. Sun YN, Wang CD, Zhang XM et al (2011) Shape dependence of gold nanoparticles on in vivo acute toxicological effects and biodistribution. J Nanosci Nanotechnol 11(2):1210–1216
22. de Ven HV, Van Dyck L, Weyenberg W et al (2010) Nanosuspensions of chemically modified saponins: reduction of hemolytic side effects and potential tool in drug targeting strategy. J Control Release 148(1):e122–e123
23. Shelma CP, Sharma R (2011) Development of lauroyl sulfated chitosan for enhancing hemocompatibility of chitosan. Colloids Surf B Biointerfaces 84(2):561–570
24. Ilinskaya AN, Dobrovolskaia MA (2013) Nanoparticles and the blood coagulation system. Part II: safety concerns. Nanomedicine (Lond) 8(6):969–981
25. Ilinskaya AN, Dobrovolskaia MA (2013) Nanoparticles and the blood coagulation system. Part I: benefits of nanotechnology. Nanomedicine (Lond) 8(5):773–784

26. McGuinnes C, Duffin R, Brown S et al (2011) Surface derivatization state of polystyrene latex nanoparticles determines both their potency and their mechanism of causing human platelet aggregation in vitro. Toxicol Sci 119(2):359–368
27. Oslakovic C, Cedervall T, Linse S et al (2012) Polystyrene nanoparticles affecting blood coagulation. Nanomedicine 8(6):981–986
28. Malik N, Wiwattanapatapee R, Klopsch R et al (2000) Dendrimers: relationship between structure and biocompatibility in vitro, and preliminary studies on the biodistribution of 125I-labelled polyamidoamine dendrimers in vivo. J Control Release 65(1-2):133–148
29. Stasko NA, Johnson CB, Schoenfisch MH et al (2007) Cytotoxicity of polypropylenimine dendrimer conjugates on cultured endothelial cells. Biomacromolecules 8(12):3853–3859
30. Greish K, Thiagarajan G, Herd H et al (2012) Size and surface charge significantly influence the toxicity of silica and dendritic nanoparticles. Nanotoxicology 6(7):713–723
31. Dobrovolskaia MA, Patri AK, Potter TM et al (2012) Dendrimer-induced leukocyte procoagulant activity depends on particle size and surface charge. Nanomedicine (Lond) 7(2):245–256
32. Dobrovolskaia MA, Patri AK, Simak J et al (2012) Nanoparticle size and surface charge determine effects of PAMAM dendrimers on human platelets in vitro. Mol Pharm 9(3):382–93
33. Jones CF, Campbell RA, Franks Z et al (2012) Cationic PAMAM dendrimers disrupt key platelet functions. Mol Pharm 9(6):1599–1611
34. Bertram JP, Williams CA, Robinson R et al (2009) Intravenous hemostat: nanotechnology to halt bleeding. Sci Transl Med 1(11):11ra22
35. Okamura Y, Fukui Y, Kabata K et al (2009) Novel platelet substitutes: disk-shaped biodegradable nanosheets and their enhanced effects on platelet aggregation. Bioconjug Chem 20(10):1958–1965
36. Kona S, Dong JF, Liu Y et al (2011) Biodegradable nanoparticles mimicking platelet binding as a targeted and controlled drug delivery system. Int J Pharm 423(2):516–524
37. Shafir G, Galperin S, Margel A (2009) Synthesis and characterization of recombinant factor VIIa-conjugated magnetic iron oxide nanoparticles for hemophilia treatment. J Biomed Mater Res A 91(4):1056–1064
38. Yatuv R, Robinson M, Dayan-Tarshish I et al (2010) The use of PEGylated liposomes in the development of drug delivery applications for the treatment of hemophilia. Int J Nanomedicine 5:581–591
39. Spira J, Plyushch O, Zozulya N et al (2010) Safety, pharmacokinetics and efficacy of factor VIIa formulated with PEGylated liposomes in haemophilia A patients with inhibitors to factor VIII—an open label, exploratory, cross-over, phase I/II study. Haemophilia 16(6):910–918
40. Spira J, Plyushch O, Andreeva T et al (2012) Safety and efficacy of a long-acting liposomal formulation of plasma-derived factor VIII in haemophilia A patients. Br J Haematol 158(1):149–152
41. Heeremans JL, Prevost R, Bekkers ME et al (1995) Thrombolytic treatment with tissue-type plasminogen activator (t-PA) containing liposomes in rabbits: a comparison with free t-PA. Thromb Haemost 73(3):488–494
42. Leach JK, O'Rear EA, Patterson E et al (2003) Accelerated thrombolysis in a rabbit model of carotid artery thrombosis with liposome-encapsulated and microencapsulated streptokinase. Thromb Haemost 90(1):64–70
43. Leach JK, Patterson E, O'Rear EA (2004) Distributed intraclot thrombolysis: mechanism of accelerated thrombolysis with encapsulated plasminogen activators. J Thromb Haemost 2(9):1548–1555
44. Chung TW, Wang SS, Tsai WJ (2008) Accelerating thrombolysis with chitosan-coated plasminogen activators encapsulated in poly-(lactide-co-glycolide) (PLGA) nanoparticles. Biomaterials 29(2):228–237
45. Nguyen PD, O'Rear EA, Johnson AE et al (1990) Accelerated thrombolysis and reperfusion in a canine model of myocardial infarction by liposomal encapsulation of streptokinase. Circ Res 66(3):875–878
46. Kim IS, Choi HG, Choi HS et al (1998) Prolonged systemic delivery of streptokinase using liposome. Arch Pharm Res 21(3):248–252
47. Perkins WR, Vaughan DE, Plavin SR et al (1997) Streptokinase entrapment in interdigitation-fusion liposomes improves thrombolysis in an experimental rabbit model. Thromb Haemost 77(6):1174–1178

48. Vaidya B, Agrawal GP, Vyas SP (2011) Platelets directed liposomes for the delivery of streptokinase: development and characterization. Eur J Pharm Sci 44(5):589–594
49. Elbayoumi TA, Torchilin VP (2008) Liposomes for targeted delivery of antithrombotic drugs. Expert Opin Drug Deliv 5(11):1185–1198
50. Merkel OM, Urbanics R, Bedocs P et al (2011) In vitro and in vivo complement activation and related anaphylactic effects associated with polyethylenimine and polyethylenimine-graft-poly(ethylene glycol) block copolymers. Biomaterials 32(21):4936–4942
51. Szebeni J, Alving CR, Rosivall L et al (2007) Animal models of complement-mediated hypersensitivity reactions to liposomes and other lipid-based nanoparticles. J Liposome Res 17(2):107–117
52. Pham CT, Mitchell LM, Huang JL et al (2011) Variable antibody-dependent activation of complement by functionalized phospholipid nanoparticle surfaces. J Biol Chem 286(1):123–130
53. Allard-Vannier E, Cohen-Jonathan S, Gautier J et al (2012) Pegylated magnetic nanocarriers for doxorubicin delivery: a quantitative determination of stealthiness in vitro and in vivo. Eur J Pharm Biopharm 81(3):498–505
54. Huynh NT, Morille M, Bejaud J et al (2011) Treatment of 9L gliosarcoma in rats by ferrociphenol-loaded lipid nanocapsules based on a passive targeting strategy via the EPR effect. Pharm Res 28(12):3189–3198
55. Shan X, Yuan Y, Liu C et al (2009) Influence of PEG chain on the complement activation suppression and longevity in vivo prolongation of the PCL biomedical nanoparticles. Biomed Microdevices 11(6):1187–1194
56. Dobrovolskaia MA, McNeil SE (2012) In vitro assays for monitoring nanoparticle interaction with components of the immune system. In: Yarmush ML, Shi D (eds) Frontiers in nanobiomedical research. World Scientific Publishing, Singapore
57. Salvador-Morales C, Sim R (2016) Complement activation. In: Dobrovolskaia MA, McNeil SE (eds) Handbook of immunological properties of engineered nanomaterials. World Scientific Publishing, Singapore, pp 303–331
58. Mayer A, Vadon M, Rinner B et al (2009) The role of nanoparticle size in hemocompatibility. Toxicology 258(2-3):139–147
59. Nel AE, Madler L, Velegol D et al (2009) Understanding biophysicochemical interactions at the nano-bio interface. Nat Mater 8(7):543–557
60. Tenzer S, Docter D, Rosfa S et al (2011) Nanoparticle size is a critical physicochemical determinant of the human blood plasma corona: a comprehensive quantitative proteomic analysis. ACS Nano 5(9):7155–7167
61. Lundqvist M, Stigler J, Elia G et al (2008) Nanoparticle size and surface properties determine the protein corona with possible implications for biological impacts. Proc Natl Acad Sci U S A 105(38):14265–14270
62. Aggarwal P, Hall JB, McLeland CB et al (2009) Nanoparticle interaction with plasma proteins as it relates to particle biodistribution, biocompatibility and therapeutic efficacy. Adv Drug Deliv Rev 61(6):428–437
63. Sperling C, Fischer M, Maitz MF et al (2009) Blood coagulation on biomaterials requires the combination of distinct activation processes. Biomaterials 30(27):4447–4456
64. Behzadi S, Serpooshan V, Sakhtianchi R et al (2014) Protein corona change the drug release profile of nanocarriers: the "overlooked" factor at the nanobio interface. Colloids Surf B Biointerfaces 123:143–149
65. Hadjidemetriou M, Al-Ahmady Z, Mazza M et al (2015) In vivo biomolecule corona around blood-circulating, clinically used and antibody-targeted lipid bilayer nanoscale vesicles. ACS Nano 9(8):8142–8156
66. Kelly PM, Aberg C, Polo E et al (2015) Mapping protein binding sites on the biomolecular corona of nanoparticles. Nat Nanotechnol 10(5):472–479
67. Maiolo D, Bergese P, Mahon E et al (2014) Surfactant titration of nanoparticle-protein corona. Anal Chem 86(24):12055–12063
68. Monopoli MP, Pitek AS, Lynch I et al (2013) Formation and characterization of the nanoparticle-protein corona. Methods Mol Biol 1025:137–155

69. O'Connell DJ, Bombelli FB, Pitek AS et al (2015) Characterization of the bionano interface and mapping extrinsic interactions of the corona of nanomaterials. Nanoscale 7(37):15268–15276
70. Wang F, Yu L, Monopoli MP et al (2013) The biomolecular corona is retained during nanoparticle uptake and protects the cells from the damage induced by cationic nanoparticles until degraded in the lysosomes. Nanomedicine 9(8):1159–1168
71. Dobrovolskaia MA, Neun BW, Man S et al (2014) Protein corona composition does not accurately predict hematocompatibility of colloidal gold nanoparticles. Nanomedicine 10(7):1453–1463
72. Shen CC, Wang CC, Liao MH et al (2011) A single exposure to iron oxide nanoparticles attenuates antigen-specific antibody production and T-cell reactivity in ovalbumin-sensitized BALB/c mice. Int J Nanomedicine 6:1229–1235
73. Mitchell LA, Lauer FT, Burchiel SW et al (2009) Mechanisms for how inhaled multiwalled carbon nanotubes suppress systemic immune function in mice. Nat Nanotechnol 4(7):451–456
74. Mitchell LA, Gao J, Wal RV et al (2007) Pulmonary and systemic immune response to inhaled multiwalled carbon nanotubes. Toxicol Sci 100(1):203–214
75. Blank F, Gerber P, Rothen-Rutishauser B et al (2011) Biomedical nanoparticles modulate specific CD4+ T cell stimulation by inhibition of antigen processing in dendritic cells. Nanotoxicology 5(4):606–621
76. Qu G, Zhang C, Yuan L et al (2012) Quantum dots impair macrophagic morphology and the ability of phagocytosis by inhibiting the Rho-associated kinase signaling. Nanoscale 4(7):2239–2244
77. Bregoli L, Chiarini F, Gambarelli A et al (2009) Toxicity of antimony trioxide nanoparticles on human hematopoietic progenitor cells and comparison to cell lines. Toxicology 262(2):121–129
78. Gibaud S, Andreux JP, Weingarten C et al (1994) Increased bone marrow toxicity of doxorubicin bound to nanoparticles. Eur J Cancer 30A(6):820–826
79. Hanley C, Thurber A, Hanna C et al (2009) The Influences of cell type and ZnO nanoparticle size on immune cell cytotoxicity and cytokine induction. Nanoscale Res Lett 4(12):1409–1420
80. Andersson-Willman B, Gehrmann U, Cansu Z et al (2012) Effects of subtoxic concentrations of TiO2 and ZnO nanoparticles on human lymphocytes, dendritic cells and exosome production. Toxicol Appl Pharmacol 264(1):94–103
81. Kao YY, Chen YC, Cheng TJ et al (2012) Zinc oxide nanoparticles interfere with zinc ion homeostasis to cause cytotoxicity. Toxicol Sci 125(2):462–472
82. Moon EY, Yi GH, Kang JS et al (2011) An increase in mouse tumor growth by an in vivo immunomodulating effect of titanium dioxide nanoparticles. J Immunotoxicol 8(1):56–67
83. Dobrovolskaia MA (2015) Pre-clinical immunotoxicity studies of nanotechnology-formulated drugs: challenges, considerations and strategy. J Control Release 220(Pt B):571–583
84. Dobrovolskaia MA, Shurin M, Shvedova AA (2016) Current understanding of interactions between nanoparticles and the immune system. Toxicol Appl Pharmacol 299:78–89
85. Han X, Corson N, Wade-Mercer P et al (2012) Assessing the relevance of in vitro studies in nanotoxicology by examining correlations between in vitro and in vivo data. Toxicology 297(1-3):1–9
86. Dobrovolskaia MA, McNeil SE (2013) Understanding the correlation between in vitro and in vivo immunotoxicity tests for nanomedicines. J Control Release 172(2):456–466
87. Crist RM, Grossman JH, Patri AK et al (2013) Common pitfalls in nanotechnology: lessons learned from NCI's Nanotechnology Characterization Laboratory. Integr Biol (Camb) 5(1):66–73
88. Ilinskaya AN, Man S, Patri AK et al (2014) Inhibition of phosphoinositol 3 kinase contributes to nanoparticle-mediated exaggeration of endotoxin-induced leukocyte procoagulant activity. Nanomedicine (Lond) 9(9):1311–1326
89. Inoue KTakano H (2011) Aggravating impact of nanoparticles on immune-mediated pulmonary inflammation. ScientificWorldJournal 11:382–390
90. Inoue K, Takano H, Yanagisawa R et al (2009) Size effects of latex nanomaterials on lung inflammation in mice. Toxicol Appl Pharmacol 234(1):68–76
91. Inoue K (2011) Promoting effects of nanoparticles/materials on sensitive lung inflammatory diseases. Environ Health Prev Med 16(3):139–143

92. Dobrovolskaia MA, Neun BW, Clogston JD et al (2010) Ambiguities in applying traditional Limulus amebocyte lysate tests to quantify endotoxin in nanoparticle formulations. Nanomedicine (Lond) 5(4):555–562
93. Dobrovolskaia MA, Neun BW, Clogston JD et al (2014) Choice of method for endotoxin detection depends on nanoformulation. Nanomedicine (Lond) 9(12):1847–1856
94. Li YBoraschi D (2016) Endotoxin contamination: a key element in the interpretation of nanosafety studies. Nanomedicine (Lond) 11(3):269–287
95. Li Y, Italiani P, Casals E et al (2015) Optimising the use of commercial LAL assays for the analysis of endotoxin contamination in metal colloids and metal oxide nanoparticles. Nanotoxicology 9(4):462–473
96. Oostingh GJ, Casals E, Italiani P et al (2011) Problems and challenges in the development and validation of human cell-based assays to determine nanoparticle-induced immunomodulatory effects. Part Fibre Toxicol 8(1):8
97. Subbarao N (2016) Nanoparticle sterility and sterilization of nanomaterials. In: Dobrovolskaia MA, McNeil SE (eds) Handbook of immunological properties of engineered nanomaterials. World Scientific Publishing, Singapore, pp 53–77
98. Zheng J, Clogston JD, Patri AK et al (2011) Sterilization of silver nanoparticles using standard gamma irradiation procedure affects particle integrity and biocompatibility. J Nanomed Nanotechnol 2011(Suppl 5):001
99. Franca A, Pelaz B, Moros M et al (2010) Sterilization matters: consequences of different sterilization techniques on gold nanoparticles. Small 6(1):89–95
100. Patri AK (2016) Translation from finding to product. In: Clinical Nanomedicine 9. Basel, Switzerland, https://www.clinam.org/images/conference.pdf
101. Bancos S, Tyner KM, Weaver JL (2016) Immunotoxicity testing of drug-nanoparticle conjugates: regulatory considerations. In: Dobrovolskaia MA, McNeil SE (eds) Handbook of immunological properties of engineered nanomaterials. World Scientific Publishing, Singapore, pp 207–222

Chapter 5
Host Response to Synthetic Versus Natural Biomaterials

Kishor Sarkar, Yingfei Xue, and Shilpa Sant

Abstract Biomaterials have gained tremendous attention in regenerative medicine and tissue engineering applications due to their ability to enhance functional tissue regeneration. After implantation of biomaterial-based device or drug carrier, it comes in contact with surrounding cells and consequently elicits confined and/or chronic inflammatory responses. The immune responses to biomaterials do not depend only on the method of implantation such as surgery and injection but also depend on source of biomaterials and their physicochemical properties such as molecular weight, chemical composition, mechanical properties and degradation rate. Therefore, it is necessary to thoroughly understand the biological responses to the implanted biomaterials. In this chapter, a brief discussion about different natural and synthetic biomaterials and their inflammatory responses is provided. Different strategies to minimize the immune response have also been discussed.

Keywords Natural biomaterial • Synthetic biomaterial • Biocompatibility • Immune response • Immunomodulation

5.1 Introduction

Over the past several decades, biomaterial-based implants or medical devices have largely changed the scope of modern medicine [1]. With a range of applications in tissue engineering, drug delivery, medical devices and biosensors, biomaterials have

K. Sarkar • Y. Xue
Department of Pharmaceutical Sciences, School of Pharmacy, University of Pittsburgh, Pittsburgh, PA, USA

S. Sant (✉)
Department of Pharmaceutical Sciences, School of Pharmacy, University of Pittsburgh, Pittsburgh, PA, USA

Department of Bioengineering, Swanson School of Engineering, University of Pittsburgh, Pittsburgh, PA, USA

McGowan Institute for Regenerative Medicine, University of Pittsburgh, Pittsburgh, PA, USA
e-mail: shs149@pitt.edu

B. Corradetti (ed.), *The Immune Response to Implanted Materials and Devices*,
DOI 10.1007/978-3-319-45433-7_5

greatly improved the treatment for numerous patients with diseases such as cancer, diabetes, cardiovascular diseases, and tissue loss. In the USA alone, there are at least 13 million biomaterial-based implants in clinical setting annually [2].

Despite the encouraging progress achieved in the recent years in areas such as polymer science, cell biology, immunology and biotechnology, biocompatibility of biomaterials remains a pressing challenge [3]. All implanted biomaterials initiate host responses, which may lead to the limited in vivo functionality and longevity, and thereby adversely affect the intended applications of biomaterial-based implants [4]. On the other hand, host response to biomaterials is necessary and beneficial in removing cellular debris due to injury and deterring the progression of infection [5]. Indeed, preventing host response such as the infiltration of macrophages was shown to lead to more severe tissue damage and decreased tissue regeneration capacity [6]. However, the initial host response to the injury can also lead to secondary tissue damage. Historically, biomaterial-based devices were designed to be inert eliciting minimal host response. However, the definition of biomaterials has further evolved to be “substances to direct, by control of interactions with components of living systems, the course of any therapeutic or diagnostic procedure” [1]. Therefore, biomaterials with the capability to modulate host response have emerged as a new frontier for biomaterial research [7].

The key concepts, mechanisms, and processes of biomaterial-related host responses such as acute/chronic inflammation, foreign body reaction (FBR), innate and acquired immunity are thoroughly discussed in the previous chapters and therefore are not reiterated here. The goal of this chapter is to compare and summarize host responses to natural and synthetic biomaterials and we envision that such discussion will foster the rational design of next generation biomaterials with enhanced biocompatibility. Based on their sources, biomaterials can be generally classified into natural materials such as extracellular matrix (ECM) proteins, polysaccharides, and decellularized tissue matrices; and synthetic materials such as organic and inorganic polymers, metals, nanoparticles, and their derivatives [2, 8]. The scope of this chapter is mainly focused on natural and synthetic polymers. In general, natural and synthetic biomaterials are largely different in inducing host reactions following implantation, which involves a series of events including provisional matrix formation, acute and chronic inflammation, blood–material interactions, and granulation/fibrous capsule development [9].

This chapter begins with the description and summary of major commonly used natural and synthetic biomaterials followed by detailed discussion of the inflammatory and immune responses induced by them. Finally, lessons learnt from previous studies and valuable strategies to improve biomaterials biocompatibility are presented along with the approaches to endow biomaterials with the capability to modulate host responses.

5.2 Natural Biomaterials

Nature has provided us with a range of materials with remarkable functional properties. Naturally derived biomaterials can be classified into proteins, polysaccharides, and decellularized tissue matrices. Protein and polysaccharide-based biomaterials can be processed by two major methods. First, proteins or polysaccharides can be extracted from living organisms by dissolving in solvents or enzymes and reconstituted into fibrils. Alternatively, these biomaterials can be prepared by removing other components in living organisms by solvents or enzymes [10]. Decellularized tissue matrices are obtained by removing cells from native tissues/organs. Multiple decellularization protocols including physical, chemical, and enzymatic approaches have been applied to enable the effective decellularization process. Overall, one of the greatest advantages of using natural materials is that they are derived from materials already present inside the living systems [10]. Natural materials do not usually pose the problems of toxicity potentially faced by a range of synthetic materials. Also, they are bioactive with specific protein binding sites and other biochemical signals that may assist in a range of cellular activities including cell attachment, cell–cell communication, and eventually tissue regeneration [11]. Therefore, the field of biomimicry (e.g., "mimicking nature") is growing rapidly [12]. However, natural materials may pose problems of immunogenicity and possible contamination. Another problem faced by natural biomaterials is their relative instability, which might result in the tendency for mechanical failure or premature decomposition. Indeed, the biodegradation and biomechanical features of natural biomaterials are difficult to control [13].

With regard to host responses to natural biomaterials, although they are considered to have remarkable biocompatibility, natural biomaterials are also immunogenic [14]. The immunogenicity issue is especially serious in the case of xenogeneic materials, where antigens such as DNA, α-Gal epitopes (Galα1-3Galβ1-(3)4GlcNAc-R), and damage-associated molecular pattern (DAMP) molecules are presented [14]. Moreover, the manufacturing methods involved in the decellularized tissue matrices also determine host response to them [14, 15]. Incomplete decellularization process may result in a residual α-Gal epitopes or DNA and may lead to ECM rejection or acute immune responses [16]. Besides, the host response to biomaterials is also device-specific, which means that in addition to the source of the biomaterial, the intended clinical application and the site of implantation may also affect the severity of the host response. In this section, we give an account for different natural polymers and their immunogenic responses. We also discuss how the chemical composition, mechanical properties, surface chemistry and degradation time of different polymers affect the immune responses as summarized in Table 5.1.

5.2.1 Collagen

Collagen is the most abundant type of protein found in connective tissues [17]. So far, at least 29 subtypes of collagen have been identified [18]. All of them have the common triple helical structure with repeated $[\text{Gly-X-Y}]_n$ sequence, where X and Y are frequently proline and hydroxyproline, respectively [19, 20]. Among these different collagens, type I collagen has been most widely studied. The triple helical structure of collagen and its fibers are often packed into highly organized fibrillar structure, which provide tensile strength and structural integrity to various types of tissues and organs [21]. Collagen is relatively stiffer than other elastic proteins such

Table 5.1 Natural versus synthetic materials: chemical composition, mechanical properties, surface chemistry, degradation time and immune responses of different polymers

Biomaterials	Young's modulus (GPa)	Tensile strength (MPa)	Degradation time	Predominant host response	Cytokine production
Natural					
Collagen	Uncrosslinked: 0.046–1.8 Crosslinked: 0.383–0.766 [21]	Uncrosslinked: 0.91–7.2 Crosslinked: 46.8–68.8 [21]	>1 month [173]	Pro- and anti-inflammatory	IFN-γ, IL-13 [41]
Gelatin	3 [174]	20 [175]	Based on their water content [176]	Pro-inflammatory	TNF-α, IL-12, IL-6 [177]
Chitosan	0.007 [178]	2.43 [178]	<1 month [179]	Pro- and anti-inflammatory	IFN-γ, IL-2, TNFα, IL-10 [57, 58]
Hyaluronan	$(0.07–0.09) \times 10^{-3}$ [180]	0.011–0.013 [180]	Around 1 week [181]	Pro-inflammatory	TNF-α, IL-1β, IL-6 [74]
Heparin	NA	NA	NA	Anti-inflammatory	IL-10 [80]
Alginate	$(0.01–0.05) \times 10^{-3}$ [182]	0.005–0.04 [182]	Controlled by molecular weight [183]	Pro-inflammatory	TNF-α, GM-CSF, IL-12, IL-6, IL-1 [87]
Silk	10–22.6[184]	300–1100 [184]	10–24 weeks [185]	Pro-inflammatory	IFN-γ, IL-2, TNF-α, IL-1β[100, 102]
Synthetic					
PGA	6–7 [186]	60–99.7 [186]	6–12 months [187]	Pro-inflammatory	IL-1β, IL-6, GM-CSF, TNF-α[188]
PLA	0.35–3.5 [186]	21–60 [186]	12–24 months [189]	Pro- and anti-inflammatory	IL-6, IL-12/23, IL-10 [114]
PLGA	1.0–4.34 [186]	41.4–55.2 [186]	<2 months [190]	Pro- and anti-inflammatory	TNF-α, IL-6, TGF-β1 [191, 192]
PCL	0.21–0.44 [186]	20.7–42 [186]	>24 months [193]	Pro-inflammatory	TNF-α, IL-1β IL-6 [194, 195]
PTFE	0.39–2.25 [196]	10–45 [196]	Nondegradable	Pro-inflammatory	TNF-α, IL-1β IL-6 [133, 197]

as elastin, but it is an elastic material with a high resilience of nearly 90 %, and is capable of reversible deformation [22]. Biologically, collagen serves as a natural substrate for cellular activities, which makes collagen an excellent material for tissue engineering applications. Currently, there have been several FDA (Food and Drug Administration) approved collagen-containing products that have entered into the market for treating exuding diabetic ulcers, spinal dural repair, and regeneration of bone graft substitute [23, 24]. Moreover, collagen has also been explored in cardiovascular, musculoskeletal, and neuronal tissue engineering [25–28].

In general, the epitopes presented in the telopeptide regions of tropocollagen molecule are responsible for immune response [29, 30]. The immunogenic response of collagen depends upon the helical part conformation as well as the amino acid sequence of the polymerized collagen fibril [31–33]. Collagen is one of the primary initiators of the coagulation cascade and often used for attracting fibroblasts in vivo during wound healing [34]. The high thrombogenicity of collagen has led to its application as hemostatic agent. There are several collagen-based products that have already entered the market or undergoing clinical trials for surgical sealants or hemostat application [35]. The in vivo response of collagen was studied by implanting collagen sponge in rats for up to 8 weeks [36]. Scar tissue was developed within 1 week after implantation with signs of slight inflammation. Subsequently, fibrous tissue was observed at two weeks after implantation. At the same time, the collagen sponge was found to be completely degraded. Four weeks after implantation, previously observed fibrous layer had thickened to form wave-like scar tissue. The same scar tissue further matured in the following two weeks. Eventually, this wave-like scar tissue then began to be resolved after 8 weeks [36].

The mechanical property of collagen significantly decreases during extraction, scaffold fabrication, and sterilization steps [37–39]. Therefore, extra chemical cross-linking is necessary to regain the mechanical property and stability of collagen for tissue engineering application. The incorporation of such external crosslinking agents may impart cytotoxicity and host immune response to collagen [40–42]. Ye et al. [41] investigated the inflammatory response of two differently cross-linked dermal sheep collagen disks (hexamethylenediisocyanate, HDSC or glutaraldehyde cross-linked collagen, GDSC) in mice. It was observed that GDSC showed higher neutrophil infiltration at day 2 and 21 with release of high levels of interferon-gamma (IFN-γ), a cytokine related to pro-inflammatory response whereas HDSC showed little neutrophil infiltration at day 2. It was also found that GDSC completely degraded after 28 days but HDSC remained intact. HDSC increased the level of interleukin-13 (IL-13), anti-inflammatory cytokine. Therefore, it should be noted that the inflammatory response depends not only on the biomaterial type but also on their compositions.

5.2.2 *Gelatin*

Gelatin is a mixture of proteins produced from hydrolysis of collagen obtained from the connective tissues [43]. Structurally, gelatin molecules contain repeating sequences of glycine–proline/hydroxyproline–proline/hydroxyproline triplets, which then form the triple helical structure of gelatin [8]. It has good ability to form gels because the helical regions in the gelatin protein chains are able to immobilize water [17]. Gelatin possesses better biocompatibility than its precursor collagen with lower risk of host rejection or infection [2, 44]. Therefore, gelatin has been frequently used in biomedical application as wound dressing, adhesive, or absorbent pad for surgical use as well as tissue engineering scaffolds [8].

5.2.3 Chitosan

Chitosan is a cationic polysaccharide composed of D-glucosamine and *N*-acetyl-D-glucosamine repeating units. It is obtained by alkaline hydrolysis of chitin which is the second most abundant natural biopolymer derived from exoskeletons of shrimps, fungal cell wall, and insects [45–48].

In 1970, chitosan was discovered to facilitate the wound healing process and after that, it has been broadly used in biomedical applications from sutures and wound dressing material to drug/gene delivery and tissue engineering [47, 49–53]. Vande Vord et al. [54] implanted porous tubular chitosan scaffold in mice intraperitoneally. The dramatic infiltration of neutrophils was observed at the implant site after 1 week indicating chemotactic effect of chitosan on immune cells. Other groups have also reported similar chemotactic effect of chitosan to neutrophils [55].

The percent of cationic amine groups in chitosan varies with the degree of deacetylation (DDA) of chitin during alkaline hydrolysis and as a result, the immune response of chitosan depends on its DDA, molecular weight, ionic charge and solubility. Chitosan exhibited hemostatic effect and complement activation [56]. Previous reports suggest that chitosan has dual immune response, i.e., it shows both pro- and anti-inflammatory responses. Low molecular weight chitosan (3 kDa) was found to have more pro-inflammatory response through stimulation of tumor necrosis factor alpha (TNF-α), IL-6, and IFN-γ secretion compared to that of high molecular weight (50 kDa) chitosan [57]. In another study, Oliveira et al. [58] reported downregulation of TNF-α and upregulation of anti-inflammatory cytokine levels (IL-10 and tumor growth factor-beta1, TGF-β1) in macrophage cells with high molecular weight chitosan. However, the same chitosan showed opposite effects in dendritic cells, i.e., increased secretion of pro-inflammatory cytokines (TNF-α and IL-1β) and decreased IL-10 secretion, suggesting that the immune response can be highly dependent on the cell types.

Due to presence of strong hydrogen bonding in chitosan, it is not soluble in water at physiological pH. It is only soluble in acidic environment as its pKa value is 6.5 [59, 60]. Therefore, solubility of chitosan is another important parameter for macrophage activation. Chen et al. [61] prepared water soluble chitosan through incorporation of hydroxypropyl group in its structure and studied its effect on the macrophages. Water-soluble chitosan decreased the production of pro-inflammatory cytokines IL-6 and TNF-α when monocyte-derived macrophages were stimulated with dust mite allergen *Dermatophagoides farina*. In another study, Bajaj et al.

reported that zwitterionic chitosan derivative showed increased solubility in wide pH range [62]. They did not observe any abnormal change in cytokine levels of unstimulated macrophages in presence of unmodified chitosan and zwitterionic chitosan. However, the cytokine levels (TNF-α and IL-6) were significantly decreased by zwitterionic chitosan compared to unmodified chitosan in macrophages stimulated by lipopolysaccharide (LPS) [62].

5.2.4 *Hyaluronan*

Hyaluronan, also called hyaluronic acid (HA), is a linear polysaccharide consisting of repeating units of D-glucuronic acid and *N*-acetyl glucosamine. It is found mainly in ECM of connective and epithelial tissues. Due to its anionic nature and high structural homology across species, it is almost nontoxic, non-antigenic, and non-immunogenic [63, 64]. FDA has approved HA for various eye surgeries including retinal detachment, corneal transplantation and cataract removal [65]. Apart from these, HA has also been used as lubricant gels for various joint disorders, lip fillers, wound healing, drug/protein delivery, and tissue engineering applications [66–70].

Due to protein binding capability of HA, it shows inflammatory responses through binding with cell surface receptors, particularly CD44 and Toll-like receptors (TLR) 2 and 4 of inflammatory cells, although the extent of inflammation depends on molecular weight of HA [71–73]. It is reported that low molecular weight HA shows pro-inflammatory response through upregulation of TNF-α and IL-12β whereas anti-inflammatory response was obtained by high molecular weight HA which increased IL-10 levels [74]. Kajahn et al. studied the effect of sulfate functionalization and degree of substitution of HA on macrophage activation. Sulfated HA-derivative with higher degree of substitution showed anti-inflammatory effect compared to non-functionalized HA and low substituted HA-derivative [75].

5.2.5 *Heparin*

Heparin is a naturally occurring linear glycosaminoglycan that consists of repeating units of D-glucuronic and D-glucosamine (GlcN) linked with 1, 4 linkage. Heparin possesses highest negative charge among the known biomolecules due to presence of high contents of sulfonic and carboxylic acid groups in its chemical structure [76]. It is generally obtained from porcine mucosal tissues having molecular weights ranged from 5 to 40 kDa.

Heparin is mainly used as an anticoagulant of blood. It is also used for pain relief, anti-inflammatory, anticancer, angiogenesis regulation, and inhibitor of complement activation [77]. Heparin and its derivatives mainly show anti-inflammatory response [78] although it is suggested that it may show either pro- or anti-inflammatory activity. An anti-inflammatory response (inhibition of TNF-α) was observed for low molecular weight heparin in a porcine sepsis model [79].

In another study, heparin showed dose-dependent anti-inflammatory response (reducing the pro-inflammatory cytokines TNF-α, IL-6, IL-8, and IL-1β) in LPS-stimulated THP-1 cells and primary monocytes. However, in the presence of LPS binding protein, heparin showed pro-inflammatory effect [80].

5.2.6 Alginate

Alginate, obtained from brown seaweed, is a naturally occurring anionic polymer consisting of mannuronic acid and guluronic acid units in an irregular block-wise pattern. Due to its low toxicity, easy accessibility and good gelation property in presence of divalent cations such as Ca^{2+}, alginate has been extensively studied and used for many biomedical applications such as drug/protein delivery, tissue engineering, cell/micro-organism immobilization as well as food applications [48, 81–86].

Sodium alginate showed pro-inflammatory response in macrophage cells (RAW264.7) through nuclear factor-kappaB (NF-kB) pathway and produced IL-1β, IL-6, IL-12, and TNF-α in time and dose-dependent manner [87]. Thomas et al. [88] studied the inflammatory response of four different commercialized alginate dressings, Kaltostat® (Convatec), Tegagen HG® (3M Healthcare), Comfeel: Seasorb filler® (Coloplast), and Sorbsan® (Braun). Among all the dressings, Kaltostat® showed more pro-inflammatory response through increasing TNF-α cytokine.

The repeating units, mannuronic acid (M) and guluronic acid (G) in alginate exists in irregular block pattern with varying proportions of MM, GG, and MG blocks [89]. Iwamoto et al. [90] synthesized different alginate oligomers (saturated and unsaturated) consisting only M or G and mixed MG repeating units and studied the effect of alginate structure on inflammatory response in macrophages. The unsaturated alginate oligomers exhibited pro-inflammatory response (increased TNF-α level) in RAW264.7 cells while saturated oligomers produced low TNF-α level. Among the unsaturated oligomers, G8 (eight repeating units of G) and M7 (seven repeating units of M) showed potent pro-inflammatory response inducing secretion of TNF-α along with IL-1α, IL-1β, and IL-6.

5.2.7 Silk

Silk is a unique class of structural proteins obtained from silk producing glands of arthropods such as spiders, silkworms, scorpions, mites, and bees. Silk possesses bulky repetitive modular hydrophobic domains interrupted by small hydrophilic groups and having large molecular weight 200–350 kDa or more. The biomedical use of silk (silkworm silk) began with sutures in wound treatment [91]. Due to its exceptional biocompatibility, low immunogenicity, antibacterial activity, and controllable biodegradability, it has been widely used in biomedical field [92–96]. Silk fibers and films have been widely used for tissue engineering scaffold applications due to its high mechanical loads or tensile forces and slow degradation rate [97].

Native silk is composed of a core structural protein, fibroin which is surrounded by a glue-like protein, sericin [98]. It is thought that sericin is responsible for antigenicity of silk [99]. Panilaitis et al. [100] observed that whole silk fiber did not show any inflammatory response in RAW264.7 cells at short as well as long time periods. Interestingly, when macrophages were exposed to sericin recoated-fibers in presence of LPS, sericin synergistically released TNF. Meinel et al. [101] also observed similar inflammatory response with silk fiber. As the sericin of native silk is responsible for inflammatory response, when sericin was removed from the native silk fiber, the sericin-free silk fiber became non-inflammatory in vivo compared to native silk fiber [102]. However, the mechanical property of native silk fiber significantly decreased after removal of sericin because it is the binding component of silk. In some reports, silk-based scaffold showed some inflammatory responses and this may be caused due to remnant solvent in the scaffold during pre- or post-processing of the scaffold [103, 104].

5.2.8 Decellularized Tissue Matrices

Decellularized tissue matrices represent lipid-free, decellularized protein-based derivatives and purified protein extracts of previously living tissues or organs [15]. ECM plays an important role in the mechanical support, signal transduction, and nutrients/waste transportation. Decellularization is a multistep process to remove all cell components (which are the major antigens) from tissue/organ leaving the ECM intact. As a completely natural material, the ECM has been proposed to be immune-privileged and evade from a series of host reactions to foreign bodies [105]. However, host privileges such as the minimal FBR and improved implanted material performance has not been unequivocally demonstrated [15]. The host response to decellularized ECM-derived biologic materials involves both the innate and acquired immune response. A recent study examined the level of host response to five commercially available decellularized ECM-derived materials, including GraftJacket™ (human dermis), Restore™ (porcine small intestine submucosa), CuffPatch™ (porcine small intestine submucosa), TissueMend™ (fetal bovine skin), and Permacol™ (porcine dermis) [106]. It was shown that these five devices had large differences in terms of the acute and chronic host response and in their downstream tissue remodeling outcomes, respectively. The CuffPatch™ showed accumulation of dense collagenous tissue and a persistent FBR. The host response to TissueMend™ and Permacol™ showed low level of chronic inflammation and fibrous encapsulation. GraftJacket™, CuffPatch™, and Permacol™ induced the presence of multinucleated giant cells at implantation site, indicating the elevated FBR. This study showed that decellularized ECM-derived biologic scaffolds differ profoundly in inducing host response. Therefore, a more detailed investigation of the effect of various ECM constituents on the host immune response and tissue remodeling is needed.

5.3 Synthetic Biomaterials

In contrast to the natural biomaterials, synthetic biomaterials are easy and inexpensive to produce, have high batch-to-batch uniformity, and demonstrate more predictable and controllable physicochemical, mechanical, and degradation properties. Besides, synthetic materials also possess excellent processing characteristics, which can ensure their off-the-shelf availability. However, they also suffer from problems such as their "foreignness" to cells, eliciting inflammatory reactions, and their noncompliance or inability to integrate with host tissues [107]. The most popular synthetic biomaterials include polyesters such as polyglycolide (PGA), polylactide (PLA), polycaprolactone (PCL), polyurethane (PU), and polyhydroxybutyrate.

5.3.1 Polyglycolide or Polyglycolic Acid (PGA)

PGA is biodegradable, thermoplastic crystalline polyester with linear aliphatic structure. It is normally prepared by polycondensation or ring-opening polymerization using glycolic acid. Early successful study on PGA-based suture system encouraged the development of a wide range of biodegradable polymers as implants for different medical applications such as sutures and bone internal fixation device [35]. PGA has a fast degradation rate with acidic degradation products, which are thought to be responsible for the inflammatory reaction induced by PGA [108]. Meanwhile, PGA did not induce lymphocyte DNA synthesis. Therefore, PGA is immunologically inert. However, PGA induced major histocompatibility complex locus II antigen and IL-2R activation, showcasing its inflammatory response [109]. It has been shown that PGA initiates significant host reaction upon implantation in vivo. When synthetic PGA scaffolds seeded with somatic lung progenitor cells from mammalian lung tissue were implanted in immunocompetent mice, a serious cascades of FBR were observed that altered the integrity of the developing lung tissue [110]. However, there is no consensus on the immune effect of PGA to date. For example, tubular urethra made from PGA seeded with autologous muscle cells has been reported to survive for 6 years post-implantation in patients [111].

The immune response to PGA mainly occurs due to the degradation products through hydrolysis or enzymatic degradation [108]. A local inflammatory response has been reported after implantation of PGA-based sutures or orthopedic pins. Ceonzo et al. [112] studied the molecular mechanism of inflammation by PGA in vitro and in vivo. Both PGA and glycolic acid solution (degradation product of PGA) were injected intraperitoneally in genetically engineered mice and it was observed that glycolic acid was responsible for local inflammatory response.

5.3.2 Polylactic Acid (PLA)

PLA is biodegradable thermoplastic crystalline polyester with aliphatic chain. Unlike PGA, PLA has slow degradation rate and good strength and stiffness, which is suitable for load-bearing applications. Early study on PLA stent implanted in humans indicated its safe profile without inducing thrombosis and late stenosis for up to 6 months [113]. However, further study on PLA reported that PLA may induce inflammatory response when implanted in the body due to their acidic degradation products [114]. Tubular PLA constructs implanted beneath the skin of mice resulted in longer inflammatory reactions indicated by presence of epithelioid and giant cells [115]. The effect of phagocytosed PLA particle on macrophages was investigated in vivo. It was shown that upon phagocytosis of PLA particle, macrophage cell damage, cell death, and cell lysis were observed [116]. Like PGA, these host reactions have strongly limited their clinical applications.

5.3.3 Polycaprolactone (PCL)

PCL is also a biodegradable polyester. PCL has been approved by FDA for medical applications such as drug delivery devices and sutures. It has also been widely used as a material of choice for tissue engineered scaffold for a variety of tissues due to their elastic mechanical properties and slow degradation rate [117–122]. When PCL was implanted in the nervous system, microglia and astrocytes were found to be activated for up to 28 days post-implantation. However, 60 days post-implantation, no scar or FBR was observed around the scaffold [123].

5.3.4 Polyurethane (PU)

PUs share the common polymer backbone structure, which includes an aliphatic or aromatic units coming from the isocyanate monomers and a more complex moiety derived from polyether or polyester monomers. PU has been extensively investigated as a material of choice for long term cardiovascular medical devices, such as cardiac pacemakers and vascular grafts due to their moderate blood compatibility and mechanical properties [124]. However, they have been shown to elicit increase in the release of chemokines, cytokines, and growth factors in the in vivo models [125]. Subcutaneous implantation of lysine diisocyanate-based PUs in rats revealed that it did not aggravate capsule formation, accumulation of macrophages, or tissue necrosis [126].

5.3.5 *Polytetrafluoroethylene (PTFE)*

PTFE is another class of synthetic polymers consisting of tetrafluoroethylene repeating units in its chemical structure and commonly known as Teflon. Due to its inertness (insoluble in all common solvents), high thermal stability, and non-biodegradability, it has been used extensively in various commercial, industrial, and biomedical applications including large blood vessel repair material [127]. Apart from this, PTFE has also been used as a graft material such as in superficial femoral occlusion and left ventricular assist device. PTFE has been found to elicit mild to moderate inflammatory response in vivo. After implantation of expanded PTFE (ePTFE) in unilateral aorto-femoral bypass of dog, chronic inflammatory response was observed along with the presence of macrophages, myofibroblasts and deposition of complement C3 after 6 months of implantation [128].

5.4 Important Biomaterial Characteristics in the Host Response

Today, implanted systems are still facing the problem of host responses such as adverse blood–material interaction, inflammation, and immune reaction [129]. Minimizing the immune response to biomaterials may be achieved by the choice of materials that are intrinsically immune-inert [5]. Besides, it has been recognized that host response to polymers are closely associated with the physicochemical properties of material, which control the type, amount, conformation, and duration of proteins that could be adsorbed onto the polymer surface. Polymer chemistry can be actively utilized to widely tune the functional aspects of biomaterial matrices such as hydrophilicity, surface pore size, degradation rate, and degradation products. Tuning these physicochemical properties enable the alteration of protein adsorption, which consequently mediates the interactions with immune cells and their activation [5]. Specifically, the hydrophobicity of materials promotes protein adsorption and enhances monocyte adhesion because water on the surface of materials can be easily replaced by a hydrophobic surface of proteins [130]. On the contrary, hydrophilic polymer surfaces will easily allow water attachment and is not favorable for protein adsorption. With the reduced protein adsorption, material is shown to have decreased monocyte/macrophage adhesion and foreign body giant cell (FBGC) formation in vitro [131]. Hydrophilicity is not the only parameter that decides the extent of the host response to polymeric materials. As an example, it has been shown that hydrophilic but charged polymethacrylate can induce complement activation due to the electrostatic interaction between positively charged complement recognition protein C1q and negatively charged polymers [132]. When negatively charged polymethacrylate binds to blood plasma proteins including complement components or IgGs, complement activation and leukocyte response can be induced [132]. Biomaterial scaffolds, on the other hand, have multiple hierarchical structures ranging from molecular level where cross-linked or individual polymer chain form porous network to microscopic level where the topographic features of the scaffold are presented. The pore size can create steric hindrance between proteins and the material surface. Materials with smaller pore size present limited surface area for protein binding. One the contrary, a surface with large pores can allow the binding of both large and small proteins within the pores at their corresponding protein configuration [129]. The small pore size was demonstrated to decrease capsule formation in vivo, irrespective of surface chemistry [133]. Additionally, PCL scaffolds with an aligned fiber topography was shown to have significantly reduced capsule formation compared to scaffolds with randomly aligned fibers [134]. The effect of the architecture of micro-structured biomaterials on determining response of macrophages has also been demonstrated [135]. Tuning surface chemistry by grafting or coating with polymer, proteins, or specific peptide sequences on polymer chains also alter protein adsorption and the host responses. As an example, poly(*N*-isopropylacrylamide) grafted poly(ethylene terephthalate) copolymer reduced protein adsorption and monocyte adhesion and resulted in reduced inflammatory cytokine levels after implantation [136]. Osteopontin coatings on a positively charged copolymer of 2-hydroxyethyl methacrylate and 2-aminoethyl methacrylate surfaces have reduced capsule thickness around the implant [137]. In addition, heparin coatings can be used to reduce coagulation and complement activation by binding to and activating antithrombin, which then inactivates thrombin and blocks blood clotting process. The coating with non-fouling polymers such as polyethylene glycol (PEG) can also minimize protein adsorption [138, 139].

5.5 Strategies to Overcome and Modulate the Host Responses

The immune response, if not controlled properly, has the potential to cause extensive secondary damage. Therefore, different strategies have been applied to reduce the unwanted host response to the implant. Moreover, many recent approaches have attempted to modulate the immune response to achieve the more effective regeneration outcome. This section summarizes the major strategies that have emerged over past few years.

5.5.1 Surface Modification

The host immune response can be reduced by chemical or physical modification of the material surface. The functional groups presented on the biomaterial surface can interact with protein molecules and consequently activate the immune cells. It is reported that the hydrophobic biomaterials such as vinylidene fluoride-hexafluoropropylene copolymer (VFH), poly(styrene–isobutylene–styrene) copolymer (SIBS), and poly(butylmethacrylate) (PBMA) show more interaction with the monocytes and result local immune response at the implant site [130]. The monocyte adhesion and

FBGC formation significantly reduced in case of hydrophilic (phosphorylcholine, BioLinx, polyacrylamide) and neutral biomaterial surface (sodium salt of polyacrylic acid) [130, 131]. However, hydrophilic and neutral biomaterials showed more pro-inflammatory response (release of IL-1β and IL-6) compared to hydrophobic surfaces although the inflammatory response was time dependent.

Change in surface topography and roughness is another strategy for immunomodulation [140, 141]. Higher cell infiltration and reduced fibrous capsule formation were obtained with aligned PCL nanofiber topography compared to randomly aligned PCL nanofibers [142]. Chen et al. [135] modulated the macrophage activation by imprinting parallel gratings (0.25–2 μm line width) on different biopolymers such as PLA, PCL, and PDMS. It was found that the density of macrophage cell attachment decreased on 2 μm gratings. In addition to this, larger grating line width (1 μm) induced more pro-inflammatory response (TNF-α) at 24 h, but the response decreased at 48 h. In another study, in vitro monocyte/macrophage stimulation was observed with variation of PTFE scaffold topography (different intra-nodal distances) [133]. Scaffold with larger intra-nodal distance (4.4 μm) showed 15-fold higher stimulation compared to nonporous scaffolds.

5.5.2 Surface Coatings

Apart from the surface chemistry or architecture, surface coating on biomaterial is another approach to mask the immune response of implant device. The immune response to implant device arises from nonspecific protein adsorption on the surface of implant device and results in leucocyte adhesion, called "biofouling." Therefore, surface coating of biomaterial may reduce such "biofouling" and can adversely affect immune response. Pre-adsorption of less inflammatory proteins (albumin) on polystyrene and PU surface was used previously due to its simple and straightforward approach [143, 144]. FBR was also decreased after coating of osteopontin on positively charged polymer surface [137]. The coating layer provides an interface between the implant surface and the tissue fluids, enabling different protein binding and downstream signaling in the immune cells, thereby possibly minimizing the induced tissue reactions [145]. Due to the lack of stability of such protein-based coating, non-fouling polymer (that prevents protein adsorption) coating has become alternative route for immunomodulation. PEG has been extensively applied as non-fouling polymer [146]. The non-fouling activity of PEG depends on its chain length or molecular weight, PEG chain density, and conformation [147–149]. Apart from PEG, PAAm, poly(*N*-isopropyl acrylamide), and poly(2-hydroxyethyl methacrylate) have also been used to prevent protein adsorption [150–152]. Hydrogel-type coatings have emerged as an interesting type and have been applied in a broad range of biomaterial devices [153, 154]. Hydrogel system can be made of materials from natural sources including ECM proteins (such as gelatin) [155], polysaccharides (such as alginate, chitosan), and synthetic polymers (such as poly(acrylamide)) [156].

5.5.3 Delivery of Bioactive Molecules

Immunomodulation can also be controlled through systemic delivery of anti-inflammatory cytokines [5]. The flexible polymeric biomaterial structures enable the anti-inflammatory and immunomodulatory therapy by the incorporation of bioactive molecules such as cytokines, growth factors, and anti-inflammatory drugs [4, 5]. As soon as the payloads are released, the anti-inflammatory effects fade and inflammatory response will resume. Therefore, the beneficial effects of the immune response on regeneration may be retained using localized delivery systems along with biomaterial, which may not impact the entire immune system and have the potential to selectively recruit specific immune cells or create a local anti-inflammatory microenvironment.

Host response at the implant site can be controlled with the use of steroidal and nonsteroidal anti-inflammatory drugs. Glucocorticoids are potent suppressors of immune responses and have been used to inhibit the immune response by inhibiting the formation and secretion of inflammatory cytokines. Glucocorticoid treatment resulted in reduced inflammatory cells at the injury site by inhibiting inflammatory mediators, decreasing capillary permeability, and fibroblast proliferation [157]. Meanwhile, the inflammation and immune response were resolved by promoting anti-inflammatory cytokine secretion and inhibiting cellular (T helper 1, Th1) immunity in favor of humoral (Th2) immunity [158]. Biomaterial-based drug carrier such as microspheres, nanoparticles, hydrogels, microspheres-hydrogel composites have been designed to deliver drugs of interest to the implant site [159–162]. As an example, delivery of dexamethasone using PLGA microsphere-polyvinyl alcohol (PVA) hydrogel composite at the implantation site resulted in reduced implant-associated inflammatory reaction as indicated by the initial decreased levels of polymorphonuclear leukocytes and minimal macrophages and lymphocytes infiltration and fibrous capsule formation in the later stage [163]. However, an undesired effect of using dexamethasone as a therapeutic is its ability to reduce the secretion of vascular endothelial growth factor (VEGF) in the surrounding tissue, which down regulates angiogenesis and would potentially inhibit wound healing [4]. By the combination therapy of dexamethasone and VEGF delivery, this problem could potentially be overcome [159]. Similarly, delivery of nonsteroidal anti-inflammatory drugs has reduced IL-8 and polymorphonuclear leukocyte levels while not reducing significantly monocyte chemoattractant protein-1 and monocyte levels [164]. Coating of biomaterial surfaces with nitric oxide (NO)-releasing layer is another strategy suggested for long-term control of immune responses. Hetrick et al. applied NO releasing diazeniumdiolate-modified xerogel polymer coating on silicone elastomer implant, which resulted in reduced inflammatory cell recruitment and extent of inflammatory reaction at the implant site. This effect was sustained even after exhaustion of the payload release from the NO reservoir [165].

A range of signaling network of growth factors, including epidermal growth factor (EGF), fibroblast growth factor (FGF), VEGF, transforming growth factor beta (TGFβ), and platelet-derived growth factor (PDGF) control adhesion, migration, proliferation, and differentiation of fibroblasts, keratinocytes, and endothelial cells during injury [166]. Biomaterials coated with or encapsulating these bioactive molecules are envisioned to have immunomodulatory effect.

Recently, biopolymer-based, microparticles or nanoparticles-based controlled delivery of immunomodulatory proteins have been studied as novel approaches [167, 168]. Rusanova et al. [169] encapsulated synthetic thrombin receptor (PAR1) agonist peptide into biodegradable PLGA microspheres and the controlled release of PAR1 from microsphere reduced the inflammatory response and resulted wound healing in ulcer rat model. Nucleic acid delivery has also been shown to effectively reduce the inflammatory response [170, 171]. Recently, Mirandi et al. reported the modulation of macrophage response to collagen based scaffold by the controlled delivery of cytokine IL-4 from PLGA-multistage silicone vector [172]. In the presence of IL-4, rat bone marrow derived macrophage showed overexpression of anti-inflammatory and M2 associated genes such as *Il10* both in vitro and in vivo.

5.6 Conclusions

In this chapter, we discuss various natural and synthetic biomaterials and how they affect the host immune responses. The nature of immune response, whether acute or chronic, depends on various factors such as implantation techniques, biomaterial source and their composition, molecular weight, surface property, mechanical properties, and degradation rate. The implanted device first comes in contact with blood plasma and ECM proteins. The adsorbed ECM proteins on the biomaterial surface attract the neutrophils and monocytes through cellular response and consequently result in the inflammatory response by macrophage. The immune response to biomaterials can be modulated through inhibition of protein adsorption on biomaterial surface by various techniques such as surface modification, surface coating and delivery of immune modulating agents. Therefore, the biomaterial-based implants should be engineered in such a way that the materials result in no or minimum immune response and unnecessary health risks to provide the best clinical outcomes for the patients.

References

1. Williams DF (2009) On the nature of biomaterials. Biomaterials 30(30):5897–5909
2. Wang X (2013) Overview on biocompatibilities of implantable biomaterials. Adv Biomater Sci Appl Biomed. doi:10.5772/53461
3. Williams DF (2008) On the mechanisms of biocompatibility. Biomaterials 29(20):2941–2953
4. Morais JM, Papadimitrakopoulos F, Burgess DJ (2010) Biomaterials/tissue interactions: possible solutions to overcome foreign body response. AAPS J 12(2):188–196
5. Boehler RM, Graham JG, Shea LD (2011) Tissue engineering tools for modulation of the immune response. Biotechniques 51(4):239–240, 242, 244 passim

6. Tidball JG, Wehling-Henricks M (2007) Macrophages promote muscle membrane repair and muscle fibre growth and regeneration during modified muscle loading in mice in vivo. J Physiol 578(1):327–336
7. Franz S, Rammelt S, Scharnweber D et al (2011) Immune responses to implants – A review of the implications for the design of immunomodulatory biomaterials. Biomaterials 32(28):6692–6709
8. Bao Ha TL, Quan TM, Nguyen Vu D, Si DM (2013) Naturally derived biomaterials: preparation and application, regenerative medicine and tissue engineering. In: Andrades JA (ed). InTech. doi:10.5772/55668
9. Anderson JM, Rodriguez A, Chang DT (2008) Foreign body reaction to biomaterials. Semin Immunol 20(2):86–100
10. Ige OO, Umoru LE, Aribo S (2012) Natural products: a minefield of biomaterials. ISRN Mater Sci 2012:1–20
11. Lapidot S, Meirovitch S, Sharon S et al (2012) Clues for biomimetics from natural composite materials. Nanomedicine 7(9):1409–1423
12. Ruys A (2013) Biomimetic biomaterials structure and applications, vol 57, Woodhead publishing series in biomaterials. Woodhead, Cambridge, UK, p 344, p. 1 online resource
13. Fishman JM, Wiles K, Wood KJ (2015) Host response to biomaterials the impact of host response on biomaterial selection. Academic Press, Cambridge, MA, pp 151–187
14. Badylak SF, Gilbert TW (2008) Immune response to biologic scaffold materials. Semin Immunol 20(2):109–116
15. Aamodt JM, Grainger DW (2016) Extracellular matrix-based biomaterial scaffolds and the host response. Biomaterials 86:68–82
16. Parenteau-Bareil R, Gauvin R, Berthod F (2010) Collagen-based biomaterials for tissue engineering applications. Materials 3(3):1863–1887
17. Wnek GE, Bowlin GL (2008) Encyclopedia of biomaterials and biomedical engineering. Informa Healthcare, New York, p. 1 online resource (4 v. (xxviii, 3110, 3172 p.))
18. Zeugolis D, Raghunath M, Ducheyne P et al (2011) Collagen: materials analysis and implant uses. Comprehens Biomater 2:261
19. van der Rest M, Garrone R, Herbage D (1993) Collagen: a family of proteins with many facets. Adv Mol Cell Biol 6:1–67
20. Kielty CM, Grant ME (2003) The collagen family: structure, assembly, and organization in the extracellular matrix. In: Connective tissue and its heritable disorders: molecular, genetic, and medical aspects, 2nd edn. Wiley, Hoboken, NJ, p 159–221
21. Chen QZ, Liang SL, Thouas GA (2013) Elastomeric biomaterials for tissue engineering. Prog Polym Sci 38(3–4):584–671
22. Gosline J, Lillie M, Carrington E et al (2002) Elastic proteins: biological roles and mechanical properties. Philos Trans R Soc B Biol Sci 357(1418):121–132
23. Narotam PK, Jose S, Nathoo N et al (2004) Collagen matrix (DuraGen) in dural repair: analysis of a new modified technique. Spine 29(24):2861–2867, discussion 2868-2869
24. Thornton JF, Rohrich RJ (2005) Dermal substitute (Integra) for open nasal wounds. Plast Reconstr Surg 116(2):677
25. Tedder ME, Liao J, Weed B et al (2008) Stabilized collagen scaffolds for heart valve tissue engineering. Tissue Eng Part A 15(6):1257–1268
26. Yost MJ, Baicu CF, Stonerock CE et al (2004) A novel tubular scaffold for cardiovascular tissue engineering. Tissue Eng 10(1–2):273–284
27. Liu C (2015) Collagen–hydroxyapatite composite scaffolds for tissue engineering. In: Mucalo M (ed) Hydroxyapatite (Hap) for biomedical applications. Woodhead, Cambridge, pp 211–234
28. Phillips JB, Bunting SC, Hall SM et al (2005) Neural tissue engineering: a self-organizing collagen guidance conduit. Tissue Eng 11(9–10):1611–1617
29. Schmitt F, Levine L, Drake M et al (1964) The antigenicity of tropocollagen. Proc Natl Acad Sci 51(3):493–497

30. Steffen C, Timpl R, Wolff I (1968) Immunogenicity and specificity of collagen: V. Demonstration of three different antigenic determinants on calf collagen. Immunology 15(1):135
31. Michaeli D, Martin GR, Kettman J et al (1969) Localization of antigenic determinants in the polypeptide chains of collagen. Science 166(3912):1522–1523
32. Furthmayr H, Beil W, Timpl R (1971) Different antigenic determinants in the polypeptide chains of human collagen. FEBS Lett 12(6):341–344
33. Timpl R, Beil W, Furthmayr H et al (1971) Characterization of conformation independent antigenic determinants in the triple-helical part of calf and rat collagen. Immunology 21(6):1017
34. Smith QT (1975) Collagen metabolism in wound healing. In: Trauma. Springer, Heidelberg, pp 31–45
35. Nair LS, Laurencin CT (2007) Biodegradable polymers as biomaterials. Prog Polym Sci 32(8–9):762–798
36. Wang X (2006) A comparison of chitosan and collagen sponges as hemostatic dressings. J Bioact Compat Polym 21(1):39–54
37. Zeugolis D, Paul R, Attenburrow G (2008) Factors influencing the properties of reconstituted collagen fibers prior to self-assembly: animal species and collagen extraction method. J Biomed Mater Res A 86(4):892–904
38. Zeugolis DI, Khew ST, Yew ES et al (2008) Electro-spinning of pure collagen nano-fibres–just an expensive way to make gelatin? Biomaterials 29(15):2293–2305
39. Delgado LM, Pandit A, Zeugolis DI (2014) Influence of sterilisation methods on collagen-based devices stability and properties. Expert Rev Med Devices 11(3):305–314
40. Brown BN, Londono R, Tottey S et al (2012) Macrophage phenotype as a predictor of constructive remodeling following the implantation of biologically derived surgical mesh materials. Acta Biomater 8(3):978–987
41. Ye Q, Harmsen MC, van Luyn MJ et al (2010) The relationship between collagen scaffold cross-linking agents and neutrophils in the foreign body reaction. Biomaterials 31(35):9192–9201
42. Gough JE, Scotchford CA, Downes S (2002) Cytotoxicity of glutaraldehyde crosslinked collagen/poly (vinyl alcohol) films is by the mechanism of apoptosis. J Biomed Mater Res 61(1):121–130
43. Ward AG, Courts A (1977) The science and technology of gelatin. In: Ward AG, Courts A (eds) Food science and technology: a series of monographs. Academic, New York, p 564, xvi
44. Kakiuchi M, Hosoya T, Takaoka K et al (1985) Human bone matrix gelatin as a clinical alloimplant. A retrospective review of 160 cases. Int Orthop 9(3):181–188
45. Muzzarelli RA, Boudrant J, Meyer D et al (2012) Current views on fungal chitin/chitosan, human chitinases, food preservation, glucans, pectins and inulin: a tribute to Henri Braconnot, precursor of the carbohydrate polymers science, on the chitin bicentennial. Carbohydr Polym 87(2):995–1012
46. Kundu PP, Sarkar K (2011) Natural polymeric vectors in gene therapy, Biopolymers. John Wiley, New York, pp 575–603
47. Coutinho DF, Sant S, Shakiba M et al (2012) Microfabricated photocrosslinkable polyelectrolyte-complex of chitosan and methacrylated gellan gum. J Mater Chem 22(33):17262
48. Rabanel J-M, Bertrand N, Sant S et al (2006) Polysaccharide hydrogels for the preparation of immunoisolated cell delivery systems. Polysaccharides for drug delivery and pharmaceutical applications. ACS Symp Ser 934:305–339
49. Prudden JF, Migel P, Hanson P et al (1970) The discovery of a potent pure chemical wound-healing accelerator. Am J Surg 119(5):560–564
50. Muzzarelli R, Baldassarre V, Conti F et al (1988) Biological activity of chitosan: ultrastructural study. Biomaterials 9(3):247–252
51. Nakajima M, Atsumi K, Kifune K et al (1986) Chitin is an effective material for sutures. Jpn J Surg 16(6):418–424

52. Mukhopadhyay P, Sarkar K, Chakraborty M et al (2013) Oral insulin delivery by self-assembled chitosan nanoparticles: in vitro and in vivo studies in diabetic animal model. Mater Sci Eng C 33(1):376–382
53. Sarkar K, Chatterjee A, Chakraborti G et al (2013) Blood compatible N-maleyl chitosan-graft-PAMAM copolymer for enhanced gene transfection. Carbohydr Polym 98(1):596–606
54. VandeVord PJ, Matthew HW, DeSilva SP et al (2002) Evaluation of the biocompatibility of a chitosan scaffold in mice. J Biomed Mater Res 59(3):585–590
55. Usami Y, Okamoto Y, Minami S et al (1994) Migration of canine neutrophils to chitin and chitosan. J Vet Med Sci 56(6):1215–1216
56. Mathews S, Kaladhar K, Sharma CP (2006) Cell mimetic monolayer supported chitosan-haemocompatibility studies. J Biomed Mater Res A 79(1):147–152
57. Wu N, Wen Z-S, Xiang X-W et al (2015) Immunostimulative activity of low molecular weight chitosans in RAW264.7 macrophages. Mar Drugs 13(10):6210
58. Oliveira MI, Santos SG, Oliveira MJ et al (2012) Chitosan drives anti-inflammatory macrophage polarisation and pro-inflammatory dendritic cell stimulation. Eur Cell Mater 24:136–152
59. Yui T, Imada K, Okuyama K et al (1994) Molecular and crystal structure of the anhydrous form of chitosan. Macromolecules 27(26):7601–7605
60. Tømmeraas K, Köping-Höggård M, Vårum KM et al (2002) Preparation and characterisation of chitosans with oligosaccharide branches. Carbohydr Res 337(24):2455–2462
61. Chen C-L, Wang Y-M, Liu C-F et al (2008) The effect of water-soluble chitosan on macrophage activation and the attenuation of mite allergen-induced airway inflammation. Biomaterials 29(14):2173–2182
62. Urtti A, Bajaj G, Van Alstine WG et al (2012) Zwitterionic chitosan derivative, a new biocompatible pharmaceutical excipient, prevents endotoxin-mediated cytokine release. PLoS One 7(1):e30899
63. Amarnath LP, Srinivas A, Ramamurthi A (2006) In vitro hemocompatibility testing of UV-modified hyaluronan hydrogels. Biomaterials 27(8):1416–1424
64. Jansen K, Van Der Werff J, Van Wachem P et al (2004) A hyaluronan-based nerve guide: in vitro cytotoxicity, subcutaneous tissue reactions, and degradation in the rat. Biomaterials 25(3):483–489
65. Rah MJ (2011) A review of hyaluronan and its ophthalmic applications. Optometry 82(1):38–43
66. De Andres-Santos A, Velasco-Martín A, Hernández-Velasco E et al (1994) Thermal behaviour of aqueous solutions of sodium hyaluronate from different commercial sources. Thermochimica Acta 242:153–160
67. Peattie RA, Rieke ER, Hewett EM et al (2006) Dual growth factor-induced angiogenesis in vivo using hyaluronan hydrogel implants. Biomaterials 27(9):1868–1875
68. Pike DB, Cai S, Pomraning KR et al (2006) Heparin-regulated release of growth factors in vitro and angiogenic response in vivo to implanted hyaluronan hydrogels containing VEGF and bFGF. Biomaterials 27(30):5242–5251
69. Collins MN, Birkinshaw C (2013) Hyaluronic acid based scaffolds for tissue engineering—A review. Carbohydr Polym 92(2):1262–1279
70. Yamanlar S, Sant S, Boudou T et al (2011) Surface functionalization of hyaluronic acid hydrogels by polyelectrolyte multilayer films. Biomaterials 32(24):5590–5599
71. Johnson P, Maiti A, Brown KL et al (2000) A role for the cell adhesion molecule CD44 and sulfation in leukocyte–endothelial cell adhesion during an inflammatory response? Biochem Pharmacol 59(5):455–465
72. Puré E, Cuff CA (2001) A crucial role for CD44 in inflammation. Trends Mol Med 7(5):213–221
73. Termeer C, Sleeman JP, Simon JC (2003) Hyaluronan–magic glue for the regulation of the immune response? Trends Immunol 24(3):112–114
74. Rayahin JE, Buhrman JS, Zhang Y et al (2015) High and low molecular weight hyaluronic acid differentially influence macrophage activation. ACS Biomater Sci Eng 1(7):481–493

75. Kajahn J, Franz S, Rueckert E et al (2012) Artificial extracellular matrices composed of collagen I and high sulfated hyaluronan modulate monocyte to macrophage differentiation under conditions of sterile inflammation. Biomatter 2(4):226–273
76. Capila I, Linhardt RJ (2002) Heparin–protein interactions. Angew Chem Int Ed 41(3):390–412
77. Mizrahy S, Peer D (2012) Polysaccharides as building blocks for nanotherapeutics. Chem Soc Rev 41(7):2623–2640
78. Ekre H-P, Naparstek Y, Lider O et al (1992) Anti-inflammatory effects of heparin and its derivatives inhibition of complement and of lymphocyte migration. In: Lane DA, Björk I, Lindahl U (eds) Heparin and related polysaccharides. Springer, Heidelberg, pp 329–340
79. Darien BJ, Fareed J, Centgraf KS et al (1998) Low molecular weight heparin prevents the pulmonary hemodynamic and pathomorphologic effects of endotoxin in a porcine acute lung injury model. Shock 9(4):274–281
80. Hochart H, Jenkins PV, Preston RJ et al (2008) Concentration-dependent roles for heparin in modifying liopolysaccharide-induced activation of mononuclear cells in whole blood. Thromb Haemost 99(3):570–575
81. Gombotz WR, Wee SF (2012) Protein release from alginate matrices. Adv Drug Deliv Rev 64:194–205
82. Mukhopadhyay P, Sarkar K, Soam S et al (2013) Formulation of pH-responsive carboxymethyl chitosan and alginate beads for the oral delivery of insulin. J Appl Polym Sci 129(2):835–845
83. Drury JL, Mooney DJ (2003) Hydrogels for tissue engineering: scaffold design variables and applications. Biomaterials 24(24):4337–4351
84. Orive G, Ponce S, Hernandez R et al (2002) Biocompatibility of microcapsules for cell immobilization elaborated with different type of alginates. Biomaterials 23(18):3825–3831
85. Arıca MY, Arpa Ç, Ergene A et al (2003) Ca-alginate as a support for Pb (II) and Zn (II) biosorption with immobilized Phanerochaete chrysosporium. Carbohydr Polym 52(2):167–174
86. Stephen AM (1995) Food polysaccharides and their applications, 67th edn. CRC, Boca Raton, FL
87. Yang D, Jones KS (2009) Effect of alginate on innate immune activation of macrophages. J Biomed Mater Res A 90(2):411–418
88. Thomas A, Harding K, Moore K (2000) Alginates from wound dressings activate human macrophages to secrete tumour necrosis factor-α. Biomaterials 21(17):1797–1802
89. Matsumoto T, Kawai MMasuda T (1991) Influence of concentration and mannuronate/guluronate [correction of gluronate] ratio on steady flow properties of alginate aqueous systems. Biorheology 29(4):411–417
90. Iwamoto M, Kurachi M, Nakashima T et al (2005) Structure–activity relationship of alginate oligosaccharides in the induction of cytokine production from RAW264. 7 cells. FEBS Lett 579(20):4423–4429
91. Altman GH, Diaz F, Jakuba C et al (2003) Silk-based biomaterials. Biomaterials 24(3):401–416
92. MacIntosh AC, Kearns VR, Crawford A et al (2008) Skeletal tissue engineering using silk biomaterials. J Tissue Eng Regen Med 2(2-3):71–80
93. Meinel L, Fajardo R, Hofmann S et al (2005) Silk implants for the healing of critical size bone defects. Bone 37(5):688–698
94. Cassinelli C, Cascardo G, Morra M et al (2006) Physical-chemical and biological characterization of silk fibroin-coated porous membranes for medical applications. Int J Artif Organs 29(9):881
95. Zhang X, Baughman CB, Kaplan DL (2008) In vitro evaluation of electrospun silk fibroin scaffolds for vascular cell growth. Biomaterials 29(14):2217–2227
96. Wang X, Zhang X, Castellot J et al (2008) Controlled release from multilayer silk biomaterial coatings to modulate vascular cell responses. Biomaterials 29(7):894–903

97. Steins A, Dik P, Müller WH et al (2015) In vitro evaluation of spider silk meshes as a potential biomaterial for bladder reconstruction. PLoS One 10(12)
98. Gillespie DB, Viney C, Yager P (1994) Raman spectroscopic analysis of the secondary structure of spider silk fibers. In: Silk polymers: materials science and biotechnology. ACS Symposium Series, vol 544. ACS Publications, Washington, DC
99. Soong HK, Kenyon KR (1984) Adverse reactions to virgin silk sutures in cataract surgery. Ophthalmology 91(5):479–483
100. Panilaitis B, Altman GH, Chen J et al (2003) Macrophage responses to silk. Biomaterials 24(18):3079–3085
101. Meinel L, Hofmann S, Karageorgiou V et al (2005) The inflammatory responses to silk films in vitro and in vivo. Biomaterials 26(2):147–155
102. Liu H, Ge Z, Wang Y et al (2007) Modification of sericin-free silk fibers for ligament tissue engineering application. J Biomed Mater Res B Appl Biomater 82(1):129–138
103. Wang Y, Rudym DD, Walsh A et al (2008) In vivo degradation of three-dimensional silk fibroin scaffolds. Biomaterials 29(24):3415–3428
104. Ghanaati S, Orth C, Unger RE et al (2010) Fine-tuning scaffolds for tissue regeneration: effects of formic acid processing on tissue reaction to silk fibroin. J Tissue Eng Regen Med 4(6):464–472
105. Badylak SF (2014) Decellularized allogeneic and xenogeneic tissue as a bioscaffold for regenerative medicine: factors that influence the host response. Ann Biomed Eng 42(7):1517–1527
106. Valentin JE, Badylak JS, McCabe GP, Badylak SF (2006) Extracellular matrix bioscaffolds for orthopaedic applications. A comparative histologic study. J Bone Joint Surg Am 88(12):2673
107. Ravi S, Chaikof EL (2010) Biomaterials for vascular tissue engineering. Regen Med 5(1):107–120
108. Rotter N, Ung F, Roy AK et al (2005) Role for interleukin 1α in the inhibition of chondrogenesis in autologous implants using polyglycolic acid-polylactic acid scaffolds. Tissue Eng 11(1–2):192–200
109. Santavirta S, Konttinen YT, Saito T et al (1990) Immune response to polyglycolic acid implants. J Bone Joint Surg 72(4):597–600
110. Cortiella J, Nichols JE, Kojima K et al (2006) Tissue-engineered lung: anin vivoandin vitrocomparison of polyglycolic acid and pluronic F-127 hydrogel/somatic lung progenitor cell constructs to support tissue growth. Tissue Eng 12(5):1213–1225
111. Raya-Rivera A, Esquiliano DR, Yoo JJ et al (2011) Tissue-engineered autologous urethras for patients who need reconstruction: an observational study. Lancet 377(9772):1175–1182
112. Ceonzo K, Gaynor A, Shaffer L et al (2006) Polyglycolic acid-induced inflammation: role of hydrolysis and resulting complement activation. Tissue Eng 12(2):301–308
113. Tamai H, Igaki K, Kyo E et al (2000) Initial and 6-month results of biodegradable poly-l-lactic acid coronary stents in humans. Circulation 102(4):399–404
114. Bergsma JE, Bos RRM, Rozema FR et al (1996) Biocompatibility of intraosseously implanted predegraded poly(lactide): an animal study. J Mater Sci Mater Med 7(1):1–7
115. Aframian DJ, Redman RS, Yamano S et al (2002) Tissue compatibility of two biodegradable tubular scaffolds implanted adjacent to skin or buccal mucosa in mice. Tissue Eng 8(4):649–659
116. Lam KH, Schakenraad JM, Esselbrugge H et al (1993) The effect of phagocytosis of poly(L-lactic acid) fragments on cellular morphology and viability. J Biomed Mater Res 27(12):1569–1577
117. Woodruff MA, Hutmacher DW (2010) The return of a forgotten polymer—Polycaprolactone in the 21st century. Prog Polym Sci 35(10):1217–1256
118. Mukundan S, Sant V, Goenka S et al (2015) Nanofibrous composite scaffolds of poly(ester amides) with tunable physicochemical and degradation properties. Eur Polym J 68:21–35
119. Gaharwar AK, Nikkhah M, Sant S et al (2014) Anisotropic poly (glycerol sebacate)-poly (ϵ-caprolactone) electrospun fibers promote endothelial cell guidance. Biofabrication 7(1):015001

120. Eslami M, Vrana NE, Zorlutuna P et al (2014) Fiber-reinforced hydrogel scaffolds for heart valve tissue engineering. J Biomater Appl 29(3):399–410
121. Sant S, Iyer D, Gaharwar AK et al (2013) Effect of biodegradation and de novo matrix synthesis on the mechanical properties of valvular interstitial cell-seeded polyglycerol sebacate-polycaprolactone scaffolds. Acta Biomater 9(4):5963–5973
122. Tong Z, Sant S, Khademhosseini A et al (2011) Controlling the fibroblastic differentiation of mesenchymal stem cells via the combination of fibrous scaffolds and connective tissue growth factor. Tissue Eng Part A 17(21–22):2773–2785
123. Nisbet DR, Rodda AE, Horne MK et al (2009) Neurite infiltration and cellular response to electrospun polycaprolactone scaffolds implanted into the brain. Biomaterials 30(27): 4573–4580
124. Santerre JP, Woodhouse K, Laroche G et al (2005) Understanding the biodegradation of polyurethanes: from classical implants to tissue engineering materials. Biomaterials 26(35):7457–7470
125. Schutte RJ, Xie L, Klitzman B et al (2009) In vivo cytokine-associated responses to biomaterials. Biomaterials 30(2):160–168
126. Zhang J-Y, Beckman EJ, Hu J et al (2002) Synthesis, biodegradability, and biocompatibility of lysine diisocyanate–glucose polymers. Tissue Eng 8(5):771–785
127. Tressaud A, Haufe G (2008) Fluorine and health: molecular imaging, biomedical materials and pharmaceuticals. Elsevier, Amsterdam
128. Skóra J, Pupka A, Dorobisz A et al (2012) Evaluation of the humoral and cellular immune responses after implantation of a PTFE vascular prosthesis* Ocena immunologicznej odpowiedzi humoralnej i komórkowej po zabiegach wszczepienia protezy. Postepy Hig Med Dosw (Online) 66:469–474
129. Gonzalez-Simon AL, Eniola-Adefeso O (2012) Host response to biomaterials engineering. In: Bhatia SK (ed) Biomaterials for regenerative medicine: , Novel technologies for clinical applications. Springer, New York, pp 143–159
130. Hezi-Yamit A, Sullivan C, Wong J et al (2009) Impact of polymer hydrophilicity on biocompatibility: implication for DES polymer design. J Biomed Mater Res A 90A(1):133–141
131. Jones JA, Chang DT, Meyerson H et al (2007) Proteomic analysis and quantification of cytokines and chemokines from biomaterial surface-adherent macrophages and foreign body giant cells. J Biomed Mater Res A 83A(3):585–596
132. Engberg AE, Rosengren-Holmberg JP, Chen H et al (2011) Blood protein-polymer adsorption: implications for understanding complement-mediated hemoincompatibility. J Biomed Mater Res A 97A(1):74–84
133. Bota PCS, Collie AMB, Puolakkainen P et al (2010) Biomaterial topography alters healing in vivo and monocyte/macrophage activation in vitro. J Biomed Mater Res A 95A(2):649–657
134. Cao H, McHugh K, Chew SY et al (2009) The topographical effect of electrospun nanofibrous scaffolds on the in vivo and in vitro foreign body reaction. J Biomed Mater Res A 93(3):1151–1159
135. Chen S, Jones JA, Xu Y et al (2010) Characterization of topographical effects on macrophage behavior in a foreign body response model. Biomaterials 31(13):3479–3491
136. Bridges AW, Singh N, Burns KL et al (2008) Reduced acute inflammatory responses to microgel conformal coatings. Biomaterials 29(35):4605–4615
137. Liu L, Chen G, Chao T et al (2008) Reduced foreign body reaction to implanted biomaterials by surface treatment with oriented osteopontin. J Biomater Sci Polym Ed 19(6):821–835
138. Sant S, Poulin S, Hildgen P (2008) Effect of polymer architecture on surface properties, plasma protein adsorption, and cellular interactions of pegylated nanoparticles. J Biomed Mater Res A 87A(4):885–895
139. Wang S, Gupta AS, Sagnella S et al (2009) Biomimetic fluorocarbon surfactant polymers reduce platelet adhesion on PTFE/ePTFE surfaces. J Biomater Sci Polym Ed 20(5–6):619–635
140. Yim EK, Leong KW (2005) Significance of synthetic nanostructures in dictating cellular response. Nanomedicine 1(1):10–21

141. Fink J, Fuhrmann R, Scharnweber T et al (2008) Stimulation of monocytes and macrophages: possible influence of surface roughness. Clin Hemorheol Microcirc 39(1–4):205–212
142. Cao H, Mchugh K, Chew SY et al (2010) The topographical effect of electrospun nanofibrous scaffolds on the in vivo and in vitro foreign body reaction. J Biomed Mater Res A 93(3):1151–1159
143. Geelhood SJ, Horbett TA, Ward WK et al (2007) Passivating protein coatings for implantable glucose sensors: evaluation of protein retention. J Biomed Mater Res B Appl Biomater 81(1):251–260
144. Amiji M, Park H, Park K (1992) Study on the prevention of surface-induced platelet activation by albumin coating. J Biomater Sci Polym Ed 3(5):375–388
145. Wisniewski N, Reichert M (2000) Methods for reducing biosensor membrane biofouling. Colloids Surf B Biointerfaces 18(3–4):197–219
146. Kingshott P, Griesser HJ (1999) Surfaces that resist bioadhesion. Curr Opin Solid State Mater Sci 4(4):403–412
147. Unsworth LD, Sheardown H, Brash JL (2005) Polyethylene oxide surfaces of variable chain density by chemisorption of PEO-thiol on gold: adsorption of proteins from plasma studied by radiolabelling and immunoblotting. Biomaterials 26(30):5927–5933
148. Unsworth LD, Sheardown H, Brash JL (2008) Protein-resistant poly (ethylene oxide)-grafted surfaces: chain density-dependent multiple mechanisms of action. Langmuir 24(5):1924–1929
149. Michel R, Pasche S, Textor M et al (2005) Influence of PEG architecture on protein adsorption and conformation. Langmuir 21(26):12327–12332
150. Wang C, Yu B, Knudsen B et al (2008) Synthesis and performance of novel hydrogels coatings for implantable glucose sensors. Biomacromolecules 9(2):561–567
151. Nolan CM, Reyes CD, Debord JD et al (2005) Phase transition behavior, protein adsorption, and cell adhesion resistance of poly (ethylene glycol) cross-linked microgel particles. Biomacromolecules 6(4):2032–2039
152. Singh N, Bridges AW, García AJ et al (2007) Covalent tethering of functional microgel films onto poly (ethylene terephthalate) surfaces. Biomacromolecules 8(10):3271–3275
153. Ahmed EM (2015) Hydrogel: preparation, characterization, and applications: a review. J Adv Res 6(2):105–121
154. Peppas NA (1986) Hydrogels in medicine and pharmacy. CRC, Boca Raton, FL
155. Geutjes PJ, Daamen WF, Buma P et al (2006) From molecules to matrix: construction and evaluation of molecularly defined bioscaffolds. Adv Exp Med Biol 585:279–295
156. de Vos P, Hoogmoed CG, Busscher HJ (2002) Chemistry and biocompatibility of alginate-PLL capsules for immunoprotection of mammalian cells. J Biomed Mate Res 60(2):252–259
157. Coutinho AE, Chapman KE (2011) The anti-inflammatory and immunosuppressive effects of glucocorticoids, recent developments and mechanistic insights. Mol Cellular Endocrinol 335(1):2–13
158. Spellberg B, Edwards JE (2001) Type 1/Type 2 immunity in infectious diseases. Clin Infect Dis 32(1):76–102
159. Patil SD, Papadmitrakopoulos F, Burgess DJ (2007) Concurrent delivery of dexamethasone and VEGF for localized inflammation control and angiogenesis. J Control Release 117(1):68–79
160. Norton LW, Koschwanez HE, Wisniewski NA et al (2007) Vascular endothelial growth factor and dexamethasone release from nonfouling sensor coatings affect the foreign body response. J Biomed Mater Res A 81A(4):858–869
161. Galeska I, Kim T-K, Patil SD et al (2005) Controlled release of dexamethasone from PLGA microspheres embedded within polyacid-containing PVA hydrogels. AAPS J 7(1):E231–E240
162. Norton LW, Tegnell E, Toporek SS et al (2005) In vitro characterization of vascular endothelial growth factor and dexamethasone releasing hydrogels for implantable probe coatings. Biomaterials 26(16):3285–3297

163. Patil SD, Papadimitrakopoulos F, Burgess DJ (2004) Dexamethasone-loaded poly(lactic-co-glycolic) acid microspheres/poly(vinyl alcohol) hydrogel composite coatings for inflammation control. Diabetes Technol Ther 6(6):887–897
164. Lopez-Armada MJ (2002) Modulation of cell recruitment by anti-inflammatory agents in antigen-induced arthritis. Ann Rheum Dis 61(11):1027–1030
165. Hetrick EM, Prichard HL, Klitzman B et al (2007) Reduced foreign body response at nitric oxide-releasing subcutaneous implants. Biomaterials 28(31):4571–4580
166. Barrientos S, Stojadinovic O, Golinko MS et al (2008) PERSPECTIVE ARTICLE: Growth factors and cytokines in wound healing. Wound Repair Regen 16(5):585–601
167. Mundargi RC, Babu VR, Rangaswamy V et al (2008) Nano/micro technologies for delivering macromolecular therapeutics using poly (D, L-lactide-co-glycolide) and its derivatives. J Control Release 125(3):193–209
168. Luten J, van Nostrum CF, De Smedt SC et al (2008) Biodegradable polymers as non-viral carriers for plasmid DNA delivery. J Control Release 126(2):97–110
169. Rusanova A, Makarova A, Strukova S et al (2006) Thrombin receptor agonist peptide immobilized in microspheres stimulates reparative processes in rats with gastric ulcer. Bull Exp Biol Med 142(1):35–38
170. Mori R, Shaw TJ, Martin P (2008) Molecular mechanisms linking wound inflammation and fibrosis: knockdown of osteopontin leads to rapid repair and reduced scarring. J Exp Med 205(1):43–51
171. Kovacs JR, Zheng Y, Shen H et al (2005) Polymeric microspheres as stabilizing anchors for oligonucleotide delivery to dendritic cells. Biomaterials 26(33):6754–6761
172. Minardi S, Corradetti B, Taraballi F et al (2016) IL-4 release from a biomimetic scaffold for the temporally controlled modulation of macrophage response. Ann Biomed Eng 44(6):2008–2019
173. Okada T, Hayashi T, Ikada Y (1992) Degradation of collagen suture in vitro and in vivo. Biomaterials 13(7):448–454
174. Kavoosi G, Dadfar SMM, Mohammadi Purfard A et al (2013) Antioxidant and antibacterial properties of gelatin films incorporated with carvacrol. J Food Saf 33(4):423–432
175. Xing Q, Yates K, Vogt C et al (2014) Increasing mechanical strength of gelatin hydrogels by divalent metal ion removal. Sci Rep 4
176. Yoshioka S, Stella VJ (2002) Chemical stability of drug substances. Springer, New York, pp 3–137
177. Kojima T, Inamura Y, Koide T et al (2005) Activity of gelatins to induce secretion of a variety of cytokines from murine peritoneal exudate macrophages. Cancer Biother Radiopharm 20(4):417–425
178. Albanna MZ, Bou-Akl TH, Walters HL et al (2012) Improving the mechanical properties of chitosan-based heart valve scaffolds using chitosan fibers. J Mech Behav Biomed Mater 5(1):171–180
179. Kean T, Thanou M (2010) Biodegradation, biodistribution and toxicity of chitosan. Adv Drug Deliv Rev 62(1):3–11
180. Burdick JA, Chung C, Jia X et al (2005) Controlled degradation and mechanical behavior of photopolymerized hyaluronic acid networks. Biomacromolecules 6(1):386–391
181. Zhang Y, Rossi F, Papa S et al (2016) Non-invasive in vitro and in vivo monitoring of degradation of fluorescently labeled hyaluronan hydrogels for tissue engineering applications. Acta Biomater 30:188–198
182. Drury JL, Dennis RG, Mooney DJ (2004) The tensile properties of alginate hydrogels. Biomaterials 25(16):3187–3199
183. Kong HJ, Kaigler D, Kim K et al (2004) Controlling rigidity and degradation of alginate hydrogels via molecular weight distribution. Biomacromolecules 5(5):1720–1727
184. Koh L-D, Cheng Y, Teng C-P et al (2015) Structures, mechanical properties and applications of silk fibroin materials. Prog Polym Sci 46:86–110
185. Leal-Egana A, Scheibel T (2010) Silk-based materials for biomedical applications. Biotechnol Appl Biochem 55(3):155–167

186. Van de Velde K, Kiekens P (2002) Biopolymers: overview of several properties and consequences on their applications. Polym Test 21(4):433–442
187. Liao S, Chan CK, Ramakrishna S (2008) Stem cells and biomimetic materials strategies for tissue engineering. Mater Sci Eng C 28(8):1189–1202
188. Parks AC, Sung K, Wu BM (2014) A three-dimensional in vitro model to quantify inflammatory response to biomaterials. Acta Biomater 10(11):4742–4749
189. de Tayrac R, Chentouf S, Garreau H et al (2008) In vitro degradation and in vivo biocompatibility of poly(lactic acid) mesh for soft tissue reinforcement in vaginal surgery. J Biomed Mater Res B Appl Biomater 85B(2):529–536
190. Silva ATCR, Cardoso BCO, e Silva MESR et al (2015) Synthesis, characterization, and study of PLGA copolymer in vitro degradation. J Biomater Nanobiotechnol 6(01):8–19
191. Zhang X, Yamaoka K, Sonomoto K et al (2014) Local delivery of mesenchymal stem cells with poly-lactic-co-glycolic acid nano-fiber scaffold suppress arthritis in rats. PLoS One 9(12):e114621
192. Semete B, Booysen L, Kalombo L et al (2010) In vivo uptake and acute immune response to orally administered chitosan and PEG coated PLGA nanoparticles. Toxicol Appl Pharmacol 249(2):158–165
193. Lam CX, Savalani MM, Teoh S-H et al (2008) Dynamics of in vitro polymer degradation of polycaprolactone-based scaffolds: accelerated versus simulated physiological conditions. Biomed Mater 3(3):034108
194. Khandwekar AP, Patil DP, Shouche Y et al (2011) Surface engineering of polycaprolactone by biomacromolecules and their blood compatibility. J Biomater Appl 26(2):227–252
195. McHugh KJ, Tao SL, Saint-Geniez M (2014) Porous poly (ε-caprolactone) scaffolds for retinal pigment epithelium transplantationPorous PCL Scaffolds for RPE Transplantation. Invest Ophthalmol Vis Sci 55(3):1754–1762
196. Rae P, Brown E (2005) The properties of poly (tetrafluoroethylene)(PTFE) in tension. Polymer 46(19):8128–8140
197. Mattana J, Effiong C, Kapasi A et al (1997) Leukocyte-polytetrafluoroethylene interaction enhances proliferation of vascular smooth muscle cells via tumor necrosis factor-α secretion. Kidney Int 52(6):1478–1485

Chapter 6
Convergence of Osteoimmunology and Immunomodulation for the Development and Assessment of Bone Biomaterials

Zetao Chen, Chengtie Wu, and Yin Xiao

Abstract The traditional biological principle for the development of bone biomaterials is to directly induce osteogenic differentiation of osteoblastic lineage cells. With this principle, most of the efforts are spent on optimizing the biomechanical and physicochemical properties of biomaterials to enhance osteogenic differentiation of mesenchymal stem cells. Given the vital roles of immune cells in bone dynamics, we propose a new concept "osteoimmunomodulation" in recognition of the importance of immune response during biomaterial-mediated osteogenesis. The paradigm of bone biomaterials design is also suggested to shift to an osteoimmunomodulatory material, and the possible evaluation strategies for the osteoimmunomodulation property of bone biomaterials are summarized. It is expected that bone biomaterials with favorable osteoimmunomodulation properties will be more clinically relevant to control new bone formation and biomaterial degradation in a controllable and biologically suitable manner.

Keywords Osteoimmunology • Immunomodulation • Osteoimmunomodulation • Bone biomaterials • Bone regeneration • Mechanical properties • Physicochemical properties

---

Z. Chen • Y. Xiao (✉)
Institute of Health and Biomedical Innovation, Queensland University of Technology, Brisbane, QLD 4059, Australia

State Key Laboratory of High Performance Ceramics and Superfine Microstructure, Chinese Academy of Sciences, Shanghai Institute of Ceramics, Shanghai, China
e-mail: yin.xiao@qut.edu.au

C. Wu
State Key Laboratory of High Performance Ceramics and Superfine Microstructure, Shanghai Institute of Ceramics, Chinese Academy of Sciences, Shanghai, China

The Australia–China Centre for Tissue Engineering and Regenerative Medicine, Queensland University of Technology, Brisbane, QLD, Australia

B. Corradetti (ed.), *The Immune Response to Implanted Materials and Devices*,
DOI 10.1007/978-3-319-45433-7_6

6.1 Introduction

Biomaterial implants are by their very nature exogenous and, therefore, tend to mediate a foreign body reaction (FBR) that may lead the host body to either reject or isolate the implants by fibrous encapsulation. FBR is a host immune response towards the implanted biomaterial(s), which consists of granulation tissue components, the formation of foreign body giant cells, and in most cases a fibrous encapsulation of the implants [1]. In an effort to bypass this deleterious host response, the early design paradigm of biomaterials was to create a biologically "inert" (biocompatible) material, which had the unintended consequence of not only minimizing reconstructive healing but also promote FBR [1, 2]. With a more sophisticated understanding of the interaction between host tissues and biomaterials, there is now an appreciation of the immune response as an essential component of the biomaterials-mediated reconstruction of a functional tissue. The immune system is modulatory and can be manipulated by modifying the properties of biomaterials. Accordingly, the paradigm on the nature of biomaterials has shifted from being an "inert" (e.g., biocompatible) material to "immunomodulatory"; the focus now being to apply strategies that incorporates, modulates and even encourages an immune response rather than attempting to suppress it [2–4]. Beneficial immunomodulation should become a prime target for the development of advanced functional biomaterials.

The close relationship between the immune and skeletal systems was first reported in the 1970s [5, 6]. Immune cells involve actively in bone dynamics under both physiological and pathological conditions via secretion of a number of regulatory molecules, including cytokines, signaling molecules, and transcription factors. The term osteoimmunology was thereby proposed to cover these overlapping scientific fields of osteology and immunology [7, 8]. Due to the specialized nature of immunobiology within the bone environment, the convergence of osteoimmunology and immunomodulation seems to be necessary, and the modulation of bone immune response should be categorized as osteoimmunomodulation. Accordingly, the immunomodulatory property of bone biomaterials should be modified as the "osteoimmunomodulatory" property [9–11]. Osteoimmunomodulation should become a new paradigm for the design and assessment of novel advanced bone biomaterials [9, 12]. By modifying the bio-physiochemical and mechanical properties of bone biomaterials, it is possible to actively manipulate osteoimmunomodulation to improve the bone reconstruction outcomes.

6.2 Immune Cells Regulate Bone Dynamics

With the advancement of bone biology, it is now widely accepted that osteoclastogenesis and osteogenesis are not simply the work of osteoclastic and osteoblastic lineage cells, but a cooperation of cells from multiple systems, including skeletal, immune, circulation systems. The field of osteoimmunology reveals that immune

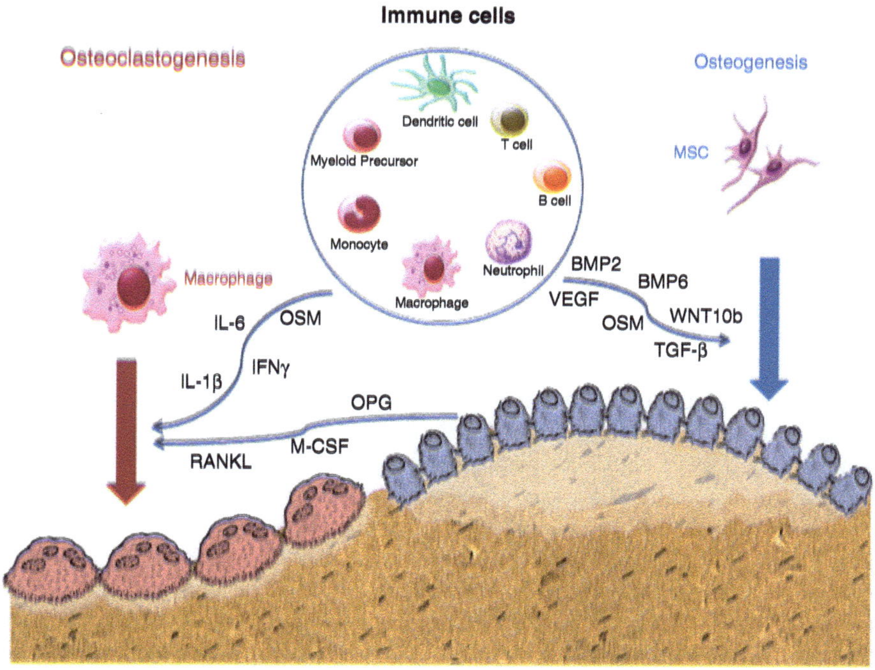

Fig. 6.1 Immune cells regulate bone metabolism. Immune cells participate actively in osteoclastogenesis and osteogenesis by releasing regulatory factors that act locally within the microenvironment

cells elicit significant effects on regulating the osteoclastogenesis and osteogenesis (Fig. 6.1). Physiological bone formation and remodeling process requires well-functioning immune cells. Abnormal functioning of immune cells could cause bone diseases such as osteolysis, osteoporosis, osteoarthritis, and rheumatoid arthritis. For example, Th17 cells, which generates a significant amount of IL-17 and have been reported to play a crucial role in the pathogenesis of rheumatoid arthritis [13, 14]. It is of great importance to be able to direct the behavior of immune cells such that they enhance bone reconstruction.

Immune cells regulate osteoclastogenesis by cytokines, such as macrophage-colony stimulating factor (M-CSF), receptor activator of NF-kB ligand (RANKL), osteoprotegerin (OPG), interleukin-6 (IL-6), and oncostatin M (OSM). The RANKL/RANK/OPG system is one vital physiological mechanism in regulating osteoclastogenesis. RANKL binds to RANK and activates the downstream cascade, thereby promoting the osteoclast differentiation, whereas osteoprotegerin (OPG), a decoy receptor derived from osteoblasts, binds to RANKL and disrupts its interaction with RANK, thereby suppressing osteoclastogenesis [15, 16]. RANKL is expressed not only by osteoblastic cells but also by activated T cells and neutrophils, indicating the involvement of these immune cells during osteoclastogenesis [17, 18]. OPG is mainly sourced from B cells [19, 20]. This implies that B cells are one of the main inhibitors of osteoclastogenesis in normal physiology. Inflammatory cytokines IL-6, IL-23, and OSM are important regulators of osteoclast formation and function, indicating the overlapping regulatory functions of inflammatory cytokines on both immune response and bone physiology. IL-6 and IL-23 is known to increase the expression of RANKL, and utilize the RANKL/RANK/OPG system to elicit indirect effects on promoting osteoclastogenesis [21–23]. IL-6 is also found to participate in the TNF-α and IL-1 induced osteoclast formation [24]. OSM can also stimulate the production of RANKL by osteoblasts and enhance the formation of osteoclasts in a dose dependent manner, which may be associated with its synergistic effects with IL-6 [25, 26]. By contrast, interferon-γ (IFN-γ) promotes the degradation of TRAF6, a key intermediate in RANKL/RANK pathway, thereby preventing massive bone destruction during inflammation [27].

As to their effects on osteogenesis, it is thought that resident macrophages are indispensable for efficient osteoblast mineralization, and the depletion of macrophages results in the loss of osteoblast-mediating bone formation in vivo [28]. In addition, many research outcomes have proved the positive roles of inflammatory cytokines on regulating osteogenesis. A combination of inflammatory cytokines TNF-α, TGF-β, IFN-γ, and IL-17 at physiological concentrations can activate MSCs mediated mineralization as effectively as dexamethasone, a commonly used osteogenic media supplement [29]. TNF-α has been found to increase alkaline phosphatase (ALP) activity and mineralization by mesenchymal stromal cells (MSCs) in a dose-dependent manner through activation of the NF-kB signaling pathway [30, 31]. The stimulatory effect of the conditioned medium from the lipopolysaccharide-activated inflammatory M1 macrophages on ALP activity is attenuated when the conditioned medium is pretreated with TNF-α neutralizing antibody [29]. Studies in IL-6 knockout mice show that the absence of IL-6 delays callus maturity, mineralization, and remodeling indicating the essential role of IL-6 in the early stages of fracture healing [32], whereas knockout of OSM in early stage leads to the reduced amount of new bone [33].

However, deleterious effects of inflammatory cytokines, especially the pro-inflammatory, on osteogenesis have also been well reported [34]. TNF-α inhibits osteogenic differentiation by reducing the release of BMP2 and elevating levels of canonical Wnt signaling pathway inhibitors dickkopf-1 (DKK-1) and sclerostin (SOST); it can also improve pro-apoptotic effects on osteoblasts [35–37]. The upregulated expression of IFN-γ and TNF-α by T lymphocytes is regarded to be a key reason for the failure of MSCs based bone tissue regeneration. Interestingly, this inhibitory effect on osteogenesis can be eliminated simply by the administration of the commonly used anti-inflammatory drug aspirin [38]. The underlying mechanism might be associated with the activation of transcription factor NF-kB in MSCs, which improves degradation of β-catenin, an important component of an osteogenic signaling pathway, the canonical Wnt signaling pathway [39]. This implies that the regulatory effects of inflammatory cytokines on bone dynamics are dose and time dependent. An adequate concentration and appropriate timing of these cytokines is of importance to induce new bone formation, which should become an immunomodulation target for immunomodulatory biomaterials.

The important role of immune cells in the bone homeostasis reveals the specialty of immunomodulation within the bone environment. The regulation of immune cell behaviors from implanted biomaterials will inevitably influence the behavior of bone cells, thereby changing bone dynamics. It is therefore important to introduce the framework of osteoimmunology into the concept of immunomodulation for the development of a new generation of bone biomaterials.

6.3 Definition of the "Osteoimmunomodulation" Property

There are quite a few studies focusing on the bone biomaterials-mediated immune responses. However, most of the efforts have focussed on whether the implants mediated foreign body reaction thereby causing excessive inflammation and rejection or encapsulation by fibrous tissue—a feature known as "biocompatibility." Although it is also a concept derived from the interaction between immune reaction and bone biomaterials, osteoimmunomodulation is different from biocompatibility. It does not simply describe the host immune reactions towards the implants, but emphasizes the influences that the local immune microenvironment has on bone cells and the subsequent osteogenesis and osteoclastogenesis. Osteoimmunomodulation is a specific biological property of bone biomaterials that describes the ability of biomaterials to modulate the local immune microenvironment, affecting the bone dynamics, thereby determining the in vivo fate of bone biomaterials in terms of de novo bone formation or fibrous encapsulation [11] (Fig. 6.2). To be specific, bone biomaterials with favorable osteoimmunomodulation properties are those that could elicit significant effects on local immune cells to generate an adequate and appropriate inflammatory response and release biomolecules that can enhance the recruitment and osteogenic differentiation of mesenchymal stromal cells, leading to new bone regeneration. Proper osteoclastogenesis and osteoclastic activities are also required, which should be able to accomplish bone functional remodeling and cell-mediating materials degradation process, while avoiding excessive bone resorption and destruction. This concept is derived from the convergence of osteoimmunology and immunomodulation, drawing on recent advances in both fields.

6.4 The Need to Include Immune Cells When Evaluating Novel Bone Biomaterials

As direct effector cells for osteogenesis, osteoblastic lineage cells are generally used to assess the in vitro osteogenic capacity of bone biomaterials, a strategy that has achieved some success in generating novel bone biomaterials. However, inconsistencies between in vitro and in vivo studies are not uncommon, and many potential bone biomaterials screened by this strategy are subsequently found to underperform in vivo

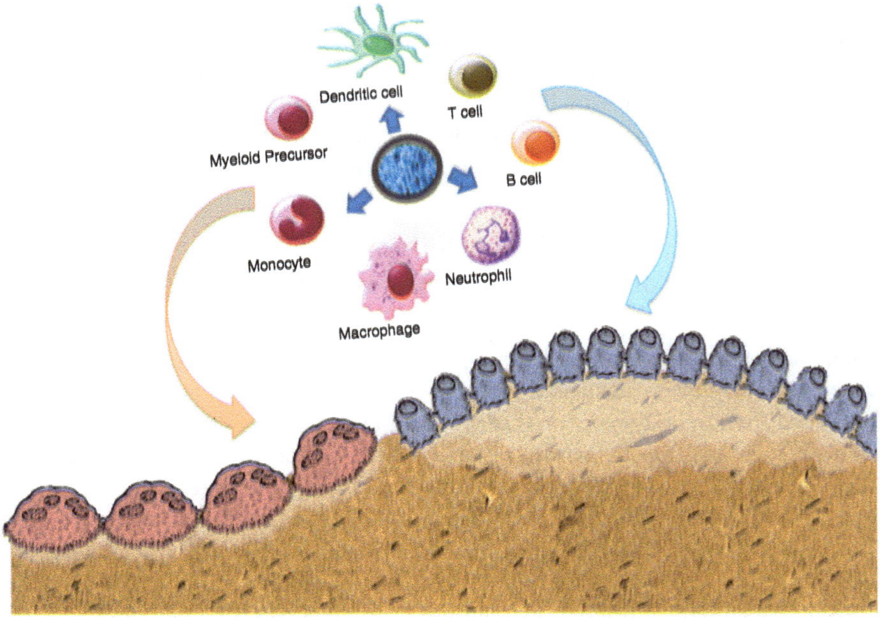

Fig. 6.2 Concept of the osteoimmunomodulation. Osteoimmunomodulation refers to the immune environment that is created by the interaction between biomaterials and immune cells, which subsequently affects the bone formation and remodeling

in term of bone regeneration. Based on the knowledge of osteoimmunomodulation, it is a logical extension that the traditional strategy, which relies only on the response of osteoblastic lineage cells, is insufficient for evaluating and developing bone biomaterials and that neglecting the role of immune cells may account for the inconsistent results between the in vitro and in vivo studies [12] (Fig. 6.3).

Immediately following its implantation, the biomaterial will elicit a number of universal inflammatory responses from the host body. These include blood clot formation, protein adsorption forming the transient matrix, and acute inflammation (neutrophil migration, mast cells degranulation, and antigen presenting). The macrophage derived osteoclasts and foreign body giant cells will commence decomposing the implants, releasing foreign components into the immediate vicinity and altering the local environment. Dendritic cells can ingest potent antigens (such as necrotic cells and tissues, proteins with conformation changes mediated by the implants) and present them to humoral immune cells (B-cells), and also release cytokines that alter the immune environment to initiate and determine further antigen-specific immune responses. Given the close relationship between the immune and skeletal systems, it should come as no surprise that stimulated immune cells can also secrete cytokines that can regulate bone dynamics. An appropriate immune reaction should release an adequate amount of osteogenic cytokines to facilitate osteogenesis, whereas an inappropriate immune reaction can lead to a chronic inflammation and the fibrous encapsulation of the implant. With these considerations in mind, we have documented the important role of immune

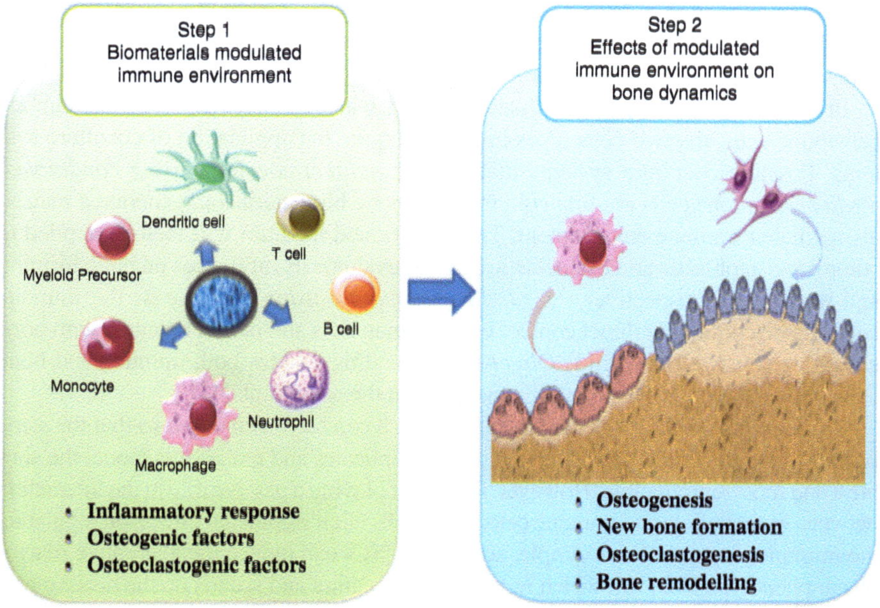

Fig. 6.3 Schematic outlining the proposed strategy to evaluate candidate bone biomaterials. All three factors (bone biomaterials, immune cells, and bone cells) must be involved in the system of evaluation. The effect of biomaterials to modulate the immune environment must be assessed to demonstrate whether the resulting immune response is favorable for bone regeneration. An immune environment that is favorable to bone regeneration and repair consists of a finely balanced mix of inflammatory cytokines, osteogenic and osteoclastogenic factors, and fibrosis enhancing factors

cells in biomaterial-stimulated osteogenesis, and the missing of immune cells in the in vitro evaluation system led to some inconsistent conclusions compared with the in vivo outcomes [12]. The traditional approach of only using osteoblastic lineage cells when evaluating such biomaterials is clearly inadequate, especially in term of the osteoimmunomodulatory properties of the materials. Immune cells, such as macrophages, dendritic cells, and T and B cells, must be included when evaluating in vitro osteogenesis mediated by bone biomaterials.

Among all the immune cells, macrophages have received the most attention in term of the interaction between biomaterials and the host response. Macrophages are innate immune cells and have a key role as effector cells in the material-induced immune response by initiating and maintaining inflammation; this process closely correlates with the formation of a fibrous capsule. Macrophages also influence bone physiology and pathology [28, 40] by acting as precursors to osteoclasts, thus participating directly in bone remodeling and material degradation. Macrophages have a direct role in osteogenesis by the expression and secretion of important regulatory molecules [41], such as bone morphogenetic protein 2 (BMP2) and transforming growth factor β (TGF-β) [42, 43]. As such, macrophages are indispensable for the efficient osteoblast mineralization and their depletion dampen the osteogenic differentiation of mesenchymal progenitors [28, 44]. This suggests that macrophages could be the key cell type determining the in vivo fate of bone substitute materials. Given this important role in the bone dynamics, macrophage can be applied as a model immune cell for the in vitro evaluation of osteogenesis with bone biomaterials [9, 10, 12, 45].

In this novel biomimicking evaluation system, which involves biomaterials/immune cells/bone cells, the two cells types involved require the application of coculture systems. Possible coculture systems include: (1) *Indirect coculture using conditioned medium:* Immune cells are first cultured with bone biomaterials to generate a materials-mediated immune environment. The conditioned medium will then be applied to stimulate osteoblastic and osteoclastic cells, to evaluate its influences on bone forming and remodeling, respectively; (2) *Indirect coculture using Transwells:* the immune cells can be cultured in direct contact with the materials and indirect contact with bone cells in a Transwell insert; (3) *Direct coculture:* this enables both immune and bone cells to be in direct contact with the materials at the same time [11].

Conditioned medium has the advantage that it can be frozen down so that the same batch of medium could be applied for several replicates and makes this model the simplest and most reproducible. However, instead of playing a passive role in the interaction between biomaterials and immune cells, bone cells elicit significant effects on regulating the immune response. For example, activated MSCs can induce the alternative macrophage phenotype M2 [46], which reduces inflammation and speeds up the healing process. They can also express TNF-α stimulated gene/protein 6 and IL-1ra to decrease the amplifying effects of pro-inflammatory cytokines (IL-1, IL-6, and TNF-α) [47, 48]. This means that the use of indirect coculture using conditioned medium can only reflect the in vivo environment to a limited extent since it focusses only on the effects the immune signaling molecules have on bone cells while ignoring the counteracting effects that bone cells have on immune cells. Indirect coculture, using transwells or direct coculture, would better mimic the in vivo environment, from the aspect of mimicking the interaction between bone cells and immune cells. However, direct coculture requires that the cells used are sourced from one individual patient, especially when using immune cells that would otherwise attack xenogenic cells. Interpatient variability and limited source of proper donor tissues makes it hard to establish this system as the standard of in vitro assessment. Indirect coculture using Transwell may, therefore, be the most suitable evaluation system for the evaluation of osteoimmunomodulation [11].

6.5 Development of "Smart" Bone Biomaterials to Manipulate Osteoimmunomodulation

Osteoimmunomodulation implies a strategy to modify the immune response, thereby manipulating the bone cells' reaction. The modulatory effects of biomaterials on the immune system have been well documented. Biomaterials elicit significant effects on regulating the immune reaction, which is determined by their bio-physicochemical and mechanical properties. This follows a strategy to develop bone biomaterials with favorable bio-physicochemical and mechanical properties that help in modulating a beneficial immune environment for bone regeneration [11].

Biological behaviors of immune cells on the surfaces of the bone biomaterials are largely determined by the surface properties, such as surface wettability [49–51], surface charge [52, 53], and topography [54, 55]. Generally, hydrophobic materials tend to improve monocyte adhesion in comparison to hydrophilic materials resulting in an immune response at the implant site [50, 56]. Positively charged particles are found to be more likely to cause inflammatory reactions than negatively charged and neutral species [52, 53]. Smooth titanium surface induces inflammatory macrophage (M1-like) activation, with increased expression of IL-1β, IL-6, and TNFα [51]. However, hydrophilic rough titanium induces M2-like macrophage activation and elevates the expression of anti-inflammatory cytokines IL-4 and IL-10 [51]. Micro-structured topography induced macrophages to an activated state that have both M1 and M2 characteristics [57], and titanium surfaces, modified by titania nanotube arrays, can reduce in vitro inflammatory response compared to the raw titanium surface [58]. Modifying surface properties therefore presents as an effective way of modulating the immune response.

A size and shape-dependent foreign body immune reaction towards implants has been observed in rodents and nonhuman primates [59], indicating the importance of size and shape properties of particles in modulating the immune response. Decreased particle size increases the surface area and improves chemical reactivity, thereby strengthening the influences on target cells, or can even elicits a different effect altogether [60, 61]. Bulk gold samples are practically inert, whereas gold nanoparticles have been reported to elicit highly reactive immune responses, including production of reactive oxidative species (ROS) [62]. Hydroxyapatite particles with the smallest size (1–30 μm) stimulate immune cells to produce the greatest amount of pro-inflammatory cytokines (TNF-α, IL-1β, IL-6) [63]. However, it does not follow that smaller sized particles necessarily mediate a more severe immune response. An in vivo study has demonstrated that reducing the size of irregularly shaped hydroxyapatite particles can inhibit the inflammatory reaction [64]. Some studies report that increasing the size of implanted materials led to larger foreign body reactions and the formation of a thicker layer of fibrocapsule surrounding the implant [65]. Spheres with a diameter of 1.5 mm and above could significantly inhibit foreign body reactions and fibrosis compared to smaller sized spheres [59]. As for the effect of particle shapes, circular rods of medical-grade polymers were found to have the weakest foreign body response, followed by pentagonal and then triangular [66]. A systematic evaluation of a broad spectrum of particle sizes and shapes within each category of bone biomaterials is therefore necessary to gain an understanding of how these parameters affect the inflammatory response.

Porosity and pore size are two important parameters for the fabrication of 3D bone tissue engineering scaffolds, which are also vital factors in determining the tissue types growing inside the scaffolds (inflammatory granuloma tissue, vascular tissue, or bone tissue) [67]. It seems that with an increase in pore size, the activity of the foreign body reaction decreases [68, 69]. The underlying mechanism may be related to macrophage polarization since there appears to be a correlation between increasing fiber/pore size and increased expression of the M2 markers, along with decreased expression of the M1 markers [70]. Small pores will severely hamper the diffusion of nutrients and oxygen supplied from blood and interstitial fluid, especially in the core area of the scaffold, resulting in a local hypoxic environment [71]. Hypoxia enhances local inflammation, leading to the formation of granuloma. This would completely block the small pores, creating a barrier between the implant and the surrounding bone cells that prevents bone tissue ingrowth from taking place [68]. Pores in the size range of 90–120 μm have been found to hamper vascularization and lead to chondrogenesis, whereas larger sized pores (350 μm) enhance vascularization and osteogenesis [72]. A better understanding of how porosity and pore size affect immune response is needed to develop advanced functional scaffolds for bone reconstructions.

Bioactive bone biomaterials normally undergo degradation to different extents following implantation, either by physicochemical dissolution, cell-medicated dissolution, hydrolysis, enzymatic decomposition, or corrosion [73]. Ions released from biomaterials during degradation can elicit significant effects by altering the local immune environment. Calcium (Ca) is involved in certain inflammatory signaling pathways, for example the noncanonical Wnt5A/Ca^{2+} signaling pathway and the calcium sensing receptor (CaSR) signaling cascade [74–76]. Si ions may also regulate immune reaction. It is thought that long-term exposure to components of silicone gel-filled breast implants may lead to autoimmune or inflammatory diseases, while Si is usually found at higher concentrations in the lesion areas and patient's blood [77]. Magnesium (Mg) can suppress pro-inflammatory cytokine production by inhibiting toll-like receptor (TLR) pathway [78], while cobalt (Co) has pro-inflammatory effects via stabilizing hypoxia-inducible factors (HIFs) [79, 80]. Zinc (Zn)-substituted ceramics can elevate the release of anti-inflammatory cytokine IL-10, while reducing the expression of TNF-α and IL-1β, the latter two which are associated with the regulation of TLR-4 pathway [81, 82]. Strontium (Sr)-substituted Ca/P materials could inhibit the release of TNFα in human primary monocytes [83, 84]. It is well known that bioactive ions also elicit significant effects on regulating osteogenesis and osteoclastogenesis, in addition to their effects on the immune system, which makes nutrient elements obvious candidates for the development of bone biomaterials with multidirectional effects.

Based on these considerations, we have conducted studies in which we combined nutrient elements to develop novel bone biomaterials with multidirectional effects on regulating immune response, osteogenesis, and osteoclastogenesis (Fig. 6.4). Three types of bioactive coating materials were successfully prepared using defined combinations of nutrient elements ($Sr_2MgSi_2O_7$, $Sr_2ZnSi_2O_7$, $MgSiO_3$) [10, 85, 86]. All three materials were found to endow the titanium substrate (Ti-6Al-4V) with favorable osteoimmunomodulation, suppressing inflammation and inhibiting osteoclastogenesis (Figs. 6.5, 6.6, and 6.7). Interestingly, we found that $Sr_2ZnSi_2O_7$ and $MgSiO_3$ modified immune environment in such a way that it enhanced osteogenesis, whereas the $Sr_2MgSi_2O_7$ material was only comparable with hydroxyapatite coating materials in terms of osteogenesis [10, 85, 86]. This implies that the osteoimmunomodulatory effects depend on the composition and concentration of the ions and requires further investigation. The strategy to manipulate osteoimmunomodulation by controlled release of defined combinations of nutrient elements is, therefore, one worthy of careful consideration.

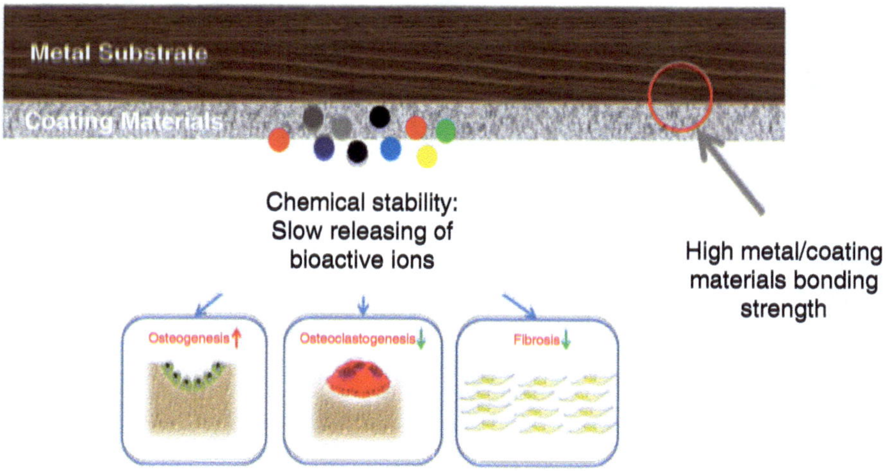

Fig. 6.4 Ideal coating materials for bone substitute biomaterials must have high bonding strength, functional ion release, and multidirectional effects for regulating immune response, and be capable of favoring osteogenesis over osteoclastogenesis

Potential strategies for the manipulations are: (1) modifying the surface properties of materials, including topography (size and roughness), and surface chemistry (hydrophilicity and electric potential); (2) changing particle size and shape; (3) for 3D structure scaffolds, optimizing their pore size and porosity; (4) incorporating bioactive nutrient elements; (5) introducing biomolecules such as macrophage inducers (e.g., IL-4, LPS) or inflammatory cytokines (e.g., IL-10, TNF-α, IFN-γ), to selectively induce a dominant macrophage phenotype or a phenotype switch pattern; and (6) coupling with immunomodulatory drugs, e.g., steroidal and nonsteroidal anti-inflammatory drugs (dexamethasone, aspirin). These strategies should all be considered when designing biomaterials, since their combination can have synergistic effects. Needless to say, a better understanding of the mechanisms underlying biomaterials mediated immune response and their effects on osteoclastogenesis and osteogenesis is essential for developing advanced bone biomaterials with favorable osteoimmunomodulation properties.

6.6 Conclusions

The convergence of osteoimmunology and immunomodulation offers a novel strategy in the development of new generation bone biomaterials. The paradigm for developing bone biomaterials has shifted to osteoimmunomodulatory biomaterials, emphasizing the importance of osteoimmunomodulation. This property is becoming the essential parameter when evaluating and developing of advanced bone biomaterials. The aim should be to develop "smart" materials capable of modulating an immune environment in a way that enhances the new bone formation. Osteoimmunomodulation,

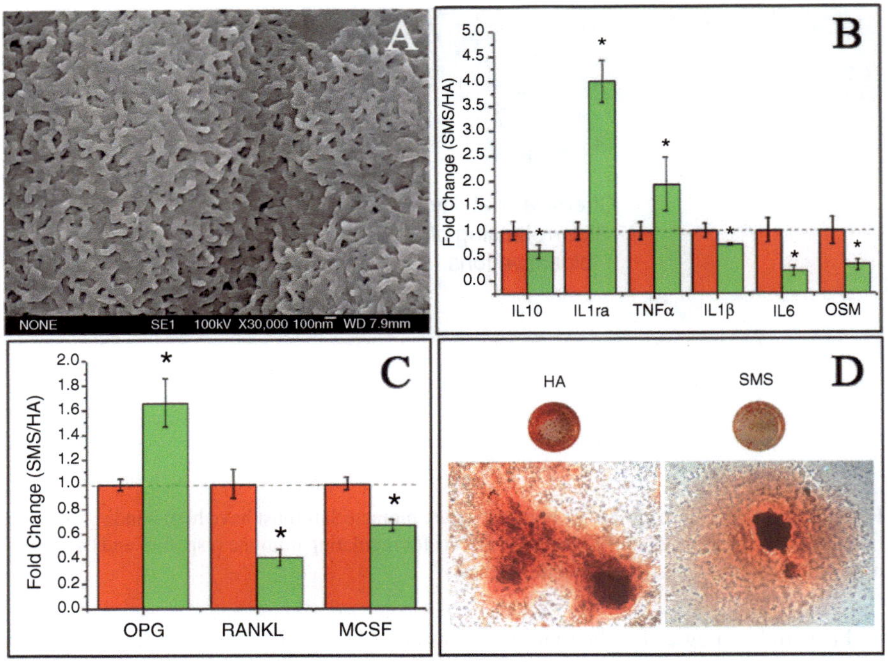

Fig. 6.5 Multidirectional effects of Sr2MgSi2O7 (SMS) coating. (**a**) Enhanced the Ca/P apatite formation on the SMS coating following immersion in simulated body fluid for 14 days; image taken by scanning electron microscopy. (**d**) Suppressing the pro-inflammatory gene expression by macrophages, cultured on the surface of the SMS coating for 3 days; gene expression was detected by RT-qPCR and hydroxyapatite (HA) coating used as a control. (**c**) Inhibiting gene expression of osteoclastogenic factors by BMSCs, stimulated by the SMS coating/macrophage conditioned medium; gene expression was detected by RT-qPCR and HA coating was used as control. (**d**) Maintaining comparable amount of mineralization nodules with hydroxyapatite coating; mineralization nodules were formed by BMSCs, stimulated by coating/macrophage conditioned medium; alizarin red was used to stain the nodules. Adapted with permission from Ref. [85], Copyright 2014 American Chemical Society

as it is defined in this chapter includes not only the modulation of inflammatory response towards the implants, but the effects the immune microenvironment has on bone dynamics, regulating the balance of osteogenesis versus osteoclastogenesis. This balance effectively determines the in vivo fate of bone biomaterials, towards either de novo bone formation or fibrosis encapsulation. Traditional strategies for the development of bone biomaterials mainly focused on manipulating osteoblastic lineage cells to the exclusion of immune cells [87–89]. This paradigm failed to reflect the in vivo condition in which the host immune reaction is activated following the implantation of a biomaterial. In this chapter we discuss a novel strategy for the development and evaluation of bone biomaterials that emphasizes osteoimmunomodulation. Future studies should focus on the components of the immune system that,

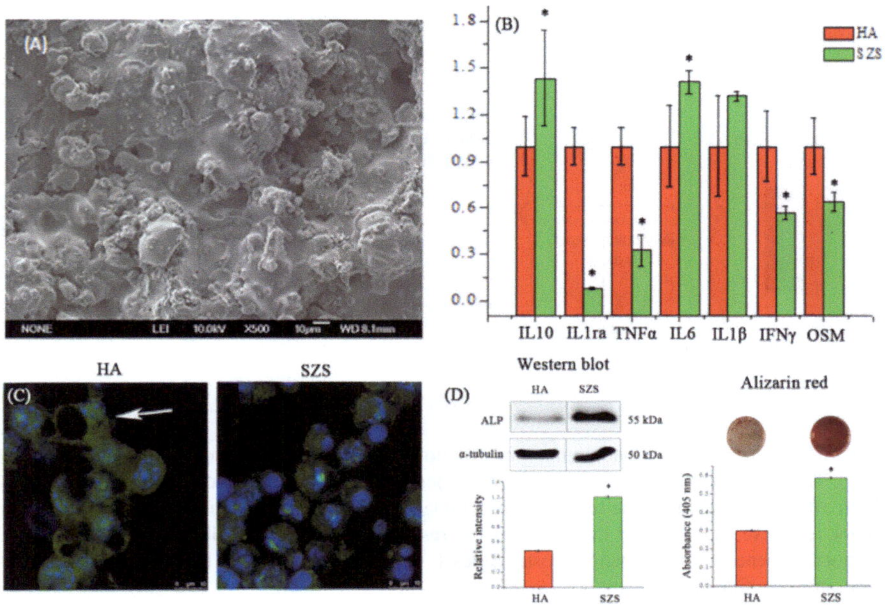

Fig. 6.6 Sr2ZnSi2O7 (SZS) coating induces osteoimmunomodulation that promotes bone formation. (**a**) The surface of SZS coatings on the Ti-6Al-4V substrate by SEM. (**b**) Gene expression of inflammatory cytokines by macrophages, cultured on the surface of the SZS coating for 7 h; gene expression was detected by RT-qPCR and HA coating was used as control. (**c**) Phagocytosis of the coating materials. Macrophages were stained by DAPI and phalloidin with FITC for the nuclei and cytoskeleton respectively. *HA* Macrophages internalized the HA particles, generating a big "bubble" intracellularly, which was observed in most of the stimulated cells, *SZS* this feature was rarely observed in the SZS coating treated macrophages. (**d**) Western blot analysis of ALP expression by BMSCs stimulated by the macrophage/SZS conditioned medium. Alizarin Red staining of mineralized nodules by BMSCs treated with the conditioned medium containing osteogenic supplements. Adapted from Ref. [10] with permission from The Royal Society of Chemistry

when activated by a biomaterials implant, can enhance osteogenesis. Knowledge such as this will inform the design paradigms for the next generation of advanced bone biomaterials.

Acknowledgements The authors would like to acknowledge the Q-CAS Biotechnology Fund (GJHZ1505), the Prince Charles Hospital Foundation, QUT Funding for Australia-China Centre of Tissue Engineering and Regenerative Medicine (ACCTERM), Recruitment Program of Global Young Talent, China (C.W.), the National High Technology Research and Development Program of China (863 Program, SS2015AA020302), Natural Science Foundation of China (Grant 31370963), Program of Shanghai Outstanding Academic Leaders (15XD1503900), Key Research Program of Chinese Academy of Sciences (Grant KGZDEW-T06), Innovative Project of SIC, CAS, and ARC (DP120103697). We thank Dr Thor Friis who proofread the draft of the manuscript.

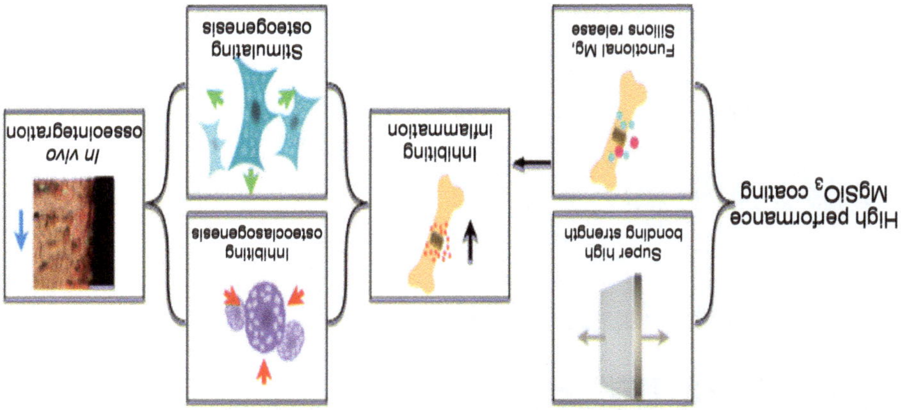

Fig. 6.7 Manipulation of osteoimmunomodulation using nutrient elements (magnesium and silicon). Titanium substrates were coated with clinoenstatite (MgSiO3), which binds with extremely high affinity to titanium. The release of functional Mg and Si ions modulates the local osteoimmune response and dampens the inflammatory response around the titanium implants, decreased osteoclastogenesis, thus enhancing osteogenesis. These effects translate into superior osseointegration of the titanium implant in vivo [86]

References

1. Bryers JD, Giachelli CM, Ratner BD (2012) Engineering biomaterials to integrate and heal: the biocompatibility paradigm shifts. Biotechnol Bioeng 109(8):1898–1911
2. Franz S, Rammelt S, Scharnweber D et al (2011) Immune responses to implants – a review of the implications for the design of immunomodulatory biomaterials. Biomaterials 32(28):6692–6709
3. Williams DF (2009) On the nature of biomaterials. Biomaterials 30(30):5897–5909
4. Mokarram N, Bellamkonda RV (2014) A perspective on immunomodulation and tissue repair. Ann Biomed Eng 42(2):338–351
5. Mundy GR, Raisz LG, Cooper RA et al (1974) Evidence for the secretion of an osteoclast stimulating factor in myeloma. N Engl J Med 291(20):1041–1046
6. Horton JE, Raisz LG, Simmons HA et al (1972) Bone resorbing activity in supernatant fluid from cultured human peripheral blood leukocytes. Science 177(4051):793–795
7. Takayanagi H (2007) Osteoimmunology: shared mechanisms and crosstalk between the immune and bone systems. Nat Rev Immunol 7(4):292–304
8. Takayanagi H (2005) Inflammatory bone destruction and osteoimmunology. J Periodontal Res 40(4):287–293
9. Chen Z, Mao X, Tan L et al (2014) Osteoimmunomodulatory properties of magnesium scaffolds coated with beta-tricalcium phosphate. Biomaterials 35(30):8553–8565
10. Chen Z, Yi D, Zheng X et al (2014) Nutrient element-based bioceramic coatings on titanium alloy stimulating osteogenesis by inducing beneficial osteoimmmunomodulation. J Mater Chem B 2(36):6030–6043
11. Chen Z, Klein T, Murray RZ et al (2016) Osteoimmunomodulation for the development of advanced bone biomaterials. Mater Today 19(6):304–321
12. Chen Z, Yuen J, Crawford R et al (2015) The effect of osteoimmunomodulation on the osteogenic effects of cobalt incorporated beta-tricalcium phosphate. Biomaterials 61:126–138
13. van den Berg WB, Miossec P (2009) IL-17 as a future therapeutic target for rheumatoid arthritis. Nat Rev Rheumatol 5(10):549–553

14. Sato K, Suematsu A, Okamoto K et al (2006) Th17 functions as an osteoclastogenic helper T cell subset that links T cell activation and bone destruction. J Exp Med 203(12):2673–2682
15. Wright HL, McCarthy HS, Middleton J et al (2009) RANK, RANKL and osteoprotegerin in bone biology and disease. Curr Rev Musculoskelet Med 2(1):56–64
16. Boyce BF, Xing L (2007) Biology of RANK, RANKL, and osteoprotegerin. Arthritis Res Ther 9(Suppl 1):S1
17. Theill LE, Boyle WJ, Penninger JM (2002) RANK-L and RANK: T cells, bone loss, and mammalian evolution. Annu Rev Immunol 20:795–823
18. Chakravarti A, Raquil MA, Tessier P et al (2009) Surface RANKL of Toll-like receptor 4-stimulated human neutrophils activates osteoclastic bone resorption. Blood 114(8):1633–1644
19. Li Y, Toraldo G, Li A et al (2007) B cells and T cells are critical for the preservation of bone homeostasis and attainment of peak bone mass in vivo. Blood 109(9):3839–3848
20. Pacifici R (2010) The immune system and bone. Arch Biochem Biophys 503(1):41–53
21. Yoshitake F, Itoh S, Narita H et al (2008) Interleukin-6 directly inhibits osteoclast differentiation by suppressing receptor activator of NF-kappaB signaling pathways. J Biol Chem 283(17):11535–11540
22. Abdel Meguid MH, Hamad YH, Swilam RS et al (2013) Relation of interleukin-6 in rheumatoid arthritis patients to systemic bone loss and structural bone damage. Rheumatol Int 33(3):697–703
23. Adamopoulos IE, Tessmer M, Chao C-C et al (2011) IL-23 is critical for induction of arthritis, osteoclast formation, and maintenance of bone mass. J Immunol 187(2):951–959
24. Devlin RD, Reddy SV, Savino R et al (1998) IL-6 mediates the effects of IL-1 or TNF, but Not PTHrP or 1,25(OH)2D3, on osteoclast-like cell formation in normal human bone marrow cultures. J Bone Miner Res 13(3):393–399
25. Richardson IG (2000) The nature of the hydration products in hardened cement pastes. Cem Concr Compos 22(2):97–113
26. Sims NA, Quinn JMW (2014) Osteoimmunology: oncostatin M as a pleiotropic regulator of bone formation and resorption in health and disease. Bonekey Rep 3
27. Arron JR, Choi Y (2000) Bone versus immune system. Nature 408(6812):535–536
28. Chang MK, Raggatt LJ, Alexander KA et al (2008) Osteal tissue macrophages are intercalated throughout human and mouse bone lining tissues and regulate osteoblast function in vitro and in vivo. J Immunol 181(2):1232–1244
29. Rifas L (2006) T-cell cytokine induction of BMP-2 regulates human mesenchymal stromal cell differentiation and mineralization. J Cell Biochem 98(4):706–714
30. Hess K, Ushmorov A, Fiedler J et al (2009) TNFalpha promotes osteogenic differentiation of human mesenchymal stem cells by triggering the NF-kappaB signaling pathway. Bone 45(2):367–376
31. Ding J, Ghali O, Lencel P et al (2009) TNF-alpha and IL-1beta inhibit RUNX2 and collagen expression but increase alkaline phosphatase activity and mineralization in human mesenchymal stem cells. Life Sci 84(15–16):499–504
32. Yang X, Ricciardi BF, Hernandez-Soria A et al (2007) Callus mineralization and maturation are delayed during fracture healing in interleukin-6 knockout mice. Bone 41(6):928–936
33. Guihard P, Boutet MA, Brounais-Le Royer B et al (2015) Oncostatin m, an inflammatory cytokine produced by macrophages, supports intramembranous bone healing in a mouse model of tibia injury. Am J Pathol 185(3):765–775
34. Walsh NC, Reinwald S, Manning CA et al (2009) Osteoblast function is compromised at sites of focal bone erosion in inflammatory arthritis. J Bone Miner Res 24(9):1572–1585
35. Gilbert L, He X, Farmer P et al (2000) Inhibition of osteoblast differentiation by tumor necrosis factor-alpha. Endocrinology 141(11):3956–3964
36. Feldmann M, Maini RN (2010) Anti-TNF therapy, from rationale to standard of care: what lessons has it taught us? J Immunol 185(2):791–794
37. Diarra D, Stolina M, Polzer K et al (2007) Dickkopf-1 is a master regulator of joint remodeling. Nat Med 13(2):156–163

38. Liu Y, Wang L, Kikuiri T et al (2011) Mesenchymal stem cell-based tissue regeneration is governed by recipient T lymphocytes via IFN-gamma and TNF-alpha. Nat Med 17(12):1594–1601
39. Chang J, Liu F, Lee M et al (2013) NF-kappaB inhibits osteogenic differentiation of mesenchymal stem cells by promoting beta-catenin degradation. Proc Natl Acad Sci U S A 110(23):9469–9474
40. Alexander KA, Chang MK, Maylin ER et al (2011) Osteal macrophages promote in vivo intramembranous bone healing in a mouse tibial injury model. J Bone Miner Res 26(7):1517–1532
41. Pettit AR, Chang MK, Hume DA et al (2008) Osteal macrophages: a new twist on coupling during bone dynamics. Bone 43(6):976–982
42. Honda Y, Anada T, Kamakura S et al (2006) Elevated extracellular calcium stimulates secretion of bone morphogenetic protein 2 by a macrophage cell line. Biochem Biophys Res Commun 345(3):1155–1160
43. Wahl SM, McCartney-Francis N, Allen JB et al (1990) Macrophage production of TGF-beta and regulation by TGF-beta. Ann N Y Acad Sci 593:188–196
44. Vi L, Baht GS, Whetstone H et al (2015) Macrophages promote osteoblastic differentiation invivo: implications in fracture repair and bone homeostasis. J Bone Miner Res 30(6):1090–1102
45. Shi M, Chen Z, Farnaghi S et al (2016) Copper-doped mesoporous silica nanospheres, a promising immunomodulatory agent for inducing osteogenesis. Acta Biomater 30:334–344
46. Kim J, Hematti P (2009) Mesenchymal stem cell-educated macrophages: a novel type of alternatively activated macrophages. Exp Hematol 37(12):1445–1453
47. Ortiz LA, DuTreil M, Fattman C et al (2007) Interleukin 1 receptor antagonist mediates the antiinflammatory and antifibrotic effect of mesenchymal stem cells during lung injury. Proc Natl Acad Sci U S A 104(26):11002–11007
48. Oh JY, Roddy GW, Choi H et al (2010) Anti-inflammatory protein TSG-6 reduces inflammatory damage to the cornea following chemical and mechanical injury. Proc Natl Acad Sci U S A 107(39):16875–16880
49. Yun JK, DeFife K, Colton E et al (1995) Human monocyte/macrophage adhesion and cytokine production on surface-modified poly(tetrafluoroethylene/hexafluoropropylene) polymers with and without protein preadsorption. J Biomed Mater Res 29(2):257–268
50. Boehler RM, Graham JG, Shea LD (2011) Tissue engineering tools for modulation of the immune response. Biotechniques 51(4):239–240
51. Hotchkiss KM, Reddy GB, Hyzy SL et al (2015) Titanium surface characteristics, including topography and wettability, alter macrophage activation. Acta Biomater 31:425–34
52. Dobrovolskaia MA, McNeil SE (2007) Immunological properties of engineered nanomaterials. Nat Nanotechnol 2(8):469–478
53. Tan Y, Li S, Pitt BR et al (1999) The inhibitory role of CpG immunostimulatory motifs in cationic lipid vector-mediated transgene expression in vivo. Hum Gene Ther 10(13):2153–2161
54. Wojciak-Stothard B, Madeja Z, Korohoda W et al (1995) Activation of macrophage-like cells by multiple grooved substrata. Topographical control of cell behaviour. Cell Biol Int 19(6):485–490
55. Rice JM, Hunt JA, Gallagher JA et al (2003) Quantitative assessment of the response of primary derived human osteoblasts and macrophages to a range of nanotopography surfaces in a single culture model in vitro. Biomaterials 24(26):4799–4818
56. Hezi-Yamit A, Sullivan C, Wong J et al (2009) Impact of polymer hydrophilicity on biocompatibility: implication for DES polymer design. J Biomed Mater Res A 90(1):133–141
57. Paul NE, Skazik C, Harwardt M et al (2008) Topographical control of human macrophages by a regularly microstructured polyvinylidene fluoride surface. Biomaterials 29(30):4056–4064
58. Smith BS, Capellato P, Kelley S et al (2013) Reduced in vitro immune response on titania nanotube arrays compared to titanium surface. Biomater Sci 1(3):322–332
59. Veiseh O, Doloff JC, Ma M et al (2015) Size- and shape-dependent foreign body immune response to materials implanted in rodents and non-human primates. Nat Mater 14(6):643–651

60. Nel AE, Madler L, Velegol D et al (2009) Understanding biophysicochemical interactions at the nano-bio interface. Nat Mater 8(7):543–557
61. Oh WK, Kim S, Choi M et al (2010) Cellular uptake, cytotoxicity, and innate immune response of silica-titania hollow nanoparticles based on size and surface functionality. ACS Nano 4(9):5301–5313
62. Cuenya BR (2010) Synthesis and catalytic properties of metal nanoparticles: size, shape, support, composition, and oxidation state effects. Thin Solid Films 518(12):3127–3150
63. Laquerriere P, Grandjean-Laquerriere A, Jallot E et al (2003) Importance of hydroxyapatite particles characteristics on cytokines production by human monocytes in vitro. Biomaterials 24(16):2739–2747
64. Malard O, Bouler JM, Guicheux J et al (1999) Influence of biphasic calcium phosphate granulometry on bone ingrowth, ceramic resorption, and inflammatory reactions: preliminary in vitro and in vivo study. J Biomed Mater Res 46(1):103–111
65. Ward WK, Slobodzian EP, Tiekotter KL et al (2002) The effect of microgeometry, implant thickness and polyurethane chemistry on the foreign body response to subcutaneous implants. Biomaterials 23(21):4185–4192
66. Matlaga BF, Yasenchak LP, Salthouse TN (1976) Tissue response to implanted polymers: the significance of sample shape. J Biomed Mater Res 10(3):391–397
67. Karageorgiou V, Kaplan D (2005) Porosity of 3D biomaterial scaffolds and osteogenesis. Biomaterials 26(27):5474–5491
68. Klinge U, Klosterhalfen B, Birkenhauer V et al (2002) Impact of polymer pore size on the interface scar formation in a rat model. J Surg Res 103(2):208–214
69. Weyhe D, Schmitz I, Belyaev O et al (2006) Experimental comparison of monofile light and heavy polypropylene meshes: less weight does not mean less biological response. World J Surg 30(8):1586–1591
70. Garg K, Pullen NA, Oskeritzian CA et al (2013) Macrophage functional polarization (M1/M2) in response to varying fiber and pore dimensions of electrospun scaffolds. Biomaterials 34(18):4439–4451
71. Laschke MW, Harder Y, Amon M et al (2006) Angiogenesis in tissue engineering: breathing life into constructed tissue substitutes. Tissue Eng 12(8):2093–2104
72. Kuboki Y, Jin Q, Kikuchi M et al (2002) Geometry of artificial ECM: sizes of pores controlling phenotype expression in BMP-induced osteogenesis and chondrogenesis. Connect Tissue Res 43(2–3):529–534
73. Bohner M, Galea L, Doebelin N (2012) Calcium phosphate bone graft substitutes: failures and hopes. J Eur Ceram Soc 32(11):2663–2671
74. De A (2011) Wnt/Ca2+ signaling pathway: a brief overview. Acta Biochim Biophys Sin 43(10):745–756
75. MacLeod RJ, Hayes M, Pacheco I (2007) Wnt5a secretion stimulated by the extracellular calcium-sensing receptor inhibits defective Wnt signaling in colon cancer cells. Am J Physiol Gastrointest Liver Physiol 293(1):G403–G411
76. Zhao Y, Wang C-L, Li R-M et al (2014) Wnt5a promotes inflammatory responses via nuclear factor kB (NF-kB) and mitogen-activated protein kinase (MAPK) pathways in human dental pulp cells. J Biol Chem 289(30):21028–21039
77. Teuber SS, Saunders RL, Halpern GM et al (1995) Elevated serum silicon levels in women with silicone gel breast implants. Biol Trace Elem Res 48(2):121–130
78. Sugimoto J, Romani AM, Valentin-Torres AM et al (2012) Magnesium decreases inflammatory cytokine production: a novel innate immunomodulatory mechanism. J Immunol 188(12):6338–6346
79. Cramer T, Yamanishi Y, Clausen BE et al (2003) HIF-1alpha is essential for myeloid cell-mediated inflammation. Cell 112(5):645–657
80. Oda T, Hirota K, Nishi K et al (2006) Activation of hypoxia-inducible factor 1 during macrophage differentiation. Am J Physiol Cell Physiol 291(1):C104–C113

81. Grandjean-Laquerriere A, Laquerriere P, Jallot E et al (2006) Influence of the zinc concentration of sol-gel derived zinc substituted hydroxyapatite on cytokine production by human monocytes in vitro. Biomaterials 27(17):3195–3200
82. Velard F, Braux J, Amedee J et al (2013) Inflammatory cell response to calcium phosphate biomaterial particles: an overview. Acta Biomater 9(2):4956–4963
83. Cardemil C, Elgali I, Xia W et al (2013) Strontium-doped calcium phosphate and hydroxyapatite granules promote different inflammatory and bone remodelling responses in normal and ovariectomised rats. PLoS One 8(12), e84932
84. Buache E, Velard F, Bauden E et al (2012) Effect of strontium-substituted biphasic calcium phosphate on inflammatory mediators production by human monocytes. Acta Biomater 8(8):3113–3119
85. Wu C, Chen Z, Yi D et al (2014) Multidirectional effects of Sr-, Mg-, and Si-containing bioceramic coatings with high bonding strength on inflammation, osteoclastogenesis, and osteogenesis. ACS Appl Mater Interfaces 6(6):4264–4276
86. Wu C, Chen Z, Wu Q et al (2015) Clinoenstatite coatings have high bonding strength, bioactive ion release, and osteoimmunomodulatory effects that enhance in vivo osseointegration. Biomaterials 71:35–47
87. Wu CT, Fan W, Zhou YH et al (2012) 3D-printing of highly uniform CaSiO3 ceramic scaffolds: preparation, characterization and in vivo osteogenesis. J Mater Chem 22(24):12288–12295
88. Wu C, Fan W, Gelinsky M et al (2011) Bioactive SrO-SiO2 glass with well-ordered mesopores: characterization, physiochemistry and biological properties. Acta Biomater 7(4):1797–1806
89. Zhang M, Wu C, Lin K et al (2012) Biological responses of human bone marrow mesenchymal stem cells to Sr-M-Si (M = Zn, Mg) silicate bioceramics. J Biomed Mater Res A 100(11):2979–2990

Chapter 7
Modulation of Innate Immune Cells to Create Transplant Tolerance

Yue Zhao, Peixiang Lan, and Xian C. Li

Abstract Graft loss to rejection remains a key impediment to transplant success, which limits the therapeutic potential of this procedure. Though adaptive immune cells are critical in rejection, recent studies have demonstrated the importance of innate immune cells in dictating transplant outcomes (rejection or survival), highlighting the necessity in therapeutically targeting innate immune cells in the induction of tolerance to organ transplants. However, there are many challenges facing the field, as innate immune system consists of diverse cell types, molecular sensors, and soluble mediators that are different from those in the adaptive system. Also, some innate immune cells mediate graft injury, while others promote transplant survival, making therapeutic targeting of innate immune cells a challenging task. In this chapter, key elements in the innate immune system, their responses to organ transplants, as well as the challenges and opportunities in targeting those elements in favor of transplant survival are reviewed.

Keywords Innate immunity • Toll-like receptors (TLRs) • Natural killer cells • Macrophages • Dendritic cells • Complement • Ischemia–reperfusion injury • Transplant rejection • Transplant tolerance

7.1 Introduction

The innate immunity is evolutionarily conserved, and it precedes the emergency of adaptive immunity. From a functional standpoint, innate immunity acts as the first line of defense, and together with the adaptive immune system, they form an intricate system that protects the hosts from countless invading pathogens. Furthermore, the innate immune system also plays an important role in immune surveillance and tissue homeostasis by eliminating cancerous and stressed cells, thus restoring tissue integrity. These responses are carried out by innate immune cells, which include

Y. Zhao • P. Lan • X.C. Li (✉)
Immunobiology and Transplant Science Center, Houston Methodist Hospital and Houston Methodist Research Institute, 6670 Bertner Avenue, R7-211, 77030 Houston, TX, USA
e-mail: xcli@houstonmethodist.org

B. Corradetti (ed.), *The Immune Response to Implanted Materials and Devices*,
DOI 10.1007/978-3-319-45433-7_7

natural killer (NK) cells, dendritic cells (DC), macrophages, monocytes, neutrophils, mast cells, and eosinophils. As compared to adaptive T and B cells, the innate immune cells do not have somatically rearranged cell surface receptors that directly recognize foreign antigens. Instead, they express a variety of pattern recognition receptors that sense conserved pathogen moieties, danger signals, and damaged cell products; they also express receptors for complement products, antibodies, as well as complex activating and inhibitory receptors that allow them to constantly sense "self" or "missing self" to control their responses. These features allow innate immune cells to respond to potential pathogens in a timely fashion.

From a transplant standpoint, graft rejection was traditionally believed to be mediated by the adaptive immune cells. This is based on the observation that tissue and organ transplants often enjoy long term survival in transplant recipients that are deficient in T-cells. Recently, the involvement of innate immune cells in the control of transplant outcomes (rejection or acceptance) has been increasingly appreciated [1–3]. In fact, transplants provide a perfect environment for the activation of innate immune cells, as tissue damage is inevitable in transplantation due to ischemia–reperfusion injury and surgical trauma. Such innate cells influence transplant outcomes through a variety of mechanisms, which include modification of the T-cell activation programs. In different transplant settings, innate immune cells are closely involved in rejecting or protecting allotransplants, depending on the context and tolerizing protocols used; they also contribute to a wide spectrum of graft injury through diverse mechanisms. Thus, innate immune cells are important players in the overall immune responses to organ transplants, and because of that, successful induction of transplant survival or tolerance requires a detailed understanding of such cells, so that they can be specifically targeted following transplantation.

7.1.1 Elements of the Innate Immunity

The innate immune system consists of diverse cell types (Fig. 7.1), molecular sensors, cell surface activating and inhibitory receptors, as well as soluble mediators. The key elements in the system are discussed below in detail.

7.1.1.1 Innate Molecular Sensors

Innate immune cells respond vigorously to pathogens and damaged tissue products, and this is mediated by a complex array of molecular sensors initially called *pattern recognition receptors or PRRs*. These molecular sensors recognize conserved pathogen-associated molecular patterns (PAMPs) including proteins, lipids, nucleic acids, and carbohydrates derived from foreign microorganisms; they also recognize structures from damaged or stressed autologous cells known as damage-associated molecular patterns (DAMPs) or alarmins, and examples of these structures include high-mobility group box chromosomal protein 1 (HMGB1), heat shock proteins,

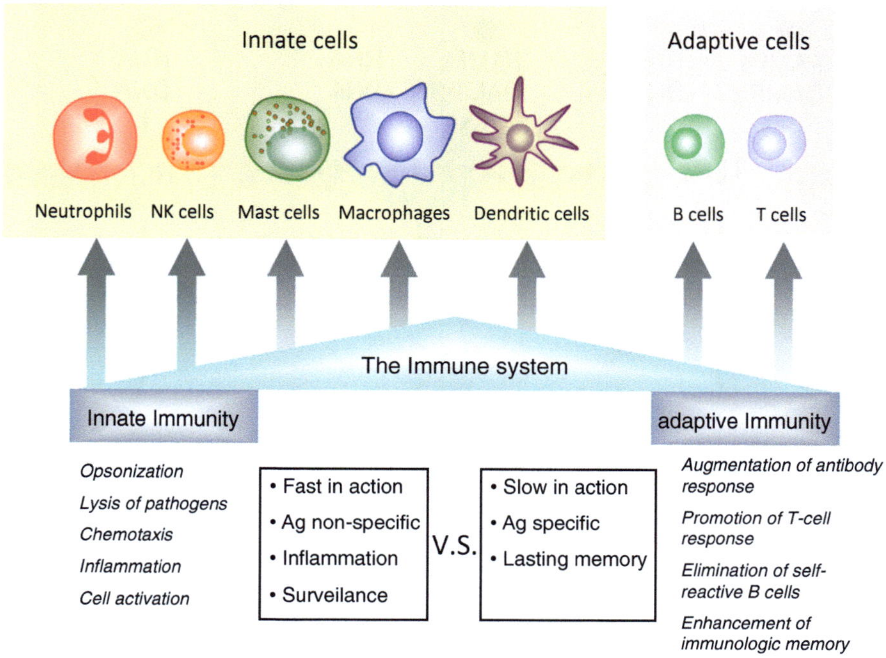

Fig. 7.1 Key cellular components in the innate system. The innate system consists of diverse cell types that form the first line of defense in the immune system. Together with the adaptive immune cells, they protect the hosts from countless invading pathogens

peptidoglycan, heparin sulfate, glucose regulated protein, fibrinogen, hyaluronic acid, and nucleotide fragments. Engagement of these innate sensors by corresponding ligands drives the activation of immune and inflammatory genes, and the products of them induce immune responses that function to limit invading pathogens or promote tissue remodeling [4].

There are three major families of innate molecular sensors identified thus far—*Toll-like receptors* or TLRs (29), *NOD-like receptors* (NLRs) [5], and *RIG-like receptors* (RLRs) [6] (Fig. 7.2). The TLR family consists of at least 13 members, and except TLR3, TLR7, TLR8, TLR9, which are intracellular structures anchored on the endosomal membrane, the rest are cell surface receptors, consisting of an extracellular domain, a transmembrane domain, and an intracellular domain. The extracellular domain contains multiple leucine-rich repeat motifs (LRRs) that are responsible for ligand binding; the intracellular domain transduces activation signals through either the MyD88-dependent pathway, which activates the NF-kB or the TRIF-dependent pathway, which stimulates Type I interferons [7]. In contrast to the membrane-anchored TLRs, NLRs are cytoplasmic proteins, consisting of multiple functional domains that are involved in ligand binding, oligomerization, and signal transduction. There are at least 23 members in the NLR family, but NOD1, NOD2, NALP1-3 and Ipaf are the best studied ones so far. NOD1 and NOD2 activate NF-kB, with subsequent induction of potent inflammatory cytokines, including

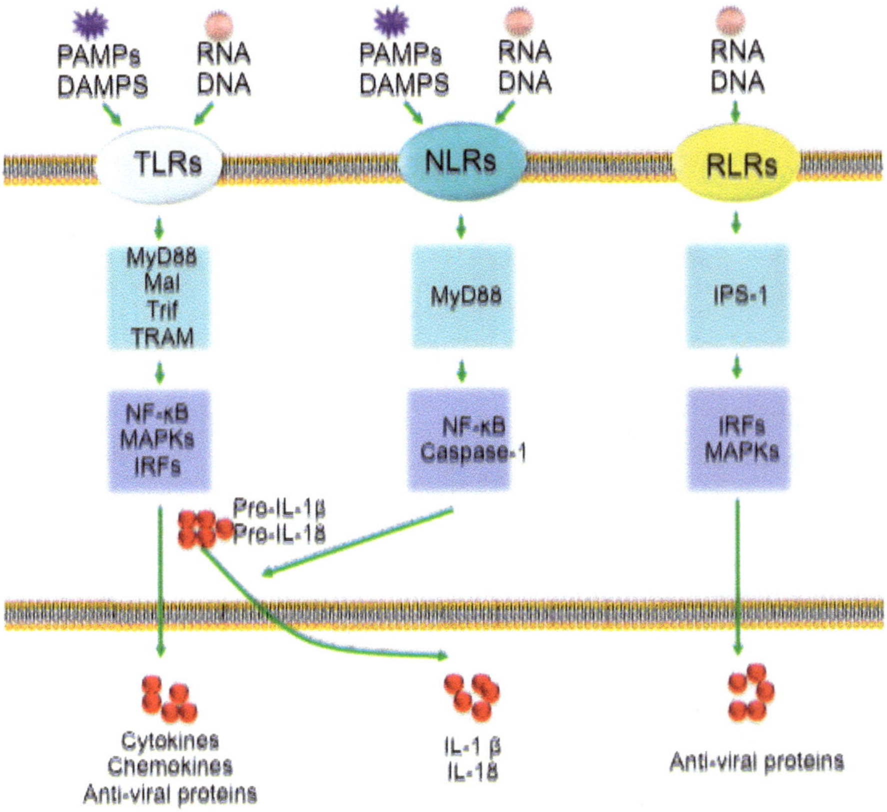

Fig. 7.2 Innate sensors and their signaling pathways. The innate sensors are broadly divided into three families, which include transmembrane receptors and intracellular ones. They respond to pathogen products and endogenous ligands from damaged cells with certain degree of specificities. Collectively, these innate sensors provide an extraordinary early warning system to detect and respond "danger signals"

IL-1β and IL-18. Furthermore, NALP1-3 and Ipaf form inflammasomes which activate caspase 1, resulting in processing and production of active IL-1β and IL-18 [5]. RLRs are an interesting family of innate sensors; they are also cytoplasmic proteins akin to NLRs. Similar to intracellular TLRs (TLR3, TLR7, TLR8, and TLR9), RLRs stimulate NF-kB activation and Type I interferon production. However, RLRs recognize viral nucleic acids in the cytoplasm, whereas intracellular TLRs recognize viral products in the endosomal compartments [6]. In addition, RLRs are widely expressed in innate immune cells and rapidly upregulated in response to Type I interferons. Together, these families of sensor molecules provide an extraordinary early warning system against pathogens and tissue injuries.

The innate molecular sensors possess multiple interesting features. TLRs, NLRs, and RLRs respond to pathogens with a certain degree of specificity. TLRs recognize extracellular and intracellular pathogens, and individual TLRs respond to different

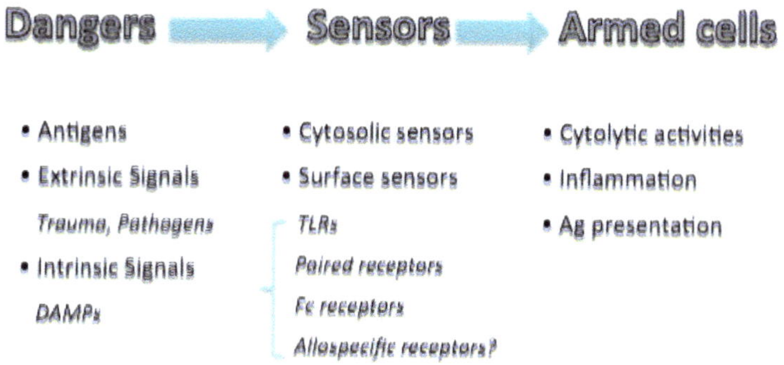

Fig. 7.3 The principle in activation of innate cells. Steps in activation of innate immune cells involve "dangers," which can be intrinsic or extrinsic; these danger signals are picked up by sensor molecules either on the surface of innate immune cells or inside the cells. The outcome of such response is the production of activated or armed innate cells to execute effector functions

molecular entities. On the other hand, NLRs sense intracellular bacteria-derived products, whereas RLRs detect intracellular viral nucleic acids. They also cooperate with each other as a flexible network to ensure elimination of invading pathogens while avoiding harmful immunopathology. The enhanced function between of PRRs could be exemplified by the interplay of NOD1-2 and NALP1-3 in the generation of active IL-1 and IL-18. After LPS stimulation, NLRC3 expression is diminished, thus attenuating TLR signaling due to a negative feedback regulation [8]. Stimulation of these receptors results in the activation of innate immune cells, production of potent pro-inflammatory cytokines, maturation of DCs, and initiation of adaptive immunity (Fig. 7.3). Clearly, the innate molecular sensors play a critical role linking the innate and adaptive responses.

7.1.1.2 Complement

Complement consists of numerous serum proteins (C1 to C9), which are produced primarily in the liver. But recent studies indicate that other tissues and cells also produce complement components in situ. In essence, complement requires activation to exert its functions and complement activation is tightly regulated by both positive and negative regulators [9].

There are three pathways whereby complement activation is initiated- *the classical pathway*, *the alternative pathway,* and *the lectin pathway* (Fig. 7.4). The classical pathway is triggered by antigen-antibody complexes, which activate C1 (made up of C1q, C1r, and C1s components). A conformational change in C1q leads to cleavage of C1s, which activates C2 and C4. When C4a and C4b fragments are produced, a sulfhydryl group on C4b is targeted for inactivation by factor I. The resultant product C4b complexes with C2a to form C3-convertase, which cleaves C3 into C3a and C3b. C3b then binds to the C4b/C2a complex to form C5-convertase. This cleaves C5 into C5a and C5b, with C5b initiating the formation of membrane-attacking complex (MAC)

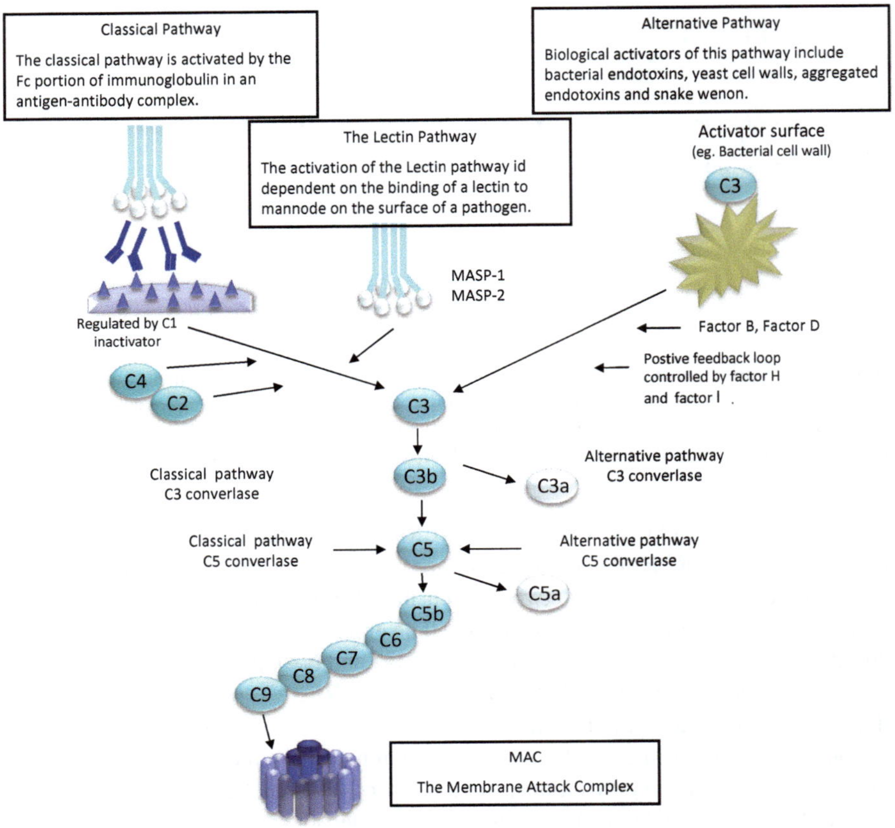

Fig. 7.4 Pathways of complement activation. There are three pathways to trigger complement activation—the classical pathway, which is initiated by antigen–antibody complexes, the alternative pathway, and the lectin pathway. All pathways converge on the activation of C3 convertase, and end with the formation of the membrane attack complex (MAC), which mediates cell lysis by breaking down cell membrane

consisting of C5b to C9. MAC then mediates lysis of target cells by disrupting cell membrane. In transplant settings, complement activation can be induced by organ ischemia during transplantation surgery, as well as by antibody binding graft alloantigens during acute and chronic rejection [10].

All three pathways converge at the production of C3 convertase, which promotes the formation of MAC and the generation of soluble complement fragments C3a and C5a that serve as powerful chemoattractants and opsonins for activation of other innate immune cells. The classical pathway of complement activation is a primary effector mechanism in antibody-mediated vascular injury. However, in other inflammatory responses including ischemia–reperfusion injury, complement products can additionally function as a "danger signal" and aid in T cell priming by acting as chemoattractants. Complement activation can be inhibited by multiple regulatory proteins. *Membrane cofactor protein* (CD46) and *Decay accelerating factor* (DAF, also known as CD55) are cell surface molecule expressed on endothelial cells, accelerates the decay of C3 convertases, thus preventing the amplification of complement cascade and formation of the downstream membrane attack complex [11]. Upon complement deposition, endothelial cells may upregulate CD59, increasing the resistance of the cells to MAC formation [12]. Also, C4-binding protein is a soluble regulatory protein that inactivates C4b and thus the classical pathway convertase [13]. CR1 (CD35) and factor H can also affect the degree of complement activation [14, 15]. Because of their regulatory properties, such inhibitory proteins have been targeted in preventing complement-mediated graft injury, including ischemia–reperfusion injury, which is discussed below.

Besides target cell lysis, chemoattraction, and innate cell activation, complement has other roles, such as costimulation of T cell activation [16, 17], reduction of induced-Treg (iTreg) generation [18], and maturation of DCs [19]. T cells and DCs can produce complement components and also express receptors for C3a and C5a. Both cell types employ the complement pathway to optimally function. In this setting, complement produced in situ, rather than systemic complement, is critically involved [20]. Similarly, graft-derived complement, not systemic ones, has been shown to contribute to graft injury following transplantation [21].

7.1.1.3 Innate Immune Cells

Natural Killer Cells

Natural killer (NK) cells are well represented in the blood, spleen, lymph nodes, and other lymphoid tissues; they also reside in non-lymphoid sites in large numbers, especially in the liver and the lungs [22]. NK cells are a major cell type in the innate immune system and represent the third largest lymphocyte population (5–15 %) in the blood (besides T cells and B cells). NK cells are present at inflammatory sites where they interact with other innate and adaptive immune cells.

Mature NK cells exhibit potent cytolytic activities; they readily kill target cells without prior antigen priming, which contrasts sharply to adaptive T cells. Target cell killing is mediated primarily by the release of perforin and granzymes, which are tightly packed inside NK cells as large preformed granules, and target cell recognition by NK cells triggers rapid degranulation and target cell apoptosis. NK cells also produce copious amounts cytokines upon stimulation, and such cytokines include both pro-inflammatory (e.g., IFN-γ, TNF-α) and anti-inflammatory cytokines (e.g., IL-10, TGF-β), thus exerting diverse impacts on the nature of the immune response.

NK cells express a multiplicity of cell surface receptors, and these receptors collectively control development, education, self tolerance, as well as effector functions of NK cells [23, 24]. From a functional standpoint, NK receptors are divided into *activating receptors* and *inhibitory receptors* (Fig. 7.5). The activating receptors include natural cytotoxicity receptors or NCR (NKp46, NKp44, and NKp30), c-type lectin-like Ly49 receptors in the mouse (e.g., Ly49H, Ly49D) and Killer Immunoglobulin-like Receptors (KIR) in humans, and NKG2 family recep-

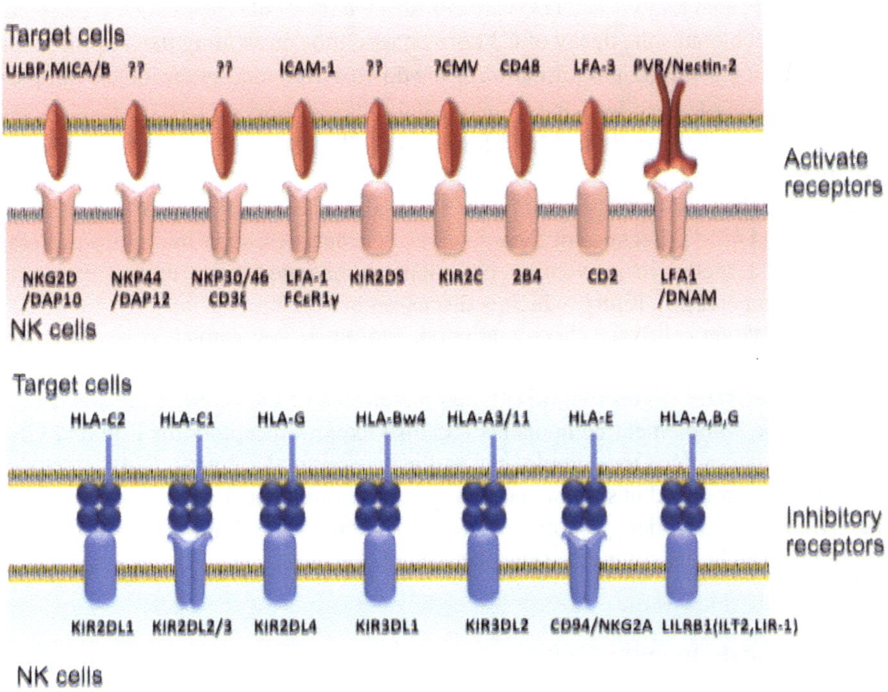

Fig. 7.5 NK cells express complex activating and inhibitory receptors. The NK receptors can be divided into activating receptors and inhibitory receptors. The ligands for the inhibitory receptors are self-MHC class I molecules, and those for the activating receptors include pathogens, some MHC molecules, and induced surface molecules on stressed or damaged cells. Signals from both activating and inhibitory receptors collectively control the functional status of NK cells

tors (NKG2C, NKG2D). Certain NK subsets also express CD16 and CD27 that function as activating receptors. In mice, the inhibitory NK receptors belong to the Ly49 family (e.g., Ly49C, Ly49G, Ly49I), whereas human NK cells express inhibitory receptors of the KIR family, which include KIR2DL1, KIR2DL2, KIR2DL3, KIR3DL1, and KIR3DL2. In both humans and mice, NK cells also express the heterodimeric inhibitory receptor NKG2A-CD94. With few exceptions, most NK activating receptors lack signaling motifs in their cytoplasmic domains. Instead, they recruit adaptor molecules DAP10, DAP12, CD3ς, or FcεRIγ to transduce activating signals. In contrast, the inhibitory NK receptors directly signal through the immunoreceptor-based tyrosine inhibitory motifs (ITIM) in their cytoplasmic tails. As can be envisioned, the ligands for such NK receptors are extremely diverse, which include pathogens, polymorphic MHC class I molecules or class I-related molecules, de novo induced molecules on damaged or stressed cells.

The NK inhibitory receptors attracted most attentions initially. As mentioned above, the ligands for such inhibitory receptors are self MHC class I molecules, and engagement of NK inhibitory receptors by self MHC class I prevent self destruction and ensures self tolerance of NK cells. During NK development, those that have their inhibitory receptors engaged are selected to undergo further maturation and become functionally competent killers against nonself targets. This feature is developmentally acquired and often called "*NK education" or "NK licensing*" [25]. As most mature NK cells in the periphery express at least one inhibitory receptor, and therefore they are fully licensed; the more inhibitory receptors individual NK cells express, the greater the killing activities toward target cells. Those that fail to express inhibitory receptors are render anergic or incompetent (*unlicensed NK cells*). Hence, upon encountering target cells that are missing or downregulated self MHC class I molecules, NK cells will unleash their potent killing activities. This also applies to MHC class I mismatched target cells in transplantation (both bone marrow and solid organ transplants) [26, 27], and NK-mediated killing in this setting is called "*missing self recognition*" [28]. However, damaged and stressed cells often express multiple ligands for NK activating receptors in addition to self MHC class I, and therefore, are often killed by NK cells, a process also called "*induced self recognition.*" It is now widely accepted that integration of signals from both activating and inhibitory receptors determines NK cells actions. Thus, if multiple activating receptors are engaged, such activating receptors can override the inhibitory signals to initiate NK mediated killing of target cells.

In essence, NK cells are controlled by a hierarchy of stimulatory and inhibitory signals, with predominance of inhibitory signals under physiological conditions. There is evidence that under conditions of viral infection, NK cells that do not express inhibitory receptors (unlicensed NK cells) respond much better than those that express the inhibitory receptors. Also, NK licensing is a dynamic and continuous process in such a way that mature NK cells can be re-educated [29]. Additionally, different regulatory mechanisms may operate in the control of cytokine production by NK cells. The diversity and plasticity of NK cells suggest that their roles in immune responses are complex.

Monocytes and Macrophages

Monocytes and macrophages are mononuclear phagocytes in the myeloid lineage. Monocytes are developed from bone marrow myeloid precursor cells, reside in the blood, and upon inflammatory triggering, they rapidly infiltrate inflammatory sites where they mature into macrophages [30]. In some organs, macrophages acquire additional tissue specific features and become integral cellular component of the organ, such as Kupffer's cells in the liver and glia cells in the brain. In humans, monocytes constitute 5–10 % of the peripheral circulating leukocytes in the blood. Although monocytes and macrophages are closely related, they can be distinguished by different surface markers. For example, monocytes are $CD11b^{high}$ and $F4/80^{-}$, they express low levels of MHC class II and CD80, CD86, CD40 costimulatory molecules. In contrast, macrophages are $CD11b^{high}$ and $F4/80^{+}$, and also have high levels of surface MHC class II, CD80, CD86, and CD40 molecules. Besides differentiating to macrophages at inflammatory sites, monocytes also differentiate into myeloid DCs in situ in selected models. Additionally, $CD14^{+}CD16^{-}$ monocytes are highly efficient at phagocytosis, can produce high levels of inflammatory cytokines, whereas CD14^{low}CD16^{+} monocytes express MHC class II molecules and are highly efficient at antigen presentation [30]. Furthermore, recent studies using intravital microscopy to directly examine blood monocytes in vivo revealed additional complexity. For instance, Ly6C^{-}GR1^{-} monocytes patrol the vessel walls, and after extravasation differentiate into Type II macrophages (see below), whereas Ly6C^{+}GR1^{+} monocytes differentiate into dendritic cell-like cells or generate an Type I macrophages [31].

Macrophages are a major cellular infiltrate at inflammatory sites with considerable diversity and plasticity; they primarily function as APCs and inflammatory cells [32]. Macrophages efficiently phagocytose foreign entities and then present the antigenic epitopes to T cells to initiate the adaptive immunity; they also produce as well as respond to a wide array of inflammatory cytokines, which further amplify their phagocytosis and APC functions. Similar to DCs, macrophages also consist of multiple phenotypic and functionally different subsets. The cytokine milieu in which macrophages are stimulated plays a decisive role in the polarization of macrophages. Activation of TLRs and/or Th1 type of immunity often polarizes macrophages to M1 cells (also called classically activated macrophages), which are potent producers of nitric oxide and pro-inflammatory cytokines that mediate acute inflammation and cellular toxicity. On the other hand, presence of IL-4 and IL-13 or activation of Th2 cells usually skews macrophage to M2 cells or alternatively activated macrophages that are involved in immune regulation, wound healing, as well as in tissue repair and regeneration. In fact, M1 and M2 cells most likely represent the extreme ends of a wide spectrum where multiple other subsets exist, especially under in vivo settings. However, the M1/M2 paradigm does highlight the dynamic nature of macrophages and the importance of environmental cues in dictating macrophage functions [33].

Under certain conditions some macrophages can become potent inducers of Foxp3^{+} Tregs, most likely via the expression of PD-L1, which indirectly contribute to immune regulation [34, 35], while others contribute to immune regulation and tolerance by differentiating into *myeloid-derived suppressor cells* (MDSC), a cell type with potent immunosuppressive properties [36]. Finally, there is emerging evidence that macrophages can be alloreactive and respond directly to allo [37] or xenoantigens [38], although the exact molecular basis for such alloreactivity remains to be defined.

Dendritic Cells

Dendritic cells (DCs) are an important cell type in the innate immune system, representing ~1 % of total PBMCs; they are rare compared to other cell types, but they are widely distributed in the body. DCs are identified morphologically by having long dendrites projecting from the body of the cells. Dendritic cells are developed from bone-marrow precursors, and based on anatomic locations, surface markers, and maturation status, they can be divided into various subsets [39, 40]. DCs in the lymphoid tissues are called lymphoid tissue resident DCs and those in other sites are often termed interstitial DCs. Phenotypically, DCs express the β integrin CD11c, and in combination with other surface markers, they can be further divided into *myeloid DCs* ($CD11c^+CD11b^+CD205^-$), *lymphoid DCs* ($CD11c^+CD11b^-CD205^+$), and *plasmacytoid DCs* ($CD11c^+B220^+PDCA^+$). Thus, DCs, though highly specialized, are extremely heterogeneous, and different DC subsets perform different functions in vivo. Earlier studies indicate that various DC subsets reside in different locations in the lymphoid tissues and exhibit striking difference in their expression of chemokines, chemokine receptors, and homing receptors. They also secrete different cytokines upon activation, which subsequently create different cytokine milieu for the differentiation of adaptive T cells. For example, plasmacytoid DCs produce copious amount of type I interferons whereas myeloid DCs secret high levels of IL-12 upon activation.

DCs represent the first line of defense through uptake, process, and present foreign antigens to adaptive immune cells. However, DCs are very dynamic; their phenotypes and functions can be profoundly modulated by a variety of mechanisms. They express an incredibly complex array of Toll-like Receptors (see below) that allow them to respond to microbial products or endogenous tissue products collectively called "danger signals"; they are capable of producing and also responding to a plethora of inflammatory cytokines. These responses often result in further proliferation and maturation of DCs, as reflected by the heightened expression of MHC molecules and T cell costimulatory molecules on the cell surface (e.g., CD80, CD86, CD40, and OX40 ligand). During this process, DCs are often transformed into potent antigen-presenting cells (APCs) that are highly efficient in the activation of adaptive immune cells. Thus, depending on the presence or absence of "danger signals," DCs can remain in a state of resting cells or become fully activated, and there is evidence that resting and activated DCs may perform different functions in vivo. While activated DCs are potent triggers of immune activation, resting DCs that expresses low levels of MHC II and low levels of costimulatory molecules, are likely required for immune tolerance by engaging T cell anergy and supporting regulatory cells [41, 42]. The recent demonstration that genetic ablation of DCs in naïve mice results in widespread autoimmune diseases suggests that DCs are essential not only for robust immunity but also for tolerance and immune homeostasis [43]. Thus, DCs provide an essential link between the innate and adaptive immune systems; they are also well positioned at the interface between immunity and immune tolerance.

Other Innate Cells

Neutrophils are generated from bone marrow and function as inflammatory cells in infection. Besides, neutrophils participate in the transplant-mediated ischemia–reperfusion injury. Upon activation, neutrophils can infiltrate to the damaged tissue within minutes and further exacerbate tissue damage through release of ROS, proteinases, and cationic peptides [44]. Tissue damage can be further exacerbated through neutrophils occlusion in capillaries, which prevents reperfusion [45]. In fact, attenuating neutrophil-mediated tissue damage during reperfusion significantly postpones T-cell infiltration and improves cardiac allografts survival [46].

Mast cells are regarded as cells in allergic responses and parasite infection immunity. Increasing evidence suggests that mast cells exhibit diverse phenotypes and constantly adjust themselves to microenvironment cues; they also interact with adaptive immune cells in transplantation. In humans, mast cells that express HLA-DR molecules present peptide antigens to antigen specific $CD4^+$ T cells, resulting in $CD4^+$ T cell activation [47]. As a consequence, the interplay between mast cells and $CD4^+$ T cells favors the upregulation of HLA-DR on mast cells in return. Aside from its role as antigen presenting cells to activate T helper cells, mast cells interact with Treg cells, highlighting a role for mast cells as regulators of tolerance. It has been documented that mast cell degranulation lead to a loss of Treg-cell mediated skin allograft tolerance, while the presence of Tregs hampers mast cell degranulation [48]. Moreover, transplant tolerance was restored after transferring mast cells into mast cell-deficient mice that graft would otherwise be easily rejected [49].

Innate lymphoid cells (ILCs) are lymphocytes lacking rearranged antigen-specific receptors and lineage markers that define conventional T and B cells. Based on the key transcription factors and cytokines they secrete, ILCs are divided into group 1, group 2, and group 3 innate lymphocytes (e.g., ILC1, ILC2 and ILC3). ILC1 express the transcription factor T-bet and produce type 1 cytokines such as IFN-γ and TNF-α. ILC2, similarly to Th2 cells, depend on GATA-3 and produce type 2 cytokines like IL-5, IL-13, and amphiregulin. ILC3 are characterized by RORγt expression and IL-17 and IL-22 secretion. ILCs played an important role in maintaining immune homeostasis at mucosal surface in the gastrointestinal tract. A significant number of ILC1 and ILC3 cells are present in human intestinal allografts and regulate the balance between rejection and tolerance [50]. Similarly, IL-22 producing ILC3 cells, which promote tissue repair, tissue homeostasis, and immunity to extracellular bacteria, can also attenuate GVHD [51].

Natural killer T (NKT) cells are CD1d-restricted cells; they belong to a subset of T lymphocytes that express an invariant T cell receptor as well as NK cell markers. NKT cells play an important role in bridging innate immunity and adaptive immunity [52]. Previous data indicated that NKT cells are required in maintaining transplant tolerance. In an islet xenograft tolerance model induced by using anti-CD4 mAbs, as well as a cardiac transplant tolerance model using costimulation blockade therapy, tolerance to grafts failed to establish when recipient mice were deficient in NKT cells, but could be restored by adoptive transfer of NKT cells to host mice [53, 54].

7.1.2 Responses of Innate Cells to Transplants

The innate immune cells use different mechanisms to respond to transplants, and such responses include ischemia–reperfusion injury, acute rejection and chronic rejection, and eventually transplant tolerance. Overall, innate cells are key contributors to rejection, but they are also key regulators of transplant tolerance.

7.1.2.1 Ischemia–Reperfusion Injury

Surgical trauma, graft preservation, ischemia, and reperfusion are integral parts of organ transplantation, which inevitably results in tissue injury. The imbalance of metabolic supply and demand during ischemia also creates an intense intragraft hypoxia and vascular dysfunction, leading to further graft damage. Damaged or stressed graft constituents first mobilize the innate immune cells via a variety of pathways that lead to tissue inflammation, production of reactive oxygen species (ROS) and pro-inflammatory cytokines, APC activation, and ultimately, greater risk of graft loss. This process involves virtually all innate immune cells, molecular sensors and pathways.

Graft IRI triggers rapid influx of monocytes, macrophages, neutrophils, NK cells as well as certain memory T cells into the grafts; IRI also mobilizes graft interstitial DCs and migration of host DCs. This massive cellular infiltration, probably in response to potent inflammatory cytokines, chemokines, and complement products, creates an ideal ground for collaboration, activation, and amplification of cellular responses. In this context, key processes include (1) stimulation of TLRs by DAMPs from damaged cells, (2) production of additional inflammatory, cytotoxic cytokines and ROS by activated innate cells, (3) maturation of APCs capable of inducing adaptive responses, and (4) induction of stress markers on otherwise healthy cells, which are recognized by NK cells and trigger NK-mediated killing. All of these responses are extensively studied in preclinical models, and certain pathways are becoming attractive therapeutic targets for further clinical development. For example, damaged cells release a large amount of the nuclear protein HMGB1, which engages TLR2, TLR4, or RAGE to activate innate immune cells. Thus, mice lacking TLR4, MyD88, and/or TRIF were protected from IRI [55]. Similarly, in liver IRI and myocardial infarction models, the degree of tissue damage was markedly reduced in TLR4 deficient mice [56, 57]. At a cellular level, depletion of macrophages with liposomal clodronate or inhibiting their trafficking to the graft ameliorates tissue damage. Furthermore, in models of renal IRI, depletion of kidney DCs in CD11c-DTR reporter mice with diphtheria toxin protected against tubular cell necrosis, leading to less renal dysfunction [58]. Additionally, NK cells readily kill renal tubular epithelial cells through recognition of an induced molecule Rae-1 after ischemia–reperfusion, and NK deficiency or antibody-mediated NK depletion protected against kidney IR injury [59]. Moreover, inflammatory macrophages played an important role in IRI injury [60]. On the other hand, alternatively activated macrophages may contribute to tissue repair after the injury [61]. These findings provide solid evidence on the importance of innate immune system in graft IRI.

Complement also contributes to ischemia–reperfusion injury, and seminal studies using cobra venom factor to inhibit complement demonstrated mitigation of IRI [62]. Other studies have since shown the importance of various complement components in IRI [63]. For instance, complement receptor 1, complement receptor-related gene Y, C1 esterase, C3 [64], C5 [65], factor B, and decay-accelerating factor (DAF) [11, 66] have all been implicated in IRI in various animal models. Indeed, treatment of mice with DAF, which inhibits C3 convertase, prior to reperfusion protects them from IR injury [67]. This protective effect was due to the decrease in complement activation and pro-inflammatory cytokine release. While, serum complement derived from the liver is responsible for antibody-initiated mediated injury, it appears that the graft-derived complement also contributes to IRI [21, 68]. It has been shown that transplant recipients with normal serum complement levels show prolonged graft survival when transplanted with C3 deficient kidneys. Additionally, transplant of these grafts into syngeneic recipients prevented IR injury [69]. Further proof of the importance of local complement to tissue injury came from studies where over-expression of a complement regulatory protein DAF on graft endothelium reduced the extent of IRI following transplantation [70].

7.1.2.2 Acute Transplant Rejection

In most cases, innate immune cells do not directly mediate acute and complete allograft rejection by themselves. However, they are involved in the control of adaptive T cell programs, thus indirectly affecting the nature of the rejection response. This aspect of the innate immune system plays important roles in transplant outcomes and has being increasingly appreciated.

The transplanted grafts provide an ideal environment for the innate immune cells. The "danger signals" generated by IRI stimulate the production of potent pro-inflammatory cytokines, such as IL-1, IL-6, TNF-α, and chemokines (e.g., MIP-1α, MIP-1β, MCP-1, IP-10, and Rantes), which mobilize graft interstitial DCs and also mediate rapid influx of host monocytes, macrophages, neutrophils, host DCs as well as adaptive T cells and B cells. Apart from danger signals, monocytes can also recognize allogeneic nonself, inducing Th1 immunity and eventually to graft rejection [71]. The innate cells infiltrating the grafts produce additional inflammatory cytokines and chemokines upon activation, further amplifying the inflammatory milieu in the graft. This rich inflammatory milieu drives proliferation and maturation of APCs, which includes upregulation of MHC class I and class II molecules on the cell surface and induction of costimulatory molecules (CD80, CD86, CD40, OX40L, etc.). Matured APCs effectively engage T cells and B cells to initiate the rejection response [72]. Thus, innate immune cells have to partner with adaptive T cells and B cells to mediate acute rejection. In other words, adaptive immune cells require activated and mature APCs to present alloantigens to trigger their activation. This interdependence in graft rejection was shown in some preclinical animal models. In a minor alloantigen mismatched mouse model (male into female skin transplantation), genetic deficiency for the adaptor molecule MyD88, which prevents TLR signaling, and therefore maturation of APCs, led to indefinite skin graft survival [73]. In a more stringent setting, deficiency of both MyD88 and TRIF prolonged the survival of MHC mismatched allografts [74]. These findings also highlight the importance of innate sensors in APC maturation and allograft rejection.

During acute rejection, there is reciprocal migration of APCs, especially DCs. Host DCs can infiltrate the transplanted graft and graft DCs can home to the host's draining lymph nodes. The relative importance of these pathways in transplantation is a matter of continuing debates, but either pathway can trigger acute rejection. Although donor DCs and host DCs present alloantigens differently to T cells, the same mechanisms (e.g., innate sensors, pro-inflammatory cytokines) are involved in driving their activation and maturation, which allow them to optimally engage T cells. Some innate immune cells such as macrophages and NK cells, once acquiring effector functions, contribute significantly to the destruction phase of an acute rejection response. Macrophages have long been identified in allograft biopsies of human kidney transplants and in animal models; they may account for 40–60 % of infiltrating leukocytes [75] as detected by immunohistochemical staining for CD68, an intracellular lysozyme-associated glycoprotein used most commonly to detect human macrophages. There are also observations that macrophage infiltration is also seen with acute vascular rejection with endothelialitis or intimal arteritis [76]. As viewed from animal models of acute rejection where macrophage depletion or antagonism [77–79] was explored, it is clear that these cells are critical components of the acute response against the allograft. This is also supported by clinical studies. In T-cell depleted patients (with alemtuzumab), renal transplant recipients still experience acute graft rejection, which is often associated with massive infiltration of monocytes [80]. Recent data indicated that infiltrated macrophage during acute rejection exhibits a pro-inflammatory phenotype [81]. Similarly, eosinophil-driven acute rejection has been observed in intestinal transplantation using alemtuzumab or thymoglobulin [82]. Along the same line, NK cells also contribute to acute rejection. For example, in CD28KO mice, antibody-mediated blockade of NKG2D, an NK-cell activating receptor, significantly prolongs cardiac graft survival [83], suggesting that NK cells facilitate rejection [84]. Additionally, if NK cells are activated by IL-15, they themselves are capable of mediating acute rejection in the absence of T or B lymphocytes [85]. Moreover, NK cells constantly interact with other cells. NK cells can augment mature DC function through the production of TNF-α or eliminating immature DCs through cytolytic activities [84]. NK cells can also promote Th1 cell differentiation through secreting IFN-γ or killing Tregs to boost adaptive immunity [86].

Among the innate molecules, complement has a unique role in allograft rejection. In transplant models where complement activation is induced, graft destruction is incredibly fast; can be within hours, and always irreversible. Graft rejection in this setting often involves vascular endothelial damage, blood coagulation, and activation innate cells without the participation of adaptive T cells. The best example is humoral rejection or acute antibody-mediated rejection (AMR) triggered by anti-donor alloantibodies [87]. The anti-donor antibodies (mostly against donor HLA molecules, ABO antigens, and endothelial antigens) form immune complexes with donor antigens in the grafts, activate complement cascade via the classical pathway. This process is robust, initiated by the activation of C1 and resulted in formation of MAC (C5b-C9 complex), generation of chemoattractants C3a and C5a, and consequently massive cell death, intragraft inflammation, and extensive blood coagulation in the grafts. Activation of complement can further amplify the humoral immune response to foreign antigens, creating a positive feedforward loop [88]. In addition, recent studies indicate that T cells and APCs themselves produce complement components, and also express receptors for selected complement elements (e.g., receptors for C3a and C5a). Thus, they can employ the locally produced complement to optimally function, through effects on maturation of APCs and costimulation of T cell activation upon alloantigen stimulation [68, 89]. Furthermore, the engagement of C3aR and C5aR on APCs induces the release of innate cytokines (IL-12, IL-23) and upregulates costimulatory molecules, again amplifying the T effector response [17, 20]. This provides another example on how innate molecules enhance adaptive immune response in the transplant settings. The role of complement in alloreactive T-cell immunity and IR injury explains, at least in part, the fact that murine kidney allografts deficient in C3 exhibit long-term survival [21], whereas those deficient in DAF have worse outcomes [90].

7.1.2.3 Chronic Transplant Rejection

One outstanding feature is that chronic allograft rejection mainly affects the graft vasculature [91, 92]. Morphologically, chronic rejection is characterized by concentric neointimal proliferation and eventual occlusion of blood vessels, and this lesion affects vessels of all sizes in the graft. Also, extensive graft interstitial fibrosis is frequently accompanied with vasculature changes. These features are unique to transplants, and therefore also called transplant vasculopathy. In contrast to acute rejection, chronic rejection takes much longer to develop and often requires years following transplantation in patients. This has become a major cause of graft loss impeding long-term transplant success in the clinic. The exact mechanisms of chronic rejection remain elusive, but the current belief is that chronic rejection is perhaps a manifestation of graft injury and remodeling over a long period of time in which both immune and nonimmune pathways are critically involved. Importantly, current studies suggest the importance of innate immune mechanisms in the pathogenesis of chronic transplant rejection.

There are distinct features in chronic rejection. In contrast to acute rejection, innate immune cells, mostly monocytes and macrophages, dominate the cellular infiltrates in the lesions, and alloantibodies and complement depositions are also frequently detected. Furthermore, molecular profiling studies often demonstrate ongoing tissue inflammation in chronic lesions, as shown by the heightened expression of multiple inflammatory cytokines. Thus, it is conceivable that tissue damages driven by innate cells, pathways, and innate molecular sensors may contribute significantly to the development of transplant vasculopathy.

It has been shown that TLR signals are strongly associated with the development of atherosclerosis, which is a hallmark of chronic allograft vasculopathy [93, 94]. Recently, TLR2, TLR4 and the adaptor proteins MyD88 and TRIF have all been found to be key mediators of chronic rejection in a fully mismatched mouse kidney transplant model [95]. Additionally, in heart transplant patients with evidence of allograft endothelial dysfunction, TLR4 expression and secretion of IL-12 and TNF, which are downstream targets of TLR signaling, were found to be at higher levels than in heart recipients without endothelial dysfunction [96], suggesting the involvement of innate sensors in cellular activation. At a cellular level, macrophages infiltrate heart allografts and contribute to transplant vasculopathy in an animal model of chronic rejection [97]. In this model, partial depletion of macrophages using carrageenan reduced the severity of chronic vasculopathy. This was independent of phagocytosis, as treatment with gadolinium, which inhibits phagocytosis, had no effect on the severity of the disease. In another study, targeting macrophage function using an adenoviral strategy ameliorated the histological features of allograft dysfunction in a rat model of interstitial fibrosis and tubular atrophy (IF/TA) [98]. Mechanistically, monocytes/macrophages, by infiltrating the damaged allograft parenchyma under the influence of chemoattractants, secrete growth factors and pro-fibrotic cytokines such TGF-β and IL-13 [99]. They can also directly differentiate into fibrocytes [100], thus promoting synthesis of extracellular matrix proteins, and stimulating transition of tubular epithelial cells into fibroblasts [99]. In human transplant recipients, presence of macrophages in early biopsy specimens is predictive of IF/TA development [101, 102]. Among macrophage subsets, many reports highlight the role of M2 cells in chronic graft loss. A strong association between the degree of M2 phenotype and fibrosis in chronic kidney rejection patients has been reported [103]. For instance, Kaul et al. reported that macrophages with M2 features infiltrated the graft during chronic rejection [81]. Additionally, upregulation of $CD163^+$ M2 macrophages by steroids and CNI correlated with increased pro-fibrotic cytokines and accelerating rejection [104].

The presence of circulating alloantibodies against HLA, MICA, autoantigens, and endothelial antigens increases the risk of long-term graft loss [105, 106]. The effector response initiated by alloantibodies in graft damage involves primarily innate pathways. Of central importance is the activation of complement. Indeed, in kidney transplantation in humans, the glomerulopathy and arteriopathy seen in chronic rejection are closely associated with C4d deposition in the graft, supporting a role of complement activation in chronic graft injury [87]. In animal models, C6 deficiency, which affects the formation of the terminal membrane attack complex of the complement cascade, reduces the severity and onset of graft arteriosclerosis [107]. Recent data also demonstrate the importance of locally derived complement (rather than systemic complement) in chronic kidney graft injury. For example, C3 deficient kidney transplants are resistant to adriamycin-induced tubular damage when transplanted into wild type recipients [108]. In C3a receptor deficient mice, adriamycin induced less kidney injury with lower expression of interstitial type 1 collagen and α-smooth muscle actin. Injury by graft-derived complement is also thought to impact long-term graft outcome in humans.

There are circumstances where alloantibody-induced chronic allograft rejection can still arise independent of complement, and a recent study identified NK cells as key effector cells in driving chronic graft damage [109]. In a mouse model of heart transplantation in which graft recipients are deficient for T cells and B cells (e.g., Rag-deficient mice), infusion of donor alloantibodies into transplant recipients induced prominent allograft vasculopathy. Similar findings were observed using either non-complement fixing alloantibodies or recipient mice deficient in complement activation. This suggests that antibody-mediated transplant vasculopathy can still occur via a *complement-independent pathway*. In fact, there are observations in the clinic that some kidney transplant patients develop arteriopathy or glomerulopathy in the absence of C4d deposition is the grafts [110], again supporting a complement independent mechanism of chronic rejection. Interestingly, NK cells are identified as key mediators in transplant vasculopathy independent of complement, as depletion of NK cells or deficiency of NK cells in transplant recipients completely prevented the incidence of chronic rejection induced by donor alloantibodies [109]. Another observation is that in a cohort of MHC compatible kidney transplant patients, KIR mismatches between donors and recipients, which generates alloreactive NK cells, are associated with the worst graft outcomes, thus indirectly suggesting a role for NK cells in clinical chronic graft loss [111]. Together, these findings call the attention on innate immune cells and mechanisms in chronic allograft rejection [112].

7.1.2.4 Transplant Tolerance

Transplant tolerance is defined as a state in which cytopathic responses to the graft are absent, while those to pathogens are well preserved. Importantly, this tolerant state should be stably maintained without broad immunosuppression. For decades, innate immune cells are thought to be effector cells, and therefore, associated with rejection. Recent studies have revealed that such cells can also be required for transplant tolerance [1].

A key requirement in transplant tolerance is the promotion of effector T cell apoptosis, followed by expansion of regulatory cells [113]. Innate immune cells can have significant impacts on both arms of the tolerant induction process. For example, DCs are required not only for robust immunity but also for tolerance, as in vivo deletion of $CD11c^+$ DCs in naïve mice breaks down tolerance and induces widespread autoimmunity [43]. This revelation led to the characterization and application of "tolerogenic DCs" in tolerance induction. In fact, DCs contribute to tolerance in several different ways. Mature DCs can drive apoptotic deletion of cytopathic effector T cells following a proliferation burst; subsets of DCs also mediate the induction of $Foxp3^+$ Tregs or homeostasis of natural Tregs. Additionally, DCs, particularly $CD8^+$ DCs, are very efficient at phagocytosis of apoptotic cells, owing to their expression of DEC205, Clec9A, CD36, Tim-1, and Tim-4 [114–116], and clearance of apoptotic cells is a critical process for maintaining tolerance [117]. Also, phagocytosis of allogeneic apoptotic cells inhibits DC maturation, which down-modulates their allostimulatory function, whilst promoting Treg cells [118]. Immature DCs induce T-cell anergy due to the engagement of TCR on T cells without delivering costimulatory signals [119]. In addition, therapeutic manipulation of DCs has also been exploited in transplantation. It has been reported that adoptive transfer of "tolerogenic dendritic cells" can prolong allograft survival and induce donor-specific transplant tolerance in small animal models [120]. Thus, "tolerogenic DCs" as a cell therapy continues to hold promise in transplant tolerance.

Other innate cell types exhibit similar features in tolerance induction. In certain settings, monocyte/macrophage can exert potent anti-inflammatory and immunosuppressive effects that help maintain peripheral tolerance. For example, the alternatively activated M2 macrophages or regulatory macrophages (Mregs) are capable of secreting anti-inflammatory cytokines, such as IL-10 and TGF-β that are involved in tapering immune responses and resolution of graft inflammation [99]. In fact, some studies demonstrate that adoptive transfer of Mregs can ameliorate the induction of experimental autoimmune encephalitis (a model of multiple sclerosis), and prevent autoimmune colitis by inducing and expanding $Foxp3^+$ Tregs [35]. Additionally, adoptive transfer of donor-derived Mregs in a cohort of human kidney transplant recipients allowed for significant reduction in the use of immunosuppressive drugs [121]. A recent study suggests that a subpopulation of macrophages induced by IL-34 is crucial in maintaining transplant tolerance, as depletion of the subset breaks down tolerance leading to graft rejection [122]. Moreover, $DC\text{-}SIGN^+$ macrophages inhibit $CD8^+$ T cell proliferation and expand $CD4^+FoxP3^+$ Treg, thus favoring transplant tolerance [123]. Similarly, NK cells also employ different mechanisms to promote transplant tolerance. NK cells, guided by "missing self recognition," can eliminate graft-derived allogeneic DCs, thus reducing T cell priming by the direct pathway of antigen presentation [27]. Killing of donor cells by NK cells favors the indirect antigen presentation, which is implicated in tolerance induction. Also, some NK cells exhibit regulatory function through IL-10 dependent mechanisms and contribute to tolerance by tipping the balance towards regulation [124]. Moreover, NK cells can interact with monocytes/macrophages and induce tolerance in a NKG2D dependent manner [125].

The striking dichotomy of innate immune cells in transplant settings (rejection versus tolerance) is most likely context dependent, representing opposite outcomes of the immune response to allotransplants. In mice rendered tolerant by costimulatory blockade, administration of various TLR ligands induces allograft rejection [126] by favoring Th1 differentiation and inhibiting Treg function [127]. In contrast, blocking TLR signaling (e.g., MyD88 deficiency) promotes tolerance induction that otherwise difficult to achieved [73, 128]. Along this line, NK cells can be tolerogenic, and further NK maturation by IL-15 mediates rejection; M1 macrophages are pro-inflammatory and M2 macrophages are immunosuppressive. Additionally, TLR signaling mediates maturation of DCs and monocytes in rejection. As Tregs also express certain TLRs, TLR stimulation can increase the suppressive properties of Tregs, which facilitate tolerance. This context-dependent function of innate pathways and context-dependent regulation of innate immune cells constitute a major challenge in manipulating the immune responses to allotransplants.

7.1.3 Future Considerations

Innate immune cells are complex and incompletely understood, especially in the context of transplant tolerance. Only recently do we realize the critical roles of innate immune cells in affecting transplant outcomes. Innate immune cells are involved in early allograft response, becoming activated following graft ischemia injury, and affect the entire spectrum of allograft responses, ranging from acute rejection to the induction of tolerance. Importantly, depending on the models, context, and tolerizing therapies, the same cell types can either promote rejection or facilitate tolerance induction. Thus, targeting innate immune cells for transplant tolerance is unlikely to be straightforward. Additionally, the dynamic interactions between innate and adaptive immunity provide further challenges in the creation of donor specific tolerance. It is likely that a combinatorial approach will be required to target both the effector mechanisms of innate cells and their involvement in activation of the adaptive immunity to further improved transplant outcomes.

As current immunosuppression drugs primarily target adaptive immune cells, development of new approaches targeting innate cells is warranted in transplant models. Therapeutic approaches to suppress the innate arm of the immune system should be directed toward: (1) inhibiting tissue injury by pro-inflammatory cytokines during ischemia–reperfusion; (2) preventing APC activation by targeting TLR signaling; (3) inhibiting complement activation; and (4) blocking innate effector mechanisms and influx of innate cells to the transplants. Since graft injury begins as early as during donor brain death and organ harvesting, strategies targeting early activation of innate immunity should ideally begin in the donor, continuing during organ procurement, and further in the recipients well after reperfusion. As described above, the innate system continues to exert its effects on adaptive alloimmune responses throughout the life of the graft. Thus, additional strategies must also be employed to limit their long-term functions in a donor specific fashion.

A significant challenge is that the complex interactions amongst diverse subsets of innate immune cells in vivo in transplant settings are poorly understood. Also, we have a limited understanding of how such interactions affect the effector and the regulatory programs of alloreactive T cells. This complexity presents both challenges and opportunities in transplant tolerance. There are several areas that deserve immediate attention, which include mechanisms that regulate various aspects of innate immune responses to allografts, how innate immune cells interact with each other and then interact with T effector cells or vice versa; the impact of innate immunity on induction and stability of immune regulatory cells. These are important but less studied areas, but the potential impact on developing better tolerizing strategies and new diagnostic and prognostic biomarkers will be significant.

References

1. Murphy SP, Porrett PM, Turka LA (2011) Innate immunity in transplant tolerance and rejection. Immunol Rev 241(1):39–48
2. Liu W, Li XC (2010) An overview on non-T cell pathways in transplant rejection and tolerance. Curr Opin Organ Transplant 15(4):422–426
3. LaRosa DF, Rahman AH, Turka LA (2007) The innate immune system in allograft rejection and tolerance. J Immunol 178(12):7503–7509
4. Kawai T, Akira S (2010) The role of pattern-recognition receptors in innate immunity: update on toll-like receptors. Nat Immunol 11(5):373–384

5. Elinav E, Strowig T, Henao-Mejia J et al (2011) Regulation of the antimicrobial response by NLR proteins. Immunity 34(5):665–679
6. Kato H, Takahasi K, Fujita T (2011) RIG-I-like receptors: cytoplasmic sensors for non-self RNA. Immunol Rev 243(1):91–98
7. O'Neill LAJ, Bowie AG (2007) The family of five: TIR-domain-containing adaptors in Toll-like receptor signalling. Nat Rev Immunol 7(5):353–364
8. Cao X (2015) Self-regulation and cross-regulation of pattern-recognition receptor signalling in health and disease. Nat Rev Immunol 16(1):35–50
9. Zipfel PF, Skerka C (2009) Complement regulators and inhibitory proteins. Nat Rev Immunol 9(10):729–740
10. Cravedi P, Heeger PS (2014) Complement as a multifaceted modulator of kidney transplant injury. J Clin Invest 124(6):2348–2354
11. Medof ME, Kinoshita T, Nussenzweig V (1984) Inhibition of complement activation on the surface of cells after incorporation of decay-accelerating factor (DAF) into their membranes. J Exp Med 160(5):1558–1578
12. Chen Song S, Zhong S, Xiang Y et al (2011) Complement inhibition enables renal allograft accommodation and long-term engraftment in presensitized nonhuman primates. Am J Transplant 11(10):2057–2066
13. Blom AM, Villoutreix BO, Dahlback B (2004) Complement inhibitor C4b-binding protein-friend or foe in the innate immune system? Mol Immunol 40(18):1333–1346
14. Collard CD, Bukusoglu C, Agah A et al (1999) Hypoxia-induced expression of complement receptor type 1 (CR1, CD35) in human vascular endothelial cells. Am J Physiol 276(2 Pt 1):C450–C458
15. Ollert MW, David K, Bredehorst R et al (1995) Classical complement pathway activation on nucleated cells. Role of factor H in the control of deposited C3b. J Immunol 155(10):4955–4962
16. Heeger PS, Lalli PN, Lin F et al (2005) Decay-accelerating factor modulates induction of T cell immunity. J Exp Med 201(10):1523–1530
17. Lalli PN, Strainic MG, Yang M et al (2008) Locally produced C5a binds to T cell expressed C5aR to enhance effector T-cell expansion by limiting antigen-induced apoptosis. Blood 112(5):1759–1766
18. Cravedi P, van der Touw W, Heeger PS (2013) Complement regulation of T-cell alloimmunity. Semin Nephrol 33(6):565–574
19. Castellano G, Woltman AM, Nauta AJ et al (2004) Maturation of dendritic cells abrogates C1q production in vivo and in vitro. Blood 103(10):3813–3820
20. Strainic MG, Liu J, Huang D et al (2008) Locally produced complement fragments C5a and C3a provide both costimulatory and survival signals to naive CD4+ T cells. Immunity 28(3):425–435
21. Pratt JR, Basheer SA, Sacks SH (2002) Local synthesis of complement component C3 regulates acute renal transplant rejection. Nat Med 8(6):582–587
22. Shi FD, Ljunggren HG, La Cava A et al (2011) Organ-specific features of natural killer cells. Nat Rev Immunol 11(10):658–671
23. Lanier LL (2005) NK cell recognition. Annu Rev Immunol 23:225–274
24. Lanier LL (2008) Up on the tightrope: natural killer cell activation and inhibition. Nat Immunol 9(5):495–502
25. Elliott JM, Yokoyama WM (2011) Unifying concepts of MHC-dependent natural killer cell education. Trends Immunol 32(8):364–372
26. Ruggeri L, Capanni M, Urbani E et al (2002) Effectiveness of donor natural killer cell alloreactivity in mismatched hematopoietic transplants. Science 295(5562):2097–2100
27. Yu G, Xu X, Vu MD et al (2006) NK cells promote transplant tolerance by killing donor antigen-presenting cells. J Exp Med 203(8):1851–1858
28. Kroemer A, Edtinger K, Li XC (2008) The innate NK cells in transplant rejection and tolerance induction. Curr Opin Organ Transplant 13:339–343
29. Sun JC (2010) Re-educating natural killer cells. J Exp Med 207(10):2049–2052
30. Gordon S, Taylor PR (2005) Monocyte and macrophage heterogeneity. Nat Rev Immunol 5(12):953–964

31. Auffray C, Fogg D, Garfa M et al (2007) Monitoring of blood vessels and tissues by a population of monocytes with patrolling behavior. Science 317(5838):666–670
32. Murray PJ, Wynn TA (2011) Protective and pathogenic functions of macrophage subsets. Nat Rev Immunol 11(11):723–737
33. Biswas SK, Mantovani A (2010) Macrophage plasticity and interaction with lymphocyte subsets: cancer as a paradigm. Nat Immunol 11(10):889–896
34. Denning TL, Wang YC, Patel SR et al (2007) Lamina propria macrophages and dendritic cells differentially induce regulatory and interleukin 17-producing T cell responses. Nat Immunol 8(10):1086–1094
35. Brem-Exner BG, Sattler C, Hutchinson JA et al (2008) Macrophages driven to a novel state of activation have anti-inflammatory properties in mice. J Immunol 180(1):335–349
36. Allavena P, Sica A, Garlanda C et al (2008) The Yin-Yang of tumor-associated macrophages in neoplastic progression and immune surveillance. Immunol Rev 222(1):155–161
37. Liu W, Xiao X, Demirci G et al (2012) Innate NK cells and macrophages recognize and reject allogeneic nonself in vivo via different mechanisms. J Immunol 188(6):2703–2711
38. Fox A, Mountford J, Braakhuis A et al (2001) Innate and adaptive immune responses to nonvascular xenografts: evidence that macrophages are direct effectors of xenograft rejection. J Immunol 166(3):2133–2140
39. Coquerelle C, Moser M (2010) DC subsets in positive and negative regulation of immunity. Immunol Rev 234(1):317–334
40. Ueno H, Schmitt N, Klechevsky E et al (2010) Harnessing human dendritic cell subsets for medicine. Immunol Rev 234(1):199–212
41. Morelli AE, Thomson AW (2007) Tolerogenic dendritic cells and the quest for transplant tolerance. Nat Rev Immunol 7:610–621
42. Manicassamy S, Pulendran B (2011) Dendritic cell control of tolerogenic responses. Immunol Rev 241(1):206–227
43. Ohnmacht C, Pullner A, King SBS et al (2009) Constitutive ablation of dendritic cells breaks self-tolerance of CD4 T cells and results in spontaneous fatal autoimmunity. J Exp Med 206(3):549–559
44. Grommes J, Soehnlein O (2011) Contribution of neutrophils to acute lung injury. Mol Med 17(3–4):293–307
45. Fialkow L, Wang Y, Downey GP (2007) Reactive oxygen and nitrogen species as signaling molecules regulating neutrophil function. Free Radic Biol Med 42(2):153–164
46. El-Sawy T, Belperio JA, Strieter RM et al (2005) Inhibition of polymorphonuclear leukocyte-mediated graft damage synergizes with short-term costimulatory blockade to prevent cardiac allograft rejection. Circulation 112(3):320–331
47. Suurmond J, van Heemst J, van Heiningen J et al (2013) Communication between human mast cells and CD4(+) T cells through antigen-dependent interactions. Eur J Immunol 43(7):1758–1768
48. de Vries VC, Wasiuk A, Bennett KA et al (2009) Mast cell degranulation breaks peripheral tolerance. Am J Transplant 9(10):2270–2280
49. Lu LF, Lind EF, Gondek DC et al (2006) Mast cells are essential intermediaries in regulatory T-cell tolerance. Nature 442(7106):997–1002
50. Talayero P, Mancebo E, Calvo-Pulido J et al (2016) Innate lymphoid cells groups 1 and 3 in the epithelial compartment of functional human intestinal allografts. Am J Transplant 16(1):72–82
51. Konya V, Mjosberg J (2015) Innate lymphoid cells in graft-versus-host disease. Am J Transplant 15(11):2795–2801
52. Taniguchi M, Harada M, Kojo S et al (2003) The regulatory role of Valpha14 NKT cells in innate and acquired immune response. Annu Rev Immunol 21:483–513
53. Ikehara Y, Yasunami Y, Kodama S et al (2000) CD4(+) Valpha14 natural killer T cells are essential for acceptance of rat islet xenografts in mice. J Clin Invest 105(12):1761–1767
54. Seino KI, Fukao K, Muramoto K et al (2001) Requirement for natural killer T (NKT) cells in the induction of allograft tolerance. Proc Natl Acad Sci U S A 98(5):2577–2581

55. Shigeoka AA, Holscher TD, King AJ et al (2007) TLR2 is constitutively expressed within the kidney and participates in ischemic renal injury through both myD88-dependent and -independent pathways. J Immunol 178(10):6252–6258
56. Zhai Y, Shen XD, O'Connell R et al (2004) Cutting edge: TLR4 activation mediates liver ischemia/reperfusion inflammatory response via IFN regulatory factor 3-dependent MyD88-independent pathway. J Immunol 173(12):7115–7119
57. Chong AJ, Shimamoto A, Hampton CR et al (2004) Toll-like receptor 4 mediates ischemia/reperfusion injury of the heart. J Thorac Cardiovasc Surg 128(2):170–179
58. Li L, Okusa MD (2010) Macrophages, dendritic cells, and kidney ischemia-reperfusion injury. Semin Nephrol 30(3):268–277
59. Zhang ZX, Wang S, Huang X et al (2008) NK cells induce apoptosis in tubular epithelial cells and contribute to renal ischemia-reperfusion injury. J Immunol 181(11):7489–7498
60. Ysebaert DK, De Greef KE, Vercauteren SR et al (2000) Identification and kinetics of leukocytes after severe ischaemia/reperfusion renal injury. Nephrol Dial Transplant 15(10):1562–1574
61. Huen SC, Cantley LG (2015) Macrophage-mediated injury and repair after ischemic kidney injury. Pediatr Nephrol 30(2):199–209
62. Maroko PR, Carpenter CB, Chiariello M et al (1978) Reduction by cobra venom factor of myocardial necrosis after coronary artery occlusion. J Clin Invest 61(3):661–670
63. Diepenhorst GMP, Van Gulik TM, Hack CE (2009) Complement-mediated ischemia-reperfusion injury: Lessons learned from animal and clinical studies. Ann Surg 249(6):889–899
64. Brown KM, Kondeatis E, Vaughan RW et al (2006) Influence of donor C3 allotype on late renal-transplantation outcome. N Engl J Med 354(19):2014–2023
65. Zhou W, Farrar CA, Abe K et al (2000) Predominant role for C5b-9 in renal ischemia/reperfusion injury. J Clin Invest 105(10):1363–1371
66. Yamada K, Miwa T, Liu J et al (2004) Critical protection from renal ischemia reperfusion injury by CD55 and CD59. J Immunol 172(6):3869–3875
67. Lu X, Li Y, Simovic MO et al (2011) Decay-accelerating factor attenuates c-reactive protein-potentiated tissue injury after mesenteric ischemia/reperfusion. J Surg Res 167(2):e103–e115
68. Zhou W, Medof ME, Heeger PS et al (2007) Graft-derived complement as a mediator of transplant injury. Curr Opin Immunol 19(5):569–576
69. Farrar CA, Zhou W, Lin T et al (2006) Local extravascular pool of C3 is a determinant of postischemic acute renal failure. FASEB J 20(2):217–226
70. Pratt JR, Jones ME, Dong J et al (2003) Nontransgenic hyperexpression of a complement regulator in donor kidney modulates transplant ischemia/reperfusion damage, acute rejection, and chronic nephropathy. Am J Pathol 163(4):1457–1465
71. Oberbarnscheidt MH, Zeng Q, Li Q et al (2014) Non-self recognition by monocytes initiates allograft rejection. J Clin Invest 124(8):3579–3589
72. Li XC, Rothstein DM, Sayegh MH (2009) Costimulatory pathways in transplantation: challenges and new developments. Immunol Rev 229:271–293
73. Goldstein DR, Tesar BM, Akira S et al (2003) Critical role of the Toll-like receptor signal adaptor protein MyD88 in acute allograft rejection. J Clin Invest 111(10):1571–1578
74. McKay D, Shigeoka A, Rubinstein M et al (2006) Simultaneous deletion of MyD88 and Trif delays major histocompatibility and minor antigen mismatch allograft rejection. Eur J Immunol 36(8):1994–2002
75. Hancock WW, Thomson NM, Atkins RC (1983) Composition of interstitial cellular infiltrate identified by monoclonal antibodies in renal biopsies of rejecting human renal allografts. Transplantation 35:458–463
76. Matheson PJ, Dittmer ID, Beaumont BW et al (2005) The macrophage is the predominant inflammatory cell in renal allograft intimal arteritis. Transplantation 79(12):1658–1662
77. Gao W, Topham PS, King JA et al (2000) Targeting of the chemokine receptor CCR1 suppresses development of acute and chronic cardiac allograft rejection. J Clin Invest 105(1):35–44
78. Jose MD, Ikezumi Y, Van Rooijen N et al (2003) Macrophages act as effectors of tissue damage in acute renal allograft rejection. Transplantation 76(7):1015–1022

79. Qi F, Adair A, Ferenbach D et al (2008) Depletion of cells of monocyte lineage prevents loss of renal microvasculature in murine kidney transplantation. Transplantation 86(9):1267–1274
80. Kirk AD, Hale DA, Mannon RB et al (2003) Results from a human renal allograft tolerance trial evaluating the humanized CD52-specific monoclonal antibody alemtuzumab (Campath-1H). Transplantation 76(1):120–129
81. Kaul AM, Goparaju S, Dvorina N et al (2015) Acute and chronic rejection: compartmentalization and kinetics of counterbalancing signals in cardiac transplants. Am J Transplant 15(2):333–345
82. Wu T, Bond G, Martin D et al (2006) Histopathologic characteristics of human intestine allograft acute rejection in patients pretreated with thymoglobulin or alemtuzumab. Am J Gastroenterol 101(7):1617–1624
83. Kim J, Chang CK, Hayden T et al (2007) The activating immunoreceptor NKG2D and its ligands are involved in allograft transplant rejection. J Immunol 179(10):6416–6420
84. Degli-Esposti MA, Smyth MJ (2005) Close encounters of different kinds: dendritic cells and NK cells take centre stage. Nat Rev Immunol 5:112–124
85. Kroemer A, Xiao X, Degauque N et al (2008) The innate NK cells, allograft rejection, and a key role for IL-15. J Immunol 180(12):7818–7826
86. Martin-Fontecha A, Thomsen LL, Brett S et al (2004) Induced recruitment of NK cells to lymph nodes provides IFN-gamma for T(H)1 priming. Nat Immunol 5(12):1260–1265
87. Colvin RB (2007) Antibody-mediated renal allograft rejection: diagnosis and pathogenesis. J Am Soc Nephrol 18(4):1046–1056
88. Fuquay R, Renner B, Kulik L et al (2013) Renal ischemia-reperfusion injury amplifies the humoral immune response. J Am Soc Nephrol 24(7):1063–1072
89. Raedler H, Heeger PS (2011) Complement regulation of T-cell alloimmunity. Curr Opin Organ Transplant 16(1):54–60
90. Pavlov V, Raedler H, Yuan S et al (2008) Donor deficiency of decay-accelerating factor accelerates murine T cell-mediated cardiac allograft rejection. J Immunol 181(7):4580–4589
91. Solez K, Colvin RB, Racusen LC et al (2008) Banff 07 classification of renal allograft pathology: updates and future directions. Am J Transplant 8(4):753–760
92. Racusen LC, Regele H (2010) The pathology of chronic allograft dysfunction. Kidney Int 78(suppl 119):S27–S32
93. Michelsen KS, Doherty TM, Shah PK et al (2004) TLR signaling: an emerging bridge from innate immunity to atherogenesis. J Immunol 173(10):5901–5907
94. Mann DL (2011) The emerging role of innate immunity in the heart and vascular system: for whom the cell tolls. Circ Res 108(9):1133–1145
95. Wang S, Schmaderer C, Kiss E et al (2010) Recipient Toll-like receptors contribute to chronic graft dysfunction by both MyD88- and TRIF-dependent signaling. Dis Model Mech 3(1–2):92–103
96. Methe H, Zimmer E, Grimm C et al (2004) Evidence for a role of toll-like receptor 4 in development of chronic allograft rejection after cardiac transplantation. Transplantation 78 (9):1324–1331
97. Kitchens WH, Chase CM, Uehara S et al (2007) Macrophage depletion suppresses cardiac allograft vasculopathy in mice. Am J Transplant 7:2675–2682
98. Yang J, Reutzel-Selke A, Steier C et al (2003) Targeting of macrophage activity by adenovirus-mediated intragraft overexpression of TNFRp55-Ig, IL-12p40, and vIL-10 ameliorates adenovirus-mediated chronic graft injury, whereas stimulation of macrophages by overexpression of IFN-(gamma) accelerates chronic graft injury in a rat renal allograft model. J Am Soc Nephrol 14(1):214–225
99. Ricardo SD, Van Goor H, Eddy AA (2008) Macrophage diversity in renal injury and repair. J Clin Invest 118(11):3522–3530
100. Niedermeier M, Reich B, Gomez MR et al (2009) CD4+ T cells control the differentiation of Gr1+ monocytes into fibrocytes. Proc Natl Acad Sci U S A 106(42):17892–17897
101. Pilmore HL, Painter DM, Bishop GA et al (2000) Early up-regulation of macrophages and myofibroblasts: a new marker for development of chronic renal allograft rejection. Transplantation 69(12):2658–2662

102. Jevnikar AM, Mannon RB (2008) Late kidney allograft loss: what we know about it, and what we can do about it. Clin J Am Soc Nephrol 3(Suppl 2):S56–S67
103. Toki D, Zhang W, Hor KL et al (2014) The role of macrophages in the development of human renal allograft fibrosis in the first year after transplantation. Am J Transplant 14(9):2126–2136
104. Ikezumi Y, Suzuki T, Yamada T et al (2015) Alternatively activated macrophages in the pathogenesis of chronic kidney allograft injury. Pediatr Nephrol 30(6):1007–1017
105. Einecke G, Sis B, Reeve J et al (2009) Antibody-mediated microcirculation injury is the major cause of late kidney transplant failure. Am J Transplant 9(11):2520–2531
106. Gaston RS, Cecka JM, Kasiske BL et al (2010) Evidence for antibody-mediated injury as a major determinant of late kidney allograft failure. Transplantation 90(1):68–74
107. Qian Z, Hu W, Liu J et al (2001) Accelerated graft arteriosclerosis in cardiac transplants. Transplantation 72(5):900–906
108. Sheerin NS, Risley P, Abe K et al (2008) Synthesis of complement protein C3 in the kidney is an important mediator of local tissue injury. FASEB J 22(4):1065–1072
109. Hirohashi T, Chase CM, Della Pelle P et al (2012) A novel pathway of chronic allograft rejection mediated by NK cells and alloantibody. Am J Transplant 12(2):313–321
110. Sis B, Campbell PM, Mueller T et al (2007) Transplant glomerulopathy, late antibody-mediated rejection and the ABCD tetrad in kidney allograft biopsies for cause. Am J Transplant 7(7) :1743–1752
111. Van Bergen J, Thompson A, Haasnoot GW et al (2011) KIR-ligand mismatches are associated with reduced long-term graft survival in HLA-compatible kidney transplantation. Am J Transplant 11(9):1959–1964
112. Uehara S, Chase CM, Kitchens WH et al (2005) NK cells can trigger allograft vasculopathy : the role of hybrid resistance in solid organ allografts. J Immunol 175(5):3424–3430
113. Li XC, Strom TB, Turka LA et al (2001) T cell death and transplantation tolerance. Immunity 14:407–416
114. Schulz O, Reis E, Sousa C (2002) Cross-presentation of cell-associated antigens by CD8(alpha)+ dendritic cells is attributable to their ability to internalize dead cells. Immunology 107(2):183–189
115. Iyoda T, Shimoyama S, Liu K et al (2002) The CD8+ dendritic cell subset selectively endocytoses dying cells in culture and in vivo. J Exp Med 195(10):1289–1302
116. Schnorrer P, Behrens GMN, Wilson NS et al (2006) The dominant role of CD8+ dendritic cells in cross-presentation is not dictated by antigen capture. Proc Natl Acad Sci U S A 103 (28):10729–10734
117. Savill J, Fadok V (2000) Corpse clearance defines the meaning of cell death. Nature 407(6805):784–788
118. Morelli AE, Larregina AT (2010) Apoptotic cell-based therapies against transplant rejection : role of recipient's dendritic cells. Apoptosis 15(9):1083–1097
119. Mueller DL, Jenkins MK, Schwartz RH (1989) Clonal expansion versus functional clonal inactivation: a costimulatory signalling pathway determines the outcome of T cell antigen receptor occupancy. Annu Rev Immunol 7:445–480
120. Hill M, Thebault P, Segovia M et al (2011) Cell therapy with autologous tolerogenic dendritic cells induces allograft tolerance through interferon-gamma and Epstein-Barr virus-induced gene 3. Am J Transplant 11(10):2036–2045
121. Hutchinson JA, Riquelme P, Sawitzki B et al (2011) Cutting edge: immunological consequences and trafficking of human regulatory macrophages administered to renal transplant recipients. J Immunol 187(5):2072–2078
122. Bezie S, Picarda E, Ossart J et al (2015) IL-34 is a Treg-specific cytokine and mediates transplant tolerance. J Clin Invest 125(10):3952–3964
123. Conde P, Rodriguez M, van der Touw W et al (2015) DC-SIGN(+) macrophages control the induction of transplantation tolerance. Immunity 42(6):1143–1158
124. Deniz G, Erten G, Kucuksezer UC et al (2008) Regulatory NK cells suppress antigen-specific T cell responses. J Immunol 180(2):850–857

125. van der Touw W, Burrell B, Lal G et al (2012) NK cells are required for costimulatory blockade induced tolerance to vascularized allografts. Transplantation 94(6):575–584
126. Adams AB, Williams MA, Jones TR et al (2003) Heterologous immunity provides a potent barrier to transplantation tolerance. J Clin Invest 111(12):1887–1895
127. Porrett PM, Yuan X, LaRosa DF et al (2008) Mechanisms underlying blockade of allograft acceptance by TLR ligands. J Immunol 181(3):1692–1699
128. Walker WE, Nasr IW, Camirand G et al (2006) Absence of innate MyD88 signaling promotes inducible allograft acceptance. J Immunol 177(8):5307–5316

Chapter 8
Inflammatory Cytokine Response to Titanium Surface Chemistry and Topography

Stephen M. Hamlet and Saso Ivanovski

Abstract Titanium continues to be one of the most widely utilized biomaterials for use in prosthetic devices to provide anchorage into bone. In particular, titanium implants have found widespread usage in dentistry for the purpose of anchoring dental prostheses providing a superior solution over conventional prostheses for the replacement of lost teeth. A major part of this success stems from its strength and favorable weight, but most importantly its outstanding biocompatibility. Research continues however on processes that can be applied to enhance and improve the clinical utility of titanium. Of the many strategies employed, surface modification to manipulate surface chemistry and topography has proven to be most effective. Recent evidence suggests that part of the success of these titanium surface modifications may be due to a subsequent modulation of the immune response to decrease inflammation and ensure a timely switch to a more reparative microenvironment. Micro-rough sandblasted and acid etched titanium has been widely used as a model of choice for assessing the effect of topographical surface modification on a variety of bone healing related biological mechanisms. Our group has used this surface and its modifications as a model to study the in vivo and in vitro effects of surface topographical modification on the cellular and molecular mechanisms associated with osseointegration, including the influence on inflammation.

Keywords Immunomodulation • Macrophage • Titanium • Implant • Cytokine • Topography • Hydrophilicity • Inflammation • Osseointegration

---

S.M. Hamlet
Menzies Health Institute Queensland, Griffith University, Nathan, QLD, Australia

S. Ivanovski (✉)
Menzies Health Institute Queensland, Griffith University, Nathan, QLD, Australia

School of Dentistry and Oral Health, Griffith University,
Gold Coast Campus Parklands Drive, Southport, QLD 4222, Australia
e-mail: s.ivanovski@griffith.edu.au

B. Corradetti (ed.), *The Immune Response to Implanted Materials and Devices*,
DOI 10.1007/978-3-319-45433-7_8

8.1 Introduction

Titanium is one of the few materials naturally suited to the requirements for implantation into the human body. The combination of a high strength to weight ratio and excellent biocompatibility has resulted in titanium being widely utilized as a biomaterial in the manufacture of medical devices to be anchored into bone. Indeed more than 1000 tonnes of titanium devices are implanted into patients worldwide every year.

Titanium is suitable for both temporary and long-term applications. For example long-term arthritic hip and knee joint replacements use prostheses that have a titanium femoral stem and head that locates into a low friction polyethylene socket. Similarly, both short- and long-term internal and external bone-fracture fixation devices such as pins, bone-plates, screws, intra-medullary nails and external fixators, pacemaker cases and defibrillators, intra-vascular stents, orthotic calipers, and artificial limbs are all further major applications for titanium prostheses due to this material's light weight, toughness, and corrosion resistance.

In particular, titanium devices have also found widespread usage in oral, maxillofacial and craniofacial surgery to anchor prostheses required to replace facial features lost through trauma or disease, thus restoring both function and aesthetics. Even more widely utilized are dental titanium implants, which support prosthesis that replace lost teeth and restore masticatory function.

Dental implants utilize the unique biocompatibility properties of titanium, which facilitates the apposition of bone directly onto the metal surface without fibrous tissue formation at the interface, in a phenomenon known as "osseointegration." Following adequate time for osseointegration following surgical implant placement, the dental restorative superstructure can then be attached onto the implant to give an effective tooth replacement. As such, dental implants provide a superior solution over conventional prostheses for the replacement of lost teeth in terms of both function and long-term predictability [1]. However, with increased clinical usage and greater acceptance and popularity of implants, there are now greater demands placed on implant systems from both clinicians and patients alike. In particular, there is an increasing demand for implant placement in sites where the quality of bone is less than ideal, such as those encountered in the posterior maxilla or in patients with systemic conditions like osteoporosis and diabetes, where the amount of mineralized tissue is reduced and or bone wound healing is compromised.

This has led to numerous attempts to enhance clinical outcomes by improving the rate and extent of bone-to-implant integration and although significant advancements in the fields of tooth replacement and bone tissue engineering have been made, it has been reported that new devices often demonstrate a worse clinical outcome for patients when compared to older available devices. Clearly an incomplete understanding of the interactions between titanium and the biological processes involved in bone repair has hampered these new clinical innovations.

In this regard it is now appreciated that more than 30 distinct cell type populations may reside in the bone marrow microenvironment adjacent to the implant. These include hematopoietic lineage cells, mesenchymal lineage cells, blood vessels, and neural tissue. Each of these cell populations, alone or in combination has the capacity to influence bone growth and regeneration following implant insertion. Notwithstanding these various important cell-biomaterial interactions, it should be appreciated that implantation of any foreign object into the body will first induce an immuno-inflammatory response by the host. In general, following implantation, biomaterials are immediately coated with proteins such as fibronectin, vitronectin, and albumin that are adsorbed onto the biomaterial surface. This activates the coagulation cascade and complement systems leading to thrombus formation and the activation of other cell populations such as neutrophils and polymorphonuclear leukocytes. Activated platelets subsequently release chemoattractants that direct the migration of monocytes to the implant where these cells differentiate into macrophages. At the wound site, macrophages bind to the biomaterial surface via integrin-mediated interactions with the adsorbed surface proteins.

If the foreign material cannot be phagocytosed and removed, the inflammatory response persists until the material becomes encapsulated in a dense layer of fibrotic connective tissue that shields it from the immune system and isolates it from the surrounding tissues. Developed as a protective mechanism to limit exposure to toxic or allergenic materials, chronic fibrous encapsulation however also compromises the efficiency of the implant and can lead to implant failure [2]. The immune response therefore is a key factor that may ultimately determine the in vivo fate of bone biomaterials [3, 4].

While macrophages are an essential component of innate immunity and play a central role in inflammation and host defense, these cells fulfill other homeostatic functions beyond defense including tissue remodeling and orchestration of metabolic functions [5]. Creating a local environment that favors bone regeneration and osseointegration by manipulating the immune response, i.e., "immunomodulation" is therefore a sound strategy for potential bone tissue engineering applications. Indeed, an analysis of the transcriptional mechanisms involved during the early stages of successful osseointegration of dental implants in the human show that inflammation and osteogenesis are inversely related, with the resolution of the early inflammatory response correlating with the ascendency of the bone formation process [6]. Moreover, modulation of cellular phenotype, including that of macrophages, by implanted biomaterials has been recognized to play a critical role in wound healing and tissue regeneration [7–9].

Since the seminal review by Anderson [2] on the "Inflammatory Response to Implants" and subsequently expanded upon by Brown and Badylak [10], the idea that biomaterial composition can influence the subsequent macrophage response by the host has highlighted the need to consider the surface structure, both topographical and chemical, of the implant in order to ensure successful integration. Other surface properties such as particle size, porosity, and ions that can be released from the biomaterial are all biomaterial-specific factors able to influence the immune response. Hydrophobic materials for example have been shown to enhance monocyte adhesion [11] whereas hydrophilic or neutral surfaces inhibit macrophage adhesion but enhance the release of pro-inflammatory cytokines and chemokines [12]. Similarly large pore size enhances angiogenesis while inhibiting inflammation [13].

8.2 Implant Surface Modification to Enhance Osseointegration: Influence of Micro-Roughness, Nano-Roughness and Hydrophilicity

Topographical modification of the titanium surface has been well established to have a major positive effect on the rate and degree of osseointegration [14]. Surface roughness at the microscale level (Ra ~ 1–2 μm) has been demonstrated to upregulate osteogenic gene expression in bone healing sites and osteoblasts [15, 16]. Indeed the use of microscale modified implant surfaces has been credited with being one of the key factors in increasing the clinical success rate of implants, especially in areas of compromised bone quality such as the posterior maxilla [17]. Animal studies using micro-rough surfaces have shown a higher level of organization in the wound as early as 4 days following implantation, leading to greater bone to implant contact 1 week post-insertion [18]. Topographical modulation of cellular phenotypic characteristics, as demonstrated by enhanced osteoblast differentiation, growth factor and cytokine production in vitro [19], increased bone-to-implant contact in vivo [20] and improved clinical rates of wound healing [21] has also been demonstrated in cells intimately involved in osteogenesis and the bone/tissue wound healing processes.

The clinically utilized micro-rough (Ra 1–2 μm) sand-blasted acid etched titanium (SLA®, Straumann, Basel, Switzerland) titanium surface has been widely studied as a model of choice for assessing the effect of topographical surface modification on a variety of bone healing related biological mechanisms. Therefore, our group has used this surface and its modifications as a model to study the in vivo and in vitro effects of surface topographical modification on the cellular and molecular mechanisms associated with osseointegration, including the influence on inflammation, which is the focus of this book chapter.

More recently, the original SLA surface has been modified to produce a micro-rough surface with reduced surface contamination, nanoscale topography, and increased surface energy, thus increasing wettability (Fig. 8.1). This modified SLA surface, named SLActive by the manufacturer (Straumann AG, Basel, Switzerland), has been demonstrated in vivo to promote enhanced bone apposition during the early stages of osseointegration compared to the SLA surface [22] resulting in superior osseointegration as measured by higher levels of bone to implant contact [23, 24] and higher torque values [25].

At the cellular level, both in vitro and in vivo studies have also shown that the SLActive surface increased osteoblast differentiation [26], increased growth factor production [27, 28] and increased osteogenic gene expression of bone-associated genes such as BMP2, alkaline phosphatase, osteocalcin, type-I-collagen, osteoprotegerin, SPP1, and RUNX2 [29–33]. These results suggest that the superior clinical performance of modified implants with a hydrophilic surface may be the result of differences induced in gene expression ultimately leading to cellular phenotypes with superior reparative characteristics.

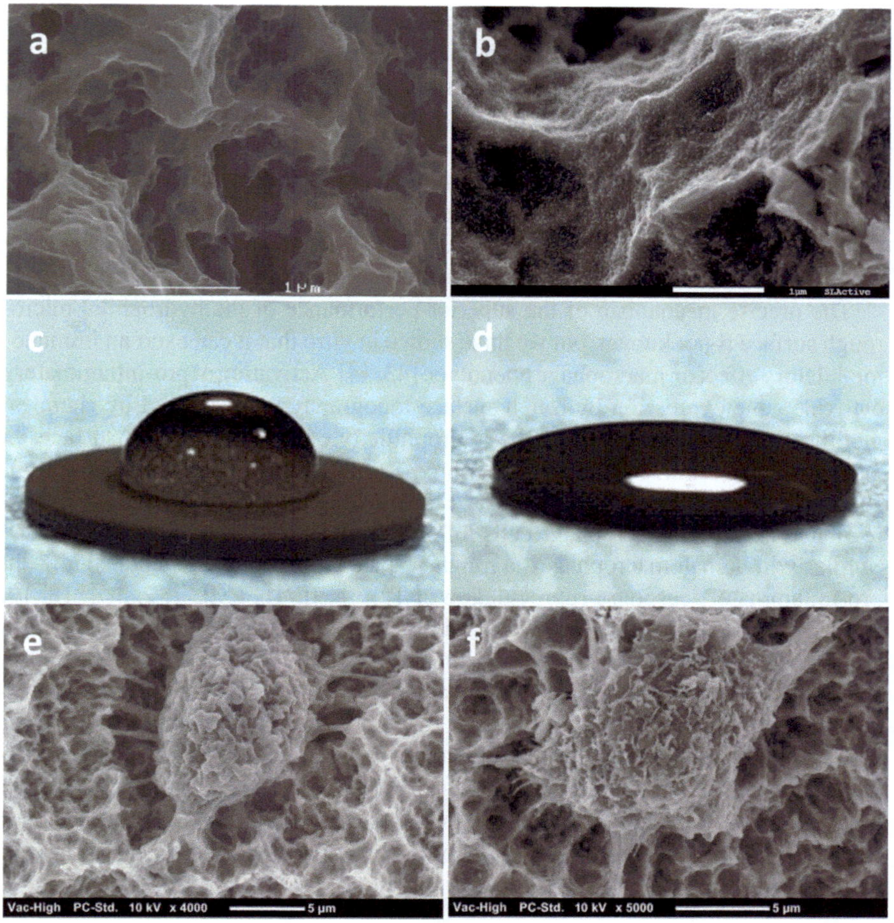

Fig. 8.1 (**a**) SEM micrograph of a micro-rough titanium surface. (**b**) SEM micrograph of micro-rough titanium surface with added nanoscale topographical features. (**c**) Hydrophobic micro-rough titanium surface and (**d**) Hydrophilic effect of increased surface energy. (**e**, **f**) Macrophages cultured on micro-rough and hydrophilic nano-rough titanium surfaces respectively

As the surface roughness of bone is around 32 nm, this allows similarly nano-structured biomaterials to be able to recapitulate the micro-architecture of the bone surface [34]. This combined with the demonstration that nanoscale surface structures can stimulate human MSCs to produce bone minerals in vitro in the absence of osteogenic supplements has generated considerable interest in the applications of nano-materials in bone tissue engineering [35]. Titanium nanostructures on the micro-rough surface, such as those identified on the SLactive surface (Fig. 8.1b), not only provide a higher surface-to-volume ratio but also have been shown to stimulate behavioral changes in both cells and tissues, such as enhanced osteoblast adhesion and functionality [36], marked osteoinduction and osteogenesis of adherent mesenchymal stem cells [37], increased bone–implant contact [38, 39], increased removal torque values of dental implants [40] and enhanced osteoconduction and bone-bonding [41]. Our in vivo human studies showed that compared with the parent SLA surface, a hydrophilic, nanostructured micro-rough surface exerted an influence not only on osteogenic- and angiogenic-related but also on inflammation-related gene expression as early as 7 days post-implantation [42], which resulted in histologically quantifiable improvement in osseointegration at 2 weeks [24]. These findings suggest that key biological events leading to the superior histomorphometric characteristics and clinical performance of modified implant surfaces occur early in the wound healing process at the tissue–implant interface.

The precise mechanism of the superior performance of the hydrophilic micro-rough surface is not known, but we have shown in vitro that it can exert an immunomodulatory effect on macrophage phenotype [43, 44]. Activation of pro-inflammatory cytokine gene expression by rough surface topography was reversed by chemical modification, i.e., by increasing the hydrophilicity of the micro-rough surface. In particular, the relative expression of three pro-inflammatory cytokines; TNF-α, IL-1α, IL-1β, and the chemokine CCL-2 all were shown to be significantly downregulated in murine macrophage cells following culture on the hydrophilic surface. Similarly in human macrophages, a rough surface topography elicited a significant pro-inflammatory response (upregulation of TNF-α, IL-1β, CCL-1, 2, 3, 4, 18, 19, and 20, CXCL-1, 5, 8, and 12, CCR7, LTB, and LTB4R gene expression) that was countered by a hydrophilic surface with downregulation of TNF-α, IL-1α and β, CCL-1, 3, 19, and 20, CXCL-1 and 8, and IL-1R1. Decreased levels of the corresponding protein secretion confirmed the cytokine gene expression changes [43].

Combined with the findings that inflammation associated whole genome expression is modulated in vivo by the nano-rough hydrophilic SLActive surface [42], it can be postulated that the modulation of the inflammatory response indeed plays a major role in the enhanced osseointegration reported with the surface modified surfaces.

8.3 Titanium and Macrophage Polarization

Although medical devices have advanced significantly in both their complexity and ultimately functionality, the host response to the biomaterial itself is still a major factor that influences the integration and ultimately the longevity of these devices. Surgical implantation of medical devices into the body stimulates a series of molecular and cellular events that can lead to encapsulation and isolation of the implant from the surrounding tissue. This series of events known as the "foreign body reaction" can limit the device's overall biocompatibility and function. The foreign body reaction generally proceeds with inflammatory cell infiltration and activation at the implant site, and ends with the formation of a fibrous capsule around the implant.

Although better known for their role in inflammatory disorders, macrophages are the major infiltrating cell population and key determinants of the overall response to the foreign material in injured tissue, whereby they act to remove damaged cells, induce inflammation and protective scar formation, support cell proliferation through the release of key growth factors and play a key role in angiogenesis. The host response to implanted biomaterials however is not simply a function of the presence and number of host inflammatory cells, but also the temporal and spatial phenotype of the responding cells. Macrophages are a highly plastic, heterogeneous cell population, exhibiting a spectrum of cell-surface markers and functions [45] linked to the type of receptor interaction on the macrophage and the presence of cytokines [46]. Following their arrival at sites of inflammation, macrophages derived from monocyte precursors become activated in response to signals present in the tissue and increase the production of cytokines, chemokines, and other molecules that contribute to the local inflammatory response. The phenotypes of the macrophages present are broadly defined by these functional properties and patterns of gene expression and are generally referred to as having either an M1 or an M2 phenotype [47, 48] mimicking the Th1–Th2 nomenclature described for T helper cells [49]. The M1 phenotype is characterized by the expression of high levels of pro-inflammatory cytokines, high production of reactive nitrogen and oxygen intermediates, promotion of a Th1 response, and strong microbicidal and tumoricidal activity. In contrast, M2 macrophages, characterized by low levels of pro-inflammatory cytokines and high expression of anti-inflammatory cytokines, play a major role in promoting cell growth and regeneration. M2 macrophages have been further subclassified (M2a, b, and c), based on the type of stimulation and the subsequent expression of surface molecules and cytokines which reflect functional and molecular specialization [48]. Successful implant integration therefore relies on a balance of classically activated (M1) macrophages to clear the wound site coupled with anti-inflammatory (M2) activated macrophages to promote wound healing and regeneration.

Cell markers alone however do not fully define the many subpopulations of macrophages [50]. The microenvironment of implanted biomaterials plays a significant role in determining the phenotype and activity of the various macrophage populations present in a regenerating wound site [51]. For example, implantation of solid biomaterials such as polycaprolactone and silicone generally lead to a typical foreign body reaction with chronic inflammatory fibrotic encapsulation. However, when these same biomaterials are fabricated with a porous structure, their implantation induces vascularized tissue–biomaterial integration rather than encapsulation [52]. Many other studies have also shown similar modulating effects of the micro- or macro-structure of biomaterials upon macrophage polarization along a reparative M2 phenotype [53, 54]. Moreover micro- and nano-structured titanium surface topographies have recently also been shown to modulate the inflammatory response of macrophages both in vitro [55, 56] and in vivo [57].

This is further illustrated by a recent transcriptome analysis of human macrophages following activation with a variety of chemical cues [58]. When macrophages were activated with known M1 or M2 chemical cues or M1- or M2-related chemical signals, they displayed biochemical behavior consistent with the M1–M2 polarization model. However activation with other chemical cues such as free fatty acids, high density lipoprotein or molecules associated with chronic inflammation, resulted in seven other distinct macrophage phenotypes, i.e., a multi-axis spectrum designated C1, C3, C4, C5, C7, C8, and C9. Clearly, the classical macrophage phenotypes M1

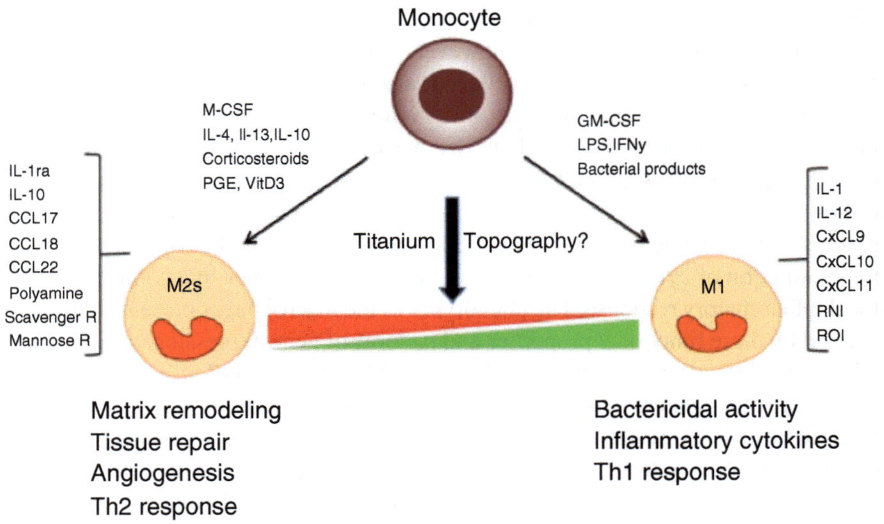

Fig. 8.2 Illustration demonstrating the classical M1 and M2 macrophage phenotypes at the ends of a possible spectrum of intervening macrophage phenotypes elicited following exposure to surface modified titanium substrates

and M2 therefore represent at best the ends of a possible spectrum of macrophage phenotypes from inflammatory to reparative (Fig. 8.2), each with distinct transcriptional profiles [45] able to modulate further downstream events at the tissue–implant surface. As such, biomaterial mediated modulation of macrophage phenotype has significant implications for improving outcomes in regenerative medicine by providing a basis for subsequent macrophage-centered diagnostic and therapeutic strategies.

While the control of macrophage phenotype with its beneficial effects upon tissue remodeling can be driven by biomaterial morphology, the precise mechanisms that mediate these responses still remain largely unknown. A network of signaling molecules, transcription factors, epigenetic mechanisms, and posttranscriptional regulators underlie the different forms of macrophage activation. Canonical IRF/STAT signaling pathways are activated by IFNs and TLR signaling to skew macrophage function toward the M1 phenotype (via STAT1) or by IL-4 and IL-13 to skew toward the M2 phenotype (via STAT6) [59]. Further regulation of STAT-mediated macrophage activation can be by members of the suppressor of cytokine signaling (SOCS) family. IL-4 and IFNγ, the latter in concert with TLR stimulation, upregulate SOCS1 and SOCS3, which in turn inhibit the action of STAT1 and STAT3, respectively [60, 61].

This functional skewing of mononuclear phagocytes occurs in vivo under both physiological and pathological conditions (e.g., chronic inflammation and tissue repair). However, in selected clinical conditions, the coexistence of cells in different activation states and unique or mixed phenotypes is often observed [62] and is a reflection of the presence of complex tissue-derived signals that result in the dynamic situations that are observed in vivo.

8.4 Macrophage Response to Titanium Surface Modification

The biocompatibility of titanium and titanium alloys is based in part on the spontaneously formed dense oxide surface layer. This 2–6 nm thick natural oxide layer is thermodynamically stable, chemically inert, and has a low solubility in body fluids. However, it has been shown that macrophages are able to liberate metal ions from solid surfaces within minutes by the membrane-bound dissolucytosis both in vitro as well as in vivo [63, 64]. Liberated metal ions may then spread from the dissolution membrane into the intercellular space between cells in the vicinity of an implant where they may be taken up and thus influence the secretion of pro-inflammatory cytokines [65].

This effect may be compounded with the use of titanium alloys, and in this context it is noteworthy that pure titanium is often alloyed with aluminum and vanadium (Ti6Al4V) [66] to improve its mechanical properties. However, further surface modification procedures such as sand-blasting and acid etching are likely to remove passive layers from the surface of the metal, exposing the less stable elements underneath. Subsequent studies have suggested potential adverse effects including stimulation of an inflammatory response and reduction in osteoblast differentiation [66, 67] as a result of vanadium and aluminum leaching from the Ti6Al4V surfaces after modification procedures.

More recently titanium has also been alloyed with zirconium (Zr) to increase implant strength similar to that seen with Ti6Al4V alloy, but without the known detrimental effects on cell function. When murine macrophages were subsequently cultured on pure titanium and titanium-zirconium alloy surfaces with a range of surface topographies and chemistries, i.e., hydrophobic micro-rough titanium (Ti-SLA) and titanium-zirconium alloy (TiZr-SLA), and hydrophilic nano-rough titanium (Ti-modSLA and TiZr-modSLA) to assess their effect on macrophage activation and cytokine production, macrophages on high energy (hydrophilic) surfaces had higher levels of anti-inflammatory gene expression that was characteristic of M2 activation and released more anti-inflammatory cytokines compared to the macrophages cultured on low energy (hydrophobic) surfaces. This effect was further enhanced on the alloy TiZr surfaces [68]. We have also shown a nanoscale-modified surface manufactured from the Ti6Al4V alloy was able to downregulate pro-inflammatory cytokine gene expression compared to the same topographical surface manufactured from commercially pure titanium, with no demonstrable effect on macrophage attachment or proliferation [44]. The potential effects of alloying elements however are still unclear, as other studies have also demonstrated differential cytokine expression in response to chemical composition. For example compared to CoCrMo metal alloy, Ti6Al4V alloy was shown to reduce TNFα secretion but increase IL-6 secretion in both murine RAW 264.7 and J774A.1 macrophages [69].

Whilst a majority of studies have demonstrated that titanium surface roughness does indeed influence and or modulate macrophage behavior, no clear consensus has been reached as yet on the actual effect that this physical cue has on macrophage phenotype (Table 8.1). For example, moderate secretion of pro-inflammatory cyto-

Table 8.1 Although the results of some studies shown here demonstrate a variability in the specifics of the cytokine response (as measured by changes in gene and/or protein secretion), in general, a micro-rough surface topography enhances the adhesion of macrophages and the secretion of inflammatory cytokines. In contrast, titanium surfaces with nanoscale topographical features and/or increased surface energy (hydrophilicity) induces macrophage polarization toward an anti-inflammatory phenotype

Reference	Titanium modification used	Macrophage response
[71]	(a) Polished, (b) sand blasted, (c) acid etched, (d) SLA	Rough surfaces (acid etched and SLA) ↑ unstimulated MΦ (J774A.1) proinflammatory cytokines (TNF-alpha). Unstimulated macrophages on the SLA surface ↓chemokines (MCP-1, MIP-1)
[84]	(a) Smooth, (b) grit-blasted/acid rough surfaces	IL-1beta ↑on grit-blasted/acid rough surfaces during the first 48 h. IL-6 ↓ on the grit-blasted/acid rough surfaces. (J774A.1)
[44]	Nanoscale CaP	↓ MΦ pro-inflammatory gene expression and protein secretion (RAW264.7)
[85]	Micro-rough ± hydrophilic	↓Monocyte and MΦ attachment
[55]	SLA and modSLA	↑Murine MΦ pro-inflammatory response to micro-rough surface ↓Murine MΦ pro-inflammatory response to nano/hydrophilic surface
[86]	SLA/Sr	↓ Osteoclast expression in vivo (rabbit femur)
[43]	SLA and modSLA	↑Human MΦ pro-inflammatory response to micro-rough surface ↓Human MΦ pro-inflammatory response to nano/hydrophilic surface
[87]	Submicron and nanometer titanium deposition	↑ NOX on submicron treated surface ↓ Pro-inflammatory cytokines by nanometer treatment
[76]	Micro- and nano-grooved	Cell elongation in grooves ↑ anti-inflammatory phenotype of murine bone marrow derived macrophages
[68]	SLA and modSLA Zr-SLA and Zr-modSLA	↑ Murine pro-inflammatory cytokines by micro-rough SLA and Zr-SLA ↑ Murine M2 gene expression by modSLA and Zr-modSLA
[88]	Nanotopography + bioactive ions	↑ M2 macrophage phenotype (J774A.1)
[73]	Hydrophobic and hydrophilic smooth, micro-rough and nano- and micro-rough Ti.	Smooth Ti induced inflammatory macrophage (M1-like) activation, Hydrophilic rough titanium induced macrophage activation similar to the anti-inflammatory M2-like state.

Key:
MΦ macrophage
↑ increased, ↓ decreased
SLA sand blasted micro-rough surface
modSLA sand blasted hydrophilic nano-rough surface
SLA/Sr sand blasted micro-rough surface containing strontium
Zr-SLA & Zr-modSLA Zirconium alloy SLA and modSLA surfaces
NOX Nitric oxide
CaP Calcium phosphate direct deposition
RAW 264.7 and *J774A.1* murine monocytic cell lines

kines and reduced production of nitric oxide by macrophages cultured on nanostructured titanium surfaces when compared to macrophages cultured on a flat titanium surface [70] is in contrast to the significant increase (an effect that was further magnified when macrophages were stimulated with LPS) in pro-inflammatory chemokines and cytokines (TNFα, IL-1β, IL-6, MCP-1, MIP-1α) that were secreted by macrophages cultured on micro-rough titanium surfaces which were also compared with macrophages cultured on a flat polished titanium surface [71]. As described earlier in our own studies using murine [55] and human macrophages [43], culture on a micro-rough surface stimulates pro-inflammatory cytokine gene expression that can be subsequently modulated (downregulated) by culturing the macrophages on a similar topographical surface but with increased surface energy (hydrophilicity). However, again in contrast, Barth et al. [72] showed that murine macrophages cultured on a micro-rough surface exhibited a dominant M2-like behavior with reduced secretion of the M1-related cytokine interferon gamma-induced protein 10 and upregulated secretion of the M2-related cytokines MCP-1 and MIP-1α when compared with macrophages cultured on a polished substrate.

In an attempt to determine which aspect(s) of titanium implant surfaces may be responsible for macrophage phenotype modulation, Hotchkiss et al. varied the surface roughness and wettability of titanium independently of each other and then assessed the effect of each property on macrophage activation and cytokine production. Macrophages were cultured on six titanium surfaces: hydrophobic and hydrophilic smooth titanium; hydrophobic and hydrophilic micro-rough titanium, and hydrophobic and hydrophilic nano-and micro-rough titanium. Increased surface wettability was shown to have a stronger immunomodulatory effect than increases in roughness with high surface wettability materials producing an anti-inflammatory microenvironment [73].

8.5 Topography-Induced Changes in Macrophage Function

While the chemistry of biomaterials dictates the behavior of infiltrating immune cells by altering cell adhesive interactions, recent evidence suggests that a biomaterial's geometry, or size and shape, may in fact dominate the overall response [57, 74, 75]. Extending this concept to titanium surface topography, surfaces that contain variable size and shape microscale features appear to enhance the adhesion of macrophages and their secretion of inflammatory cytokines when compared to macrophages on smooth surfaces although this response has been suggested to be dependent upon the actual degree of roughness [57]. The mechanism underlying topography-induced changes in macrophage function however are still not well understood although studies using precisely engineered micro-fabricated substrates have helped to elucidate the effects of topography on macrophage behavior and the host response. Titanium surfaces containing micro- and nano-patterned grooves have been shown to promote cell elongation and regulate the macrophage polarization state [76]. Surface grooves did not affect inflammatory activation but drove macrophages toward an anti-inflammatory, pro-healing phenotype [77].

It is likely that multiple cell surface mechanoreceptors may ultimately define the response of the macrophage to its mechanical environment when interacting with titanium surfaces. Integrins are candidate mechanoreceptors that couple the cell to the external environment by spanning the plasma membrane and forming attachments with the ECM. The binding of extracellular ligands to integrins may initiate intracellular signaling events, while modification of intracellular domains also regulates the binding affinity of extracellular attachments [78]. Similarly, focal adhesions are macromolecular protein complexes that create a connection between the cytoskeleton and the ECM and mediate regulatory effects of adhesion on cell behavior [79]. Focal adhesions through their connections to the F-actin cytoskeleton not only transmit force throughout the cell, but also stimulate a plethora of signaling pathways. While macrophages do not form extensive focal adhesions, a study of the actin cytoskeleton of pro-healing or micro-patterned macrophages showed an increase in actin filaments and cytoskeletal organization [77].

Macrophages have been shown to exhibit different shapes in vivo whereby pro-healing macrophages appeared more elongated and were often found within fibrous tissue architectures [80]. In contrast, we have shown that macrophages adherent to titanium surfaces (Fig. 8.1 e, f) both in vitro and in vivo appear to be more rounded in shape with many more pseudopodia-like extensions from the cell body when attached to rough surfaces [43]. Studies focusing on nuclear shape and structure have revealed strong correlations between shape change and changes in cellular phenotype. Indeed as cells differentiate, changes in cell phenotype have been correlated with the upregulation of A-type lamin [81] allowing the transmission of forces across the cytoskeleton to the nucleus, and changes in nuclear shape and stiffness [82]. Furthermore, nano-topography stimulated changes in nuclear organization have also been linked to spatially regulated phenotypic gene expression [83]. Taken together, these studies suggest that micro- and nano-patterned topographies may modulate macrophage behavior in vitro and the host response in vivo via mechanisms involving changes in cell shape and adhesion.

8.6 Conclusions

Titanium continues to be one of the most widely utilized biomaterials for use in prosthetic devices. A major part of this success stems from its strength and favorable (light) weight but most importantly its outstanding biocompatibility. Notwithstanding these attributes, research continues on possible processes and techniques that could be applied to enhance and improve the clinical utility of titanium. Of the many strategies employed, surface modification to manipulate the degree of surface roughness and topography has been proven to be most effective. Whilst originally designed to mimic the native bone microenvironment to stimulate the osteogenic activity of osteoblasts and their mesenchymal precursors, recent evidence suggests part of the success of surface modification may be due to a modulation of the immune response to decrease inflammation and ensure a timely switch to a more reparative microenvironment.

As described in this chapter, a micro-rough surface topography generally enhances the adhesion of macrophages and their secretion of inflammatory cytokines including amongst others, IL-1β, IL-6, and nitric oxide when compared to macrophages on smooth surfaces. However further modification to achieve titanium surfaces with nanoscale topographical features, increased surface energy (hydrophilicity) and a reduction in surface contamination, induces macrophage polarization toward a pro-healing (M2) phenotype. The ability therefore to control the ratio of M1 and M2 macrophages at the host–biomaterial interface is potentially a powerful new tool in regenerative medicine that will allow new bone formation and osseointegration without the prolonged immune response that leads to foreign body giant cell formation, fibrotic encapsulation, and implant failure.

References

1. Lindquist LW, Carlsson GE, Jemt T (1996) A prospective 15-year follow-up study of mandibular fixed prostheses supported by osseointegrated implants. Clinical results and marginal bone loss. Clin Oral Implants Res 7(4):329–336
2. Anderson JM (1988) Inflammatory response to implants. ASAIO Trans 34:101–107
3. Franz S, Rammelt S, Scharnweber D et al (2011) Immune responses to implants - a review of the implications for the design of immunomodulatory biomaterials. Biomaterials 32(28):6692–6709
4. Takayanagi H (2005) Inflammatory bone destruction and osteoimmunology. J Periodontal Res 40(4):287–293
5. Gordon S, Martinez FO (2010) Alternative activation of macrophages: mechanism and functions. Immunity 32(5):593–604
6. Ivanovski S, Hamlet S, Salvi GE et al (2011) Transcriptional profiling of osseointegration in humans. Clin Oral Implants Res 22(4):373–381
7. Kou PM, Babensee JE (2011) Macrophage and dendritic cell phenotypic diversity in the context of biomaterials. J Biomed Mater Res A 96(1):239–260
8. McBane JE, Sharifpoor S, Labow RS et al (2012) Tissue engineering a small diameter vessel substitute: engineering constructs with select biomaterials and cells. Curr Vasc Pharmacol 10(3):347–360
9. Weidenbusch M, Anders HJ (2012) Tissue microenvironments define and get reinforced by macrophage phenotypes in homeostasis or during inflammation, repair and fibrosis. J Innate Immun 4(5–6):463–477
10. Brown B, Badylak SF (2013) Expanded applications, shifting paradigms and an improved understanding of host–biomaterial interactions. Acta Biomaterialia 9:4948–4955
11. Hezi-Yamit A, Sullivan C, Wong J et al (2009) Impact of polymer hydrophilicity on biocompatibility: implication for DES polymer design. J Biomed Mater Res A 90(1):133–141
12. Jones JA, Chang DT, Meyerson H et al (2007) Proteomic analysis and quantification of cytokines and chemokines from biomaterial surface-adherent macrophages and foreign body giant cells. J Biomed Mater Res A 83(3):585–596
13. Laschke MW, Harder Y, Amon M et al (2006) Angiogenesis in tissue engineering: breathing life into constructed tissue substitutes. Tissue Eng 12(8):2093–2104
14. Wennerberg A, Albrektsson T (2009) Effect of titanium surface topography on bone integration: a systematic review. Clin Oral Impl Res 20:172–184
15. Brett PM, Harle J, Salih V et al (2004) Roughness response genes in osteoblasts. Bone 35(1):124–133

16. Ogawa T, Nishimura I (2003) Different bone integration profiles of turned and acid-etched implants associated with modulated expression of extracellular matrix genes. Int J Oral Maxillofac Implants 18(2):200–210
17. Ivanovski S (2010) Osseointegration—the influence of implant surface. Ann R Australas Coll Dent Surg 20:82–85
18. Abrahamsson I, Berglundh T, Linder E et al (2004) Early bone formation adjacent to rough and turned endosseous implant surfaces. An experimental study in the dog. Clin Oral Implants Res 15(4):381–392
19. Kieswetter K, Schwartz Z, Hummert TW et al (1996) Surface roughness modulates the local production of growth factors and cytokines by osteoblast-like MG-63 cells. J Biomed Mater Res 32(1):55–63
20. Cochran DL, Schenk RK, Lussi A et al (1998) Bone response to unloaded and loaded titanium implants with a sandblasted and acid-etched surface: a histometric study in the canine mandible. J Biomed Mater Res 40(1):1–11
21. Cochran DL, Buser D, ten Bruggenkate CM et al (2002) The use of reduced healing times on ITI implants with a sandblasted and acid-etched (SLA) surface: early results from clinical trials on ITI SLA implants. Clin Oral Implants Res 13(2):144–153
22. Bornstein MM, Valderrama P, Jones AA et al (2008) Bone apposition around two different sandblasted and acid-etched titanium implant surfaces: a histomorphometric study in canine mandibles. Clin Oral Implants Res 19(3):233–241
23. Buser D, Broggini N, Wieland M et al (2004) Enhanced bone apposition to a chemically modified SLA titanium surface. J Dent Res 83(7):529–533
24. Lang NP, Salvi GE, Huynh-Ba G et al (2011) Early osseointegration to hydrophilic and hydrophobic implant surfaces in humans. Clin Oral Implants Res 22(4):349–356
25. Ferguson SJ, Broggini N, Wieland M et al (2006) Biomechanical evaluation of the interfacial strength of a chemically modified sandblasted and acid-etched titanium surface. J Biomed Mater Res A 78(2):291–297
26. Rausch-fan X, Qu Z, Wieland M et al (2008) Differentiation and cytokine synthesis of human alveolar osteoblasts compared to osteoblast-like cells (MG63) in response to titanium surfaces. Dent Mater 24(1):102–110
27. Eriksson C, Nygren H, Ohlson K (2004) Implantation of hydrophilic and hydrophobic titanium discs in rat tibia: cellular reactions on the surfaces during the first 3 weeks in bone. Biomaterials 25(19):4759–4766
28. Zhao G, Raines AL, Wieland M et al (2007) Requirement for both micron- and submicron scale structure for synergistic responses of osteoblasts to substrate surface energy and topography. Biomaterials 28(18):2821–2829
29. Chakravorty N, Hamlet S, Jaiprakash A et al (2014) Pro-osteogenic topographical cues promote early activation of osteoprogenitor differentiation via enhanced TGFbeta, Wnt, and Notch signaling. Clin Oral Implants Res 25(4):475–486
30. Masaki C, Schneider GB, Zaharias R et al (2005) Effects of implant surface microtopography on osteoblast gene expression. Clin Oral Implants Res 16(6):650–656
31. Qu Z, Rausch-Fan X, Wieland M et al (2007) The initial attachment and subsequent behavior regulation of osteoblasts by dental implant surface modification. J Biomed Mater Res A 82(3):658–668
32. Vlacic-Zischke J, Hamlet SM, Friis T et al (2011) The influence of surface microroughness and hydrophilicity of titanium on the up-regulation of TGFbeta/BMP signalling in osteoblasts. Biomaterials 32(3):665–671
33. Wall I, Donos N, Carlqvist K et al (2009) Modified titanium surfaces promote accelerated osteogenic differentiation of mesenchymal stromal cells in vitro. Bone 45(1):17–26
34. Mendonca G, Mendonca DB, Aragao FJ et al (2008) Advancing dental implant surface technology—from micron- to nanotopography. Biomaterials 29(28):3822–3835
35. Dalby MJ, Gadegaard N, Tare R et al (2007) The control of human mesenchymal cell differentiation using nanoscale symmetry and disorder. Nat Mater 6(12):997–1003
36. Dalby MJ, McCloy D, Robertson M et al (2006) Osteoprogenitor response to semi-ordered and random nanotopographies. Biomaterials 27(15):2980–2987

37. Popat KC, Leoni L, Grimes CA et al (2007) Influence of engineered titania nanotubular surfaces on bone cells. Biomaterials 28(21):3188–3197
38. Goene RJ, Testori T, Trisi P (2007) Influence of a nanometer-scale surface enhancement on de novo bone formation on titanium implants: a histomorphometric study in human maxillae. Int J Periodontics Restorative Dent 27(3):211–219
39. Orsini G, Piattelli M, Scarano A et al (2007) Randomized, controlled histologic and histomorphometric evaluation of implants with nanometer-scale calcium phosphate added to the dual acid-etched surface in the human posterior maxilla. J Periodontol 78(2):209–218
40. Meirelles L, Currie F, Jacobsson M et al (2008) The effect of chemical and nanotopographical modifications on the early stages of osseointegration. Int J Oral Maxillofac Implants 23(4):641–647
41. Mendes VC, Moineddin R, Davies JE (2009) Discrete calcium phosphate nanocrystalline deposition enhances osteoconduction on titanium-based implant surfaces. J Biomed Mater Res A 90(2):577–585
42. Donos N, Hamlet S, Lang NP et al (2011) Gene expression profile of osseointegration of a hydrophilic compared with a hydrophobic microrough implant surface. Clin Oral Implants Res 22(4):365–372
43. Alfarsi MA, Hamlet SM, Ivanovski S (2014) Titanium surface hydrophilicity modulates the human macrophage inflammatory cytokine response. J Biomed Mater Res A 102(1):60–67
44. Hamlet S, Ivanovski S (2011) Inflammatory cytokine response to titanium chemical composition and nanoscale calcium phosphate surface modification. Acta Biomater 7(5):2345–2353
45. Mosser DM, Edwards JP (2008) Exploring the full spectrum of macrophage activation. Nat Rev Immunol 8(12):958–969
46. Taylor PR, Martinez-Pomares L, Stacey M et al (2005) Macrophage receptors and immune recognition. Annu Rev Immunol 23:901–944
47. Gordon S, Taylor PR (2005) Monocyte and macrophage heterogeneity. Nat Rev Immunol 5(12):953–964
48. Mantovani A, Sica A, Sozzani S et al (2004) The chemokine system in diverse forms of macrophage activation and polarization. Trends Immunol 25(12):677–686
49. Mills CD, Kincaid K, Alt JM et al (2000) M-1/M-2 macrophages and the Th1/Th2 paradigm. J Immunol 164(12):6166–6173
50. Geissmann F, Gordon S, Hume DA et al (2010) Unravelling mononuclear phagocyte heterogeneity. Nat Rev Immunol 10(6):453–460
51. Brown BN, Ratner BD, Goodman SB et al (2012) Macrophage polarization: an opportunity for improved outcomes in biomaterials and regenerative medicine. Biomaterials 33(15):3792–3802
52. Underwood RA, Usui ML, Zhao G et al (2011) Quantifying the effect of pore size and surface treatment on epidermal incorporation into percutaneously implanted sphere-templated porous biomaterials in mice. J Biomed Mater Res A 98(4):499–508
53. Bota PC, Collie AM, Puolakkainen P et al (2010) Biomaterial topography alters healing in vivo and monocyte/macrophage activation in vitro. J Biomed Mater Res A 95(2):649–657
54. Wang Z, Cui Y, Wang J et al (2014) The effect of thick fibers and large pores of electrospun poly(ε-caprolactone) vascular grafts on macrophage polarization and arterial regeneration. Biomaterials 35:5700–5710
55. Hamlet S, Alfarsi M, George R et al (2012) The effect of hydrophilic titanium surface modification on macrophage inflammatory cytokine gene expression. Clin Oral Implants Res 23(5):584–590
56. Neacsu P, Mazare A, Cimpean A et al (2014) Reduced inflammatory activity of RAW 264.7 macrophages on titania nanotube modified Ti surface. Int J Biochem Cell Biol 55:187–195
57. Ma QL, Zhao LZ, Liu RR et al (2014) Improved implant osseointegration of a nanostructured titanium surface via mediation of macrophage polarization. Biomaterials 35(37):9853–9867
58. Xue J, Schmidt SV, Sander J et al (2014) Transcriptome-based network analysis reveals a spectrum model of human macrophage activation. Immunity 40(2):274–288
59. Sica A, Bronte V (2007) Altered macrophage differentiation and immune dysfunction in tumor development. J Clin Invest 117(5):1155–1166

60. Liu Y, Stewart KN, Bishop E et al (2008) Unique expression of suppressor of cytokine signaling 3 is essential for classical macrophage activation in rodents in vitro and in vivo. J Immunol 180(9):6270–6278
61. Whyte CS, Bishop ET, Ruckerl D et al (2011) Suppressor of cytokine signaling (SOCS)1 is a key determinant of differential macrophage activation and function. J Leukoc Biol 90(5):845–854
62. Sica A, Mantovani A (2012) Macrophage plasticity and polarization: in vivo veritas. J Clin Invest 122(3):787–795
63. Larsen A, Kolind K, Pedersen DS et al (2008) Gold ions bio-released from metallic gold particles reduce inflammation and apoptosis and increase the regenerative responses in focal brain injury. Histochem Cell Biol 130(4):681–692
64. Locht LJ, Larsen A, Stoltenberg M et al (2009) Cultured macrophages cause dissolucytosis of metallic silver. Histol Histopathol 24(2):167–173
65. Catelas I, Petit A, Zukor DJ et al (2003) TNF-alpha secretion and macrophage mortality induced by cobalt and chromium ions in vitro-qualitative analysis of apoptosis. Biomaterials 24(3):383–391
66. Saulacic N, Bosshardt DD, Bornstein MM et al (2012) Bone apposition to a titanium-zirconium alloy implant, as compared to two other titanium-containing implants. Eur Cell Mater 23:273–286, discussion 286–278
67. Hallab NJ, Vermes C, Messina C et al (2002) Concentration- and composition-dependent effects of metal ions on human MG-63 osteoblasts. J Biomed Mater Res 60(3):420–433
68. Hotchkiss KM, Ayad NB, Hyzy SL et al (2016) Dental implant surface chemistry and energy alter macrophage activation in vitro. Clin Oral Implants Res. doi:10.1111/clr.12814
69. Jakobsen SS, Larsen A, Stoltenberg M et al (2007) Effects of as-cast and wrought cobalt-chrome-molybdenum and titanium-aluminium-vanadium alloys on cytokine gene expression and protein secretion in J774A.1 macrophages. Eur Cell Mater 14:45–54, discussion 54–45
70. Lee S, Choi J, Shin S et al (2011) Analysis on migration and activation of live macrophages on transparent flat and nanostructured titanium. Acta Biomater 7(5):2337–2344
71. Refai AK, Textor M, Brunette DM et al (2004) Effect of titanium surface topography on macrophage activation and secretion of proinflammatory cytokines and chemokines. J Biomed Mater Res A 70(2):194–205
72. Barth KA, Waterfield JD, Brunette DM (2013) The effect of surface roughness on RAW 264.7 macrophage phenotype. J Biomed Mater Res A 101:2679–2688
73. Hotchkiss KM, Reddy GB, Hyzy SL et al (2016) Titanium surface characteristics, including topography and wettability, alter macrophage activation. Acta Biomater 31:425–434
74. Sridharan R, Cameron AR, Kelly DJ et al (2015) Biomaterial based modulation of macrophage polarization: a review and suggested design principles. Mater Today 18(6):313–325
75. Veiseh O, Doloff JC, Ma M et al (2015) Size- and shape-dependent foreign body immune response to materials implanted in rodents and non-human primates. Nat Mater 14(6): 643–651
76. Luu TU, Gott SC, Woo BW et al (2015) Micro- and nanopatterned topographical cues for regulating macrophage cell shape and phenotype. ACS Appl Mater Interfaces 7(51):28665–28672
77. McWhorter FY, Wang T, Nguyen P et al (2013) Modulation of macrophage phenotype by cell shape. Proc Natl Acad Sci U S A 110(43):17253–17258
78. Thompson WR, Rubin CT, Rubin J (2012) Mechanical regulation of signaling pathways in bone. Gene 503(2):179–193
79. Chen CS, Alonso JL, Ostuni E et al (2003) Cell shape provides global control of focal adhesion assembly. Biochem Biophys Res Commun 307(2):355–361
80. Waldo SW, Li Y, Buono C et al (2008) Heterogeneity of human macrophages in culture and in atherosclerotic plaques. Am J Pathol 172(4):1112–1126
81. Constantinescu D, Gray HL, Sammak PJ et al (2006) Lamin A/C expression is a marker of mouse and human embryonic stem cell differentiation. Stem Cells 24(1):177–185
82. Pajerowski JD, Dahl KN, Zhong FL et al (2007) Physical plasticity of the nucleus in stem cell differentiation. Proc Natl Acad Sci U S A 104(40):15619–15624

83. Tsimbouri P, Gadegaard N, Burgess K et al (2014) Nanotopographical effects on mesenchymal stem cell morphology and phenotype. J Cell Biochem 115(2):380–390
84. Tan KS, Qian L, Rosado R et al (2006) The role of titanium surface topography on J774A.1 macrophage inflammatory cytokines and nitric oxide production. Biomaterials 27(30):5170–5177
85. Milleret V, Tugulu S, Schlottig F et al (2011) Alkali treatment of microrough titanium surfaces affects macrophage/monocyte adhesion, platelet activation and architecture of blood clot formation. Eur Cell Mater 21:430–444, discussion 444
86. Park JW, Kim YJ, Jang JH et al (2013) Positive modulation of osteogenesis- and osteoclastogenesis-related gene expression with strontium-containing microstructured Ti implants in rabbit cancellous bone. J Biomed Mater Res A 101(1):298–306
87. Lu J, Webster TJ (2015) Reduced immune cell responses on nano and submicron rough titanium. Acta Biomater 16:223–231
88. Lee CH, Kim YJ, Jang JH et al (2016) Modulating macrophage polarization with divalent cations in nanostructured titanium implant surfaces. Nanotechnology 27(8):085101

Chapter 9
The Biomechanical Environment and Impact on Tissue Fibrosis

Wayne Carver, Amanda M. Esch, Vennece Fowlkes, and Edie C. Goldsmith

Abstract The implantation of materials into the body elicits a foreign body response (FBR) that includes formation of a fibrous capsule around the implanted material. The formation of the fibrous capsule has many similarities to fibrotic responses to other insults or stressors. A number of biochemical factors are known to promote a fibrotic response including growth factors, cytokines, and hormones. Much less is known regarding the role of biomechanical forces in tissue fibrosis. The biomechanical environment plays a fundamental role in embryonic development, tissue maintenance, and pathogenesis. Mechanical forces play particularly important roles in the regulation of connective tissues including not only bone and cartilage but also the interstitial tissues of most organs. In vivo studies have correlated changes in mechanical load to modulation of the extracellular matrix and have indicated that increased mechanical force contributes to the enhanced expression and deposition of extracellular matrix components or fibrosis. A variety of in vitro models have been utilized to evaluate the effects of mechanical force on extracellular matrix-producing cells. In general, application of mechanical stretch, fluid flow, and compression results in enhanced expression and deposition of extracellular matrix components. More recent studies have indicated that tissue rigidity also provides profibrotic signals to cells. This is particularly relevant to implants as the implanted material generally alters the local biomechanical environment, which may promote fibrosis or the formation of the fibrous capsule. The

W. Carver
Department of Cell Biology and Anatomy, University of South Carolina, School of Medicine, Columbia, SC, USA

A.M. Esch
Department of Molecular and Medical Genetics, Oregon Health & Science University, School of Medicine, Portland, OR, USA

V. Fowlkes
Department of Mathematics and Science, Hagerstown Community College, Hagerstown, MD, USA

E.C. Goldsmith (✉)
Department of Cell Biology and Anatomy, University of South Carolina, School of Medicine, 6439 Garners Ferry Road, Building 1, Room B19, Columbia, SC, USA
e-mail: Edie.Goldsmith@uscmed.sc.edu

B. Corradetti (ed.), *The Immune Response to Implanted Materials and Devices*,
DOI 10.1007/978-3-319-45433-7_9

mechanisms whereby cells detect mechanical signals and transduce them into biochemical responses have received considerable attention. Cell surface receptors for extracellular matrix components and intracellular signaling pathways are instrumental in the mechanotransduction process. Understanding the effects of the biomechanical environment and the mechanisms, whereby mechanical forces are transduced into biochemical and molecular signals in the cell, will provide important insight into tissue fibrosis and fibrous capsule formation.

Keywords Extracellular matrix • Mechanotransduction • Profibrotic signals • Biomechanical environment • Fibrosis • Fibrous capsule formation

9.1 Introduction

All cells in multicellular organisms are exposed to mechanical forces to some degree. As such, physical forces play important roles in embryonic development, tissue homeostasis, and pathogenesis. The concept that mechanical forces drive embryonic development has been around for more than a century. Classical experiments have illustrated essential roles for mechanical forces in morphogenetic movements, cell shape changes, differentiation, and apoptosis associated with embryogenesis [1–4]. The effects of mechanical forces on mature cells and tissues have begun to receive more attention, as models have been developed to systematically analyze these effects. Many of the early studies in these regards were focused on cells and tissues that are influenced by obvious mechanical force including the cardiovascular and musculoskeletal systems. Early investigations in the mechanobiology field relied on relatively simple and imprecise systems. For instance, studies have utilized a hanging-drop culture system to examine the effects of tensile forces on connective tissue cells [5]. As interest grew in the mechanobiology field, innovative systems were developed to apply diverse types of forces that were more precisely quantifiable on cells and tissues [6].

The mechanobiology field began to move forward rapidly as in vitro model systems were developed to more quantitatively dissect the effects of mechanical forces on cellular processes. Various systems were engineered to apply uniaxial or multiaxial distension or stretch to cells grown on deformable substrata. These systems date back several decades to studies conducted on smooth muscle cells that were cultured on deformable elastin matrices [7, 8]. Among other responses, these studies illustrated a role for mechanical force in the growth and maintenance of musculoskeletal and cardiovascular cells [9, 10]. It has become increasingly clear that many aspects of cell behavior can be modulated by mechanical force including cell proliferation, differentiation, migration, gene expression, and even survival. The realization that most cells respond to mechanical stimuli has resulted in enhanced interest in the contribution of these forces to pathogenesis including tissue fibrosis and in the mechanisms whereby cells detect and respond to these forces.

9.2 Tissue Fibrosis

Fibrosis or accumulation of excessive extracellular matrix (ECM) components is a consequence of many stresses or pathological stimuli. The ECM is a dynamic network composed primarily of fibrillar collagens, noncollagenous glycoproteins, proteoglycans, and other components. The ECM plays many roles in tissues including functioning as a three-dimensional scaffold essential for the development and maintenance of organ structure. Alterations in ECM composition, organization, or accumulation can deleteriously impact embryonic development and organ homeostasis in adults. For instance, developmental deficits in collagen production result in vascular weakness and aneurysms [11]. On the other extreme, excessive accumulation of ECM components or fibrosis results in dysfunction of many organs. Fibrosis is associated with a number of disease conditions including hypertension, diabetes, myocardial infarction, and some immune conditions. Fibrosis can also be a consequence of exposure to many toxins and abuse of drugs like alcohol. Accumulation of excessive ECM components alters tissue biomechanical properties, which can compromise organ function and often contributes to organ failure.

A key step in the progression of fibrosis is thought to be the activation of ECM-producing cells into a myofibroblast or myofibroblast-like phenotype. This is characterized by enhanced contractile activity, formation of stress fibers, and expression of α-smooth muscle actin. Myofibroblasts are responsible for alterations in connective tissues including increased synthesis of ECM components. In addition, these cells produce cytokines and growth factors that promote the fibrotic response in an autocrine/paracrine manner. Myofibroblasts are derived from a variety of cells in response to tissue damage and stress including quiescent fibroblasts, blood-derived fibrocytes, mesenchymal stem cells, stellate cells of the liver and others [12, 13]. Regardless of their origin, myofibroblasts likely arise as an acute and beneficial response to repair damaged tissue. Continued myofibroblast contraction and production of ECM components becomes deleterious and in many cases yields to stiff fibrotic tissue that obstructs and negatively impacts organ function [14].

9.3 Implants and Fibrous Capsules

The implantation of biomaterials and medical devices into the body results in a series of events collectively recognized as the "foreign body reaction." This response can limit the functionality and biocompatibility of the implanted materials. The details of this response are more thoroughly described in other chapters of this book; however, briefly, the response includes the activation of multiple inflammatory and immune cell types including neutrophils, macrophages, and lymphocytes. Several of these cell types secrete growth factors and cytokines that promote fibrosis and serve as fibroblast chemoattractants [15]. Activated fibroblasts produce a dense fibrous capsule around the implanted material. This capsule consists largely of type I and III collagens; however, other ECM constituents including fibronectin and tenascin may also be components of the fibrous capsule. The fibrous capsule isolates the implanted material from the surrounding tissue and can often contract (constrictive fibrosis), which may impact implant structure and function. The process of fibrous capsule formation appears to be very similar to fibrosis in other tissues resulting from chronic inflammatory states. There is speculation that the implantation of biomaterials and medical devices alters local biomechanical properties of tissues and creates mechanical stress that contributes to the fibrotic response [16].

9.4 Biomechanical Forces and Tissue Fibrosis

The expression of ECM components is clearly regulated by diverse biochemical factors including growth factors, cytokines, and hormones [17]. As the importance of mechanical forces in developmental and pathological processes was realized, the impact of such forces on ECM production and organization began to receive attention. This has coincided with the development of novel in vitro systems that can be used to directly evaluate the effects of diverse types of mechanical forces on ECM-producing cells. Cells can be exposed to a variety of mechanical forces in the body including mechanical stretch (tension), compression, shear stress, and others. Studies by Leung et al. [7] were among the first to illustrate that cyclic mechanical loading promotes the production of ECM components by vascular smooth muscle cells. Since that time, a large body of literature has developed evaluating the effects of mechanical forces on ECM-producing cells and the mechanisms of these effects. Below we review the effects of several types of mechanical forces on production of ECM and activation of ECM-producing cells.

9.5 Effects of Tissue Mechanical Properties on Fibrosis

Two-dimensional in vitro systems have been invaluable in elucidating the effects of mechanical forces on cells and the mechanisms of mechanotransduction; however, cells function within a three-dimensional environment whose mechanical properties may change during development [18] or various pathological conditions including fibrosis [19, 20], cancer [21–26], and atherosclerosis [27]. Alterations in tissue biomechanical properties and stiffness are a common feature of fibrosis due to the accumulation of ECM components and their crosslinking [28, 29]. For instance, pathological scars are stiffer relative to unwounded normal skin and typically consist of thicker collagen bundles [30]. Alterations in the tissue mechanical properties can deleteriously impact cell and organ function [31]. Component cells sense and respond to ECM rigidity, which can regulate cell growth [32], shape [33], migration [34], and differentiation [35]. Stiffening of the ECM was once viewed as an endpoint to fibrosis; however, with refined techniques for measuring mechanical properties of tissue it is now clear that tissue compliance is altered early in the fibrotic process [36]. Because of the response of cells to tissue biomechanical properties, ECM stiffness can have a self-perpetuating response on fibrosis.

The effects of the biomechanical properties of the microenvironment on the expression of ECM components have been examined in a number of systems and cell types. Initial studies by Mauch and colleagues compared the expression of ECM components between cells cultured on tissue culture plastic, a rigid substratum, and three-dimensional collagen gels, a more flexible substratum [37]. These studies illustrated that collagen expression is markedly decreased in fibroblasts cultured in three-dimensional collagen scaffolds compared to cells grown on more rigid tissue culture plastic. Collagenase activity is enhanced by culture in three-dimensional scaffolds promoting a collagenolytic phenotype in the less rigid environment of the collagen gels [37]. A number of studies have subsequently supported the concept that increased matrix rigidity or stiffness promotes a fibrotic response. Culture of human colon fibroblasts on matrices that mimic the mechanical properties of the normal colon or the pathologically stiff colon of Crohn's disease patients demonstrated enhanced expression of ECM components and increased proliferation of fibroblasts on the stiffer matrix [38]. Generation of polyacrylamide scaffolds with stiffness gradients that span the range of normal and fibrotic lung tissue (0.1–50 kPa) has been utilized to evaluate the response of lung fibroblasts to alterations in tissue biomechanical properties [39]. In this system, proliferation of lung fibroblasts and expression of collagens were induced by increased scaffold stiffness. In contrast, the expression of prostaglandin, which is an endogenous anti-fibrotic factor, was inversely related to matrix rigidity. These studies and others indicate that the biomechanical properties of the microenvironment can direct the expression of ECM components and ECM-modifying enzymes with stiffer tissue properties contributing to enhanced ECM production or fibrosis. This could be an underlying property that promotes fibrous capsule formation around implants, which tend to have more rigid mechanical properties than the surrounding tissue.

Studies using three-dimensional ECM scaffolds illustrated that ECM compliance is also inversely related to the transformation of cells into a myofibroblast phenotype [40–43]. Culture of cells on plastic coated with thin films of collagen (minimal compliance and maximal generation of intracellular tension) resulted in the highest levels of α-smooth muscle actin expression. Culture of cells in free-floating collagen gels (maximal compliance and least generation of intracellular tension) yielded the lowest relative level of α-smooth muscle actin expression. Similar results have been obtained in experiments examining matrix rigidity and differentiation of bronchial fibroblasts to a myofibroblast phenotype [44]. Culture of bronchial fibroblasts on polydimethylsiloxane substrates of variable stiffness (1–50 kPa) was performed to evaluate the effects of matrix mechanical properties on myofibroblast formation. Increased scaffold stiffness promoted myofibroblast formation and increased α-smooth muscle actin and interstitial collagen expression. The conversion of hepatic stellate cells to a myofibroblast phenotype is a critical step in liver fibrosis and is part of the pathway to cirrhosis in chronic liver disease. Culture of hepatic stellate cells on tissue culture plastic and in high levels of serum results in their spontaneous conversion to a myofibroblast phenotype [45]. Culture of hepatic stellate cells on Matrigel, a relatively soft basement membrane-like matrix, retains the quiescent nature of these cells [46]. Furthermore, culture of differentiated hepatic myofibroblasts on Matrigel results in loss of myofibroblast characteristics [47]. The mechanisms of the dedifferentiation of these cells are not well understood, but these data illustrate that the myofibroblast phenotype may not be permanent. Similar studies with heart valve interstitial cells and a novel photodegradable crosslinker-polyethylene glycol scaffold in which exposure to ultraviolet light can modulate the mechanical properties of the substratum have illustrated that increased elastic modulus of the scaffold yields an enhanced proportion of myofibroblasts [48]. Interestingly, and of potential therapeutic significance, the proportion of myofibroblasts in the scaffolds diminished when the elastic modulus was decreased. In contrast to the above studies, culturing fibroblasts for prolonged periods on matrices of different mechanical properties suggest the conversion to a myofibroblast phenotype is a more "permanent" condition [49]. Understanding the plasticity of the myofibroblast phenotype is critical to development of novel therapeutic approaches to fibrosis. The above studies suggest that alterations in the ECM biomechanical properties may be an important therapeutic target that is able to modulate myofibroblast formation and fibrosis.

In addition to observed changes in cellular response based on the mechanical properties of the ECM, cellular response may be impacted by age as well. Work comparing the remodeling behavior of adult and neonatal cardiac fibroblasts revealed that while the ability of these cell populations to deform monolayer collagen substrates was nearly identical, neonatal fibroblasts were significantly better at deforming free-floating, three-dimensional collagen gels compared to adult fibroblasts [50]. This observed difference in contractile ability was paralleled by increased expression of α-smooth muscle actin in neonatal fibroblasts compared to adult fibroblasts, suggesting a larger degree of fibroblast to myofibroblast transformation in the neonatal cells. Neonatal fibroblasts in three-dimensional gels also exhibited greater expression of β1 integrin, known to be involved in collagen gel contraction, compared to their adult counterparts. Age-dependent gel contraction is not unique to collagen matrices. As shown in Fig. 9.1, neonatal cardiac fibroblasts embedded in three-dimensional fibrin gels exhibit enhanced contractile capacity compared to adult fibroblasts. The role of cellular age in response to mechanical environment may have significant clinical implications when considering fibrotic response to implantable materials.

Tissue biomechanical properties also affect cell survival. It has long been known that interactions with the ECM are necessary for survival of nontransformed cells. However, the effects of the mechanical properties of the ECM on cell survival are only recently being addressed. Using polyacrylamide gels of varying rigidity coated with type I collagen, Wang and colleagues illustrated that proliferation of NIH 3T3 cells is enhanced on stiffer scaffolds [51]. These studies also illustrated that apoptosis of NIH 3T3 cells was increased by almost twofold on less rigid collagen-coated polyacrylamide gels. A similar increase in apoptosis was seen in cells from the rat

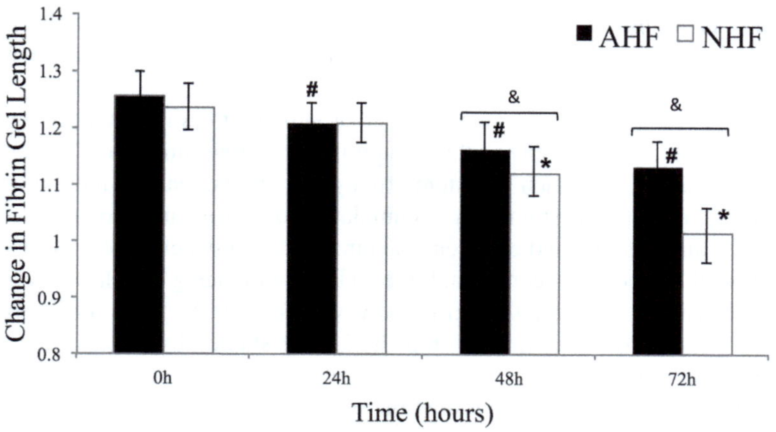

Fig. 9.1 Fibroblast-mediated contraction of free-floating fibrin gels is age dependent. Fibrin gels were seeded with equal numbers of either neonatal or adult cardiac fibroblasts and imaged at the time intervals shown. While both cell types significantly contracted fibrin gels over the study period compared to the initial gel length (#adult; *neonatal), neonatal fibroblasts exhibited a greater capacity for contraction compared to adult fibroblasts (&). Statistical significance ($p<0.05$) was determined using ANOVA and Tukey post-hoc test

annulus fibrosis when cultured on softer polyacrylamide scaffolds [52]. These studies suggest that decreasing local matrix stiffness will result in apoptosis, potentially of matrix-producing myofibroblasts or other cells.

9.6 Effects of Mechanical Stretch/Tension on Tissue Fibrosis

A number of early studies utilized cells cultured on deformable membranes to examine the cellular effects of mechanical stretch. These studies illustrated that mechanical stretch of isolated cells mimics many of the responses that had been characterized to increased load in vivo. For instance, mechanical stretch of skeletal myotubes results in reorientation of the cells and a hypertrophic response that includes increased general protein synthesis and enhanced accumulation of contractile proteins [53, 54].

Alterations in mechanical forces in vivo, such as increased cardiovascular load, have been known for some time to impact synthesis and deposition of the ECM. Increased mechanical load as seen during aortic stenosis and during hypertension promotes ventricular hypertrophy and fibrosis in the heart [55, 56]. In general, physical stretch of ECM-producing cells (largely fibroblasts and smooth muscle cells) results in increased production of ECM components or a pro-fibrotic response [57–60]. However, it has become increasingly clear that variations exist in the response of seemingly identical cells to mechanical and biochemical stimuli. Along these lines, recent studies have illustrated that dermal fibroblasts from different regions of the body have distinct responses to mechanical stretch [61]. It is likely that the in vivo physiological and pathological environments influence the response of cells, including fibroblasts, to mechanical forces.

To more accurately mimic the in vivo environment, investigators have turned to stretching cells in model systems that recapitulate the three-dimensional structure of tissues [62–64]. Using such a system, Tokuyama and colleagues have illustrated that mechanical stretch stimulates accumulation of basement membrane components including laminin and collagens in human skin equivalents [65]. Most studies examining the effects of mechanical force on ECM-producing cells have focused on simple, consistent stretching patterns; however, cells in the body are exposed to variable patterns of mechanical force. Intriguing studies by Imsirovic and colleagues have illustrated that greater variability in mechanical force elicits an enhanced response on ECM production by cells cultured in three-dimensional Gelfoam constructs [66]. These studies suggest that cells may detect not just mechanical forces, but are responsive to alterations in the magnitude, periodicity, and directionality of these forces as well. In this paradigm, cells in different tissues may have specific mechanical set points. Exposure to greater or lesser mechanical force may elicit a response.

9.7 Shear Stress and Fibrosis

Fluid movement across solid surfaces generates shear stresses, which have been shown to significantly impact cell behavior and gene expression [67–69]. The effects of shear stress generated by fluid flow have been well studied in endothelial cells and are essential for maintaining the integrity of vascular endothelium. This is due in part to the transmission of mechanical signals via PECAM-1 on the surface of endothelial cells. In contrast, shear stresses can also have deleterious effects on cells including the endothelium, depending in part on the amplitude of the stress and the cellular microenvironment.

Developmentally, fluid flow appears to be important in promoting normal development of many cells and tissues. Obstruction of blood flow in the embryonic heart alters heart morphogenesis including heart looping and valve formation [70, 71]. Studies carried out with valve primordia and in vitro bioreactors have illustrated a role for fluid flow in not only valve development but also ECM expression and deposition within the valve [72]. These studies utilized a ramping protocol to simulate physiological and pathological flow conditions. Over 7 days, ramping the flow to pathological levels resulted in enhanced expression of ECM components including collagen type I, periostin, and tenascin C. Alterations in ECM expression correlated with activation of RhoA and incubation of valve primordia with the RhoA inhibitor Y-27632 attenuated the effects of fluid flow on collagen I and tenascin C expression. The cellular mechanisms of these effects, e.g., which cells of the valve primordia are responding to alterations in flow, are not clear. Exposure of endothelial cells to flow-induced shear stresses results in expression and secretion of cytokines and growth factors [73, 74], which may modulate ECM production by valve fibroblasts. Alternatively, alterations in fluid flow may promote the conversion of endothelial cells into a mesenchymal or myofibroblast phenotype, which then participate in ECM production.

Under particular conditions, endothelial cells can acquire a fibro-proliferative phenotype through an endothelial (or epithelial)-to-mesenchyme transition. This transition is characterized by the loss of cell-to-cell contacts and reduction in endothelial markers such as VE-cadherin and PECAM. Concurrently, there is an increase in mesenchymal or myofibroblast-like markers including α-smooth muscle actin and calponin. Such a transition may be an important component of normal development [75], but may also contribute to disease processes including fibrosis of the heart, kidney, and other organs [76, 77]. Studies in vivo and in vitro have been carried out to elucidate the differences between "protective" or healthy fluid flow and patterns of flow that contribute to pathological situations such as fibrosis. Studies in animal models and correlative studies in human patients have clearly illustrated that areas of "disturbed flow," for instance at vascular branch points, are more susceptible to damage including neointima formation and transition of endothelial cells into a myofibroblast phenotype. Recent in vitro studies illustrated that laminar or uniform flow is protective against the transformation of endothelial cells into a myofibroblast phenotype [78] and that this is dependent upon the prolonged activation of the Erk5 signaling pathway by laminar flow. This is consistent with studies illustrating that uniform shear stress "strengthens" the epithelial phenotype of kidney tubule cells [79]. This includes increased tight junction formation between neighboring cells, decreased migratory capacity, and resistance to TGF-β-induced epithelial-to-mesenchymal transformation.

9.8 Mechanical Forces and Inflammation

The promotion of fibrosis by mechanical forces may be due to the direct effect of these forces on ECM-producing cells as discussed above; however, the fibrotic process may also be impacted by inflammation. Fewer studies have been performed to analyze the potential effects of mechanical forces on the inflammatory response. This could be particularly relevant to implant success as the foreign body response includes infiltration of immune/inflammatory cells and enhanced production of inflammatory cytokines. Activation of fibroblasts or myofibroblast formation not only results in enhanced ECM production, but also increased secretion of growth factors and hormones. Exposure of fibroblasts to centrifugal force (67.1 g/cm^2 for 1 h) results in elevated levels of inflammatory/oxidative stress markers including prostaglandin E2, nitric oxide, interleukin-1α, and interleukin-1β [80]. Promotion of inflammation by alterations in the biomechanical environment could be very important clinically. Studies have illustrated that orthodontic forces can be important in either the progression or resolution of clinical problems in the oral cavity. Chronic inflammation in the oral cavity can result in destruction of the periodontal ligament and alveolar bone. Recent studies have illustrated that orthodontic force results in increased production of interleukin-1β and tumor necrosis factor-α and subsequent breakdown of alveolar bone [81]. Using isolated periodontal fibroblasts in an in vitro system, Jacobs and colleagues illustrated that different levels of static tensile strain elicit diverse inflammatory responses [82]. Exposure to 10 % static strain results in an inflammatory response including the increased expression of interleukin-6, cyclooxygenase-2 (COX-2), and prostaglandin E2. In contrast, exposure to 1 % static strain yields decreased interleukin-6 expression, suggesting that moderate mechanical force may be anti-inflammatory.

Mechanical ventilation generates biomechanical forces that exacerbate lung inflammation [83, 84]. In vitro studies with lung epithelial cells have demonstrated that cyclic stretch or cyclic oscillatory pressure results in activation of NF-kB and elevated expression of inflammatory cytokines including interleukin-6, interleukin-8, and tumor necrosis factor-α [42]. These studies also illustrated that cyclic mechanical forces promote expression of miRNA-146a. This miRNA has been shown to play important modulatory roles in inflammation and innate immunity [85]. Molecular perturbation experiments illustrated that the functional role of miRNA-146a in the inflammatory response to cyclic mechanical force by lung cells involves modulation of components of the toll-like receptor-signaling pathway [86].

Little is currently known regarding the response of "classical" inflammatory/immune cells to mechanical forces. Mast cells are traditionally described for their role in hypersensitivity reactions; however, recent studies have made it clear that these cells have broader functions including modulation of tissue remodeling and fibrosis [87–89]. Mast cells are typically activated by allergen binding to Immunoglobulin E antibodies and activation of high affinity FcεR1 receptors. This results in rapid release of biochemical mediators via degranulation. More recently, other mechanisms of mast cell activation have been described including diverse biochemical factors (bacterial toxins, endothelin-1, and others) and physical forces including osmotic pressure, shear stress, and mechanical stretch. Recent studies have illustrated that cyclic stretch promotes mast cell degranulation and increased interleukin-4 expression [90, 91]. Perturbation-studies illustrated that this response involves cell surface receptors of the integrin family, a common mechanism of transduction of mechanical forces to cells as discussed in the next section.

9.9 Transduction of Mechanical Signals

Studies utilizing in vitro systems have provided fundamental information regarding the molecular mechanisms whereby cells detect and respond to mechanical forces. During the past two decades, extensive progress has been made in understanding "mechanotransduction" or the mechanisms whereby physical stimuli are converted into chemical or molecular signals by cells [92, 93]. Despite the fact that the types of mechanical forces cells experience are variable, the molecular mechanisms whereby this information is transduced appear to have similarities. Alterations in the three-dimensional conformation of mechanosensitive proteins or adhesion structures are often fundamental to this process. Studies utilizing diverse in vitro systems were essential in implicating cell surface integrins as central components of cell adhesion complexes and fundamental to mechanotransduction [94]. Integrins are heterodimers composed of an α and a β chain that serve as the primary family of receptors for ECM components [95, 96]. There are over 20 different α/β heterodimer combinations and specific α/β heterodimers serve as receptors for particular ECM ligand(s). The response of cells to mechanical force varies depending upon the ECM substratum suggesting a role for specific integrin heterodimers [94, 97]. Utilizing function-blocking antibodies to specific integrins (α4 and α5 chains) or arginine-glycine-aspartic acid (RGD) peptides to prevent integrin-ECM interactions, MacKenna and colleagues [94] were among the first to show roles for specific integrins in the response of fibroblasts to mechanical force.

These early studies set the stage for extensive research focused on the mechanisms whereby cells detect mechanical changes in the microenvironment and transduce these into biochemical and molecular alterations in the cytoplasm and nucleus. The cell-ECM linkage involving integrins and a myriad of associated proteins is a critical component of this process. It has become increasingly clear that integrin-based adhesions are dynamic and complex structures that transmit information from the ECM to the cell and vice versa [98]. Integrins, which lack intrinsic enzyme activity, provide a physical linkage from the ECM to the actin cytoskeleton and to a wide array of signaling proteins. In fact, integrin complexes can contain over a hundred different proteins, many of which bind in a force-dependent manner [99, 100]. The characterization of the ECM-integrin-cytoskeletal linkage has contributed to the concept of tensegrity in which signals can be transmitted from the ECM to the cytoplasm and nucleus via these physical connections [101, 102]. Several proteins can simultaneously bind integrins and actin and are thus thought to participate in mechanotrasduction via the physical ECM-integrin-cytoskeleton linkage including vinculin, talin, and α-actinin [103, 104].

A number of signaling molecules also associate directly or indirectly with the integrin cytoplasmic domain including focal adhesion kinase (FAK). FAK was initially identified as a Src kinase substrate [24, 105]. As integrins do not have intrinsic enzyme activity, FAK is a critical mediator of integrin-induced signaling events. The activation of FAK is initiated by autophosphorylation of tyrosine at position 397 and can be induced by clustering of integrins [106]. In turn, FAK can activate integrins, which strengthens cell adhesions with the ECM [107]. Activated FAK can act independently or as part of a Src-containing complex to phosphorylate other signaling proteins or act as a scaffold in the recruitment of additional proteins to cell adhesions.

Exposure of cells to mechanical force results in activation of numerous intracellular signaling pathways including protein kinases such as protein kinase C, c-Jun N-terminal kinases (JNK), extracellular signal-regulated kinases (ERK), and others [108]. Activation of these pathways ultimately leads to activation of transcription factors and cell activities that comprise the response of a given cell to mechanical events.

While there appear to commonalities in signaling pathways induced by various types of mechanical forces, in vitro studies illustrate that cells respond differently to diverse types of mechanical perturbations. The type of mechanical force can modulate differentiation of connective tissue cells. The ratio between tensile and compression type forces can promote either differentiation into cartilage or bone [109]. Exposing vascular endothelial cells to cyclic stretch results in differences in growth factor expression and branch formation compared to constant stretch [110]. Application of steady mechanical force on aortas results in more pronounced FAK activation compared to pulsatile stretch [111]. These studies suggest that while generalities may be developed regarding the response of cells to mechanical force, the details of this response likely vary depending on the type of force and in a cell- or tissue-specific manner.

9.10 Integration of Mechanical and Biochemical Signals

Cells are continually exposed to changes in their microenvironment or niche including alterations in mechanical forces and the biochemical milieu. Tissue development, homeostasis, and regeneration require the integration of and appropriate response to these diverse signals. The integration of signals derived from mechanical forces, biochemical factors, cell-cell interactions, and other aspects of the microenvironment are being intensely studied as artificial modulation of these may provide a means to direct stem cell differentiation and enhance tissue regeneration [112].

In regards to the fibrotic response, in vitro mechanical stretch of heart fibroblasts in the absence of serum or growth factors had little effect on the expression of collagen α1(I) mRNA [58]. However, mechanical stretch in the presence of fetal bovine serum substantially increased collagen mRNA expression and protein synthesis, illustrating that biochemical factors contained in serum are needed for the response to mechanical stretch. In these studies, mechanical stretch also had a synergistic effect with transforming growth factor-β1 (TGF-β1) and insulin-like growth factor-1 (IGF-1) on collagen production. Similar results have been described between mechanical stretch and platelet-derived growth factor (PDGF) in arterial fibroblasts [113]. We have carried out studies in our labs to evaluate the combined effects of mechanical stretch and IGF-1 on the activation of signal transduction pathways. In these experiments, heart fibroblasts were treated with IGF-1 alone (50 ng/ml), mechanical stretch alone (10 % constant equibiaxial stretch), or combined IGF-1 and mechanical stretch. The activation of signal transduction pathways was assayed by immunoblotting. Immunoblotting for phosphorylated extracellular signal-related kinases (ERK 1/2), also known as classical mitogen-activated protein kinases (MAPKs), illustrated this pathway to be activated by both IGF-1 and mechanical stretch alone (Fig. 9.2). Combined treatment with IGF-1 and mechanical stretch resulted in enhanced ERK 1/2 phosphorylation compared to either treatment alone. Similar experiments were carried out to evaluate the activation of Akt, which is an important signaling pathway activated by IGF-1. In contrast to ERK 1/2 activity,

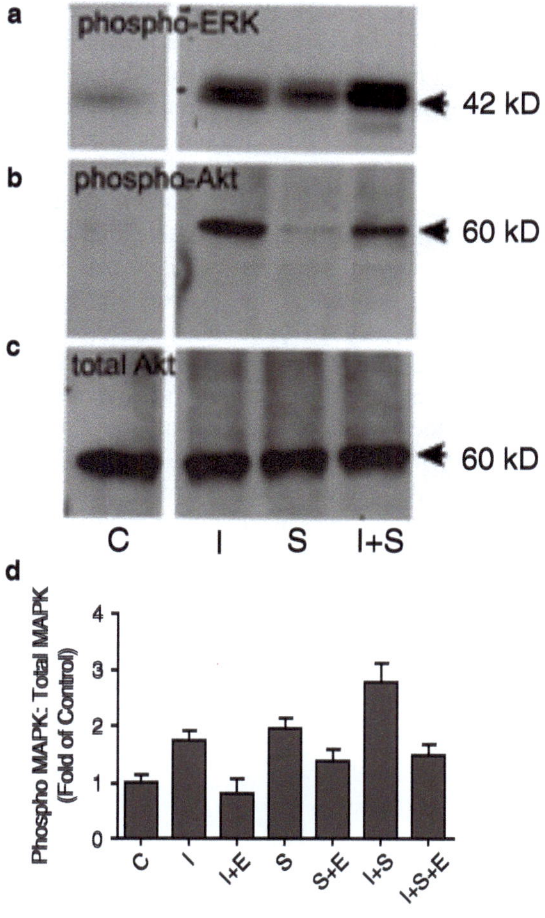

Fig. 9.2 (**a–c**) Cardiac fibroblasts were exposed to no treatment (C), IGF-1 (I), mechanical stretch (S), or IGF-1 and mechanical stretch combined (I + S) for 15 min. Immunoblots of cell lysates were performed with antisera specific for phosphorylated ERK 1/2, phosphorylated Akt, or total Akt (as a loading control). Note that IGF-1 or mechanical stretch alone both activate ERK 1/2 and activation is enhanced even further by combined treatment. Activation of Akt is only seen when IGF-1 is present and mechanical stretch has no apparent effect on this pathway. (**d**) Graphic representation of phosphorylated ERK 1/2 to total ERK 1/2 ratio. In these experiments, fibroblasts were pretreated with echistatin (E), which competes with extracellular ligands for binding to the αvβ3 integrin. Treatment with echistatin reduced activation of ERK 1/2 by IGF-1 and mechanical stretch alone, as well as combined IGF-1 and mechanical stretch treatment

Akt was phosphorylated by treatment with IGF-1 and not mechanical stretch alone. Treatment with both IGF-1 and mechanical stretch did not result in enhanced Akt activity compared to IGF-1 alone. Further experiments were carried out to investigate the role of the αvβ3 integrins in the response to mechanical stretch and IGF-1. Fibroblasts were pretreated with echistatin, which competitively inhibits binding of the αvβ3 integrin to extracellular ligand. Pretreatment with echistatin resulted in decreased ERK 1/2 activation by IGF-1 and by mechanical stretch (Fig. 9.2d). These experiments suggest that cell surface integrins provide a point of convergence between growth factor and mechanical signaling pathways. Identifying the underlying mechanisms whereby mechanical signals are transduced in cells may present new therapeutic targets for modification of the fibrotic response.

Myofibroblast formation is also regulated by the combined effects of mechanical and biochemical signaling. Mechanical tension and TGF-β1 are both critical to the formation of myofibroblasts [31, 114]. TGF-β1 is able to induce myofibroblast formation in cells grown in three-dimensional collagen gels that are mechanically restrained but not in collagen gels that are "relaxed" by free-floating in culture medium [40, 115]. Similarly, TGF-β1 promotes myofibroblast formation in cells cultured on two-dimensional substrata with stiffness that mimics that seen in fibrotic or granulation tissues but not on substrata that mimics the mechanical properties of normal dermis [116]. On the other hand, mechanical force is unable to induce myofibroblast formation in the absence of active TGF-β1 [40]. TGF-β1 is released by cells as a latent precursor that is stored as part of a complex within the ECM. Studies have illustrated that mechanical forces including generation of tension by cells can activate latent TGF-β1 and promote myofibroblast formation and that this is mediated by integrins of the αv family [114, 117, 118]. The precise molecular mechanisms of this process are currently under investigation; however, this presents the possibility that αv integrins or other proteins involved in this process could become therapeutic targets for modulation of myofibroblast activation.

9.11 Conclusions

It has become increasingly clear that most cells in the vertebrate body are exposed to varying degrees to mechanical forces. These forces impact embryonic development, homeostasis, and pathological conditions including fibrosis. The effects of the mechanical environment on implant success are particularly relevant as alterations in tissue properties by implants are thought to, at least in part, drive fibrous capsule formation. Historically, most of the studies that focused on mechanical force as a pro-fibrotic stimulus utilized two-dimensional stretch or compression models with isolated matrix-producing cells. These studies have provided substantial knowledge regarding the responses of cells to mechanical force and the underlying mechanisms of this response. However, these systems do not adequately mimic the in vivo three-dimensional environment. This has led development of three-dimensional models to evaluate the effects of mechanical forces in a more in vivo-like environment. The realization that the biomechanical properties of the microenvironment can promote fibrosis and other responses has led to renewed interest in the effects of mechanical forces on cell and tissue behavior.

While extensive knowledge has been gained regarding the effects of the mechanical environment on cells and tissues, many questions remain regarding the molecular mechanisms of these effects. Identification of novel mechano-responsive proteins will provide new therapeutic targets to modulate the deleterious effects of the biomechanical environment. As it is becomingly increasing clear that tissue stiffness may precede fibrosis or at least contribute to ongoing fibrosis, identifying methods to modulate the mechanical properties of the microenvironment may also yield novel therapeutic approaches.

References

1. Beloussov LV, Grabovsky VI (2006) Morphomechanics: goals, basic experiments and models. Int J Dev Biol 50(2-3):81–92
2. Benjamin M, Hillen B (2003) Mechanical influences on cells, tissues and organs—'mechanical morphogenesis'. Eur J Morphol 41(1):3–7
3. Farge E (2011) Mechanotransduction in development. Curr Top Dev Biol 95:243–265
4. Jones EA (2011) Mechanical factors in the development of the vascular bed. Respir Physiol Neurobiol 178(1):59–65
5. Bassett CA, Herrmann I (1961) Influence of oxygen concentration and mechanical factors on differentiation of connective tissues in vitro. Nature 190:460–461
6. Rodan GA, Mensi T, Harvey A (1975) A quantitative method for the application of compressive forces to bone in tissue culture. Calcif Tissue Res 18(2):125–131
7. Leung DY, Glagov S, Mathews MB (1976) Cyclic stretching stimulates synthesis of matrix components by arterial smooth muscle cells in vitro. Science 191(4226):475–477
8. Leung DY, Glagov S, Mathews MB (1977) A new in vitro system for studying cell response to mechanical stimulation. Different effects of cyclic stretching and agitation on smooth muscle cell biosynthesis. Exp Cell Res 109(2):285–298
9. Dzau VJ (1993) Local contractile and growth modulators in the myocardium. Clin Cardiol 16(5 Suppl 2):II5–II9
10. Samuel JL, Dubus I, Contard F et al (1990) Biological signals of cardiac hypertrophy. Eur Heart J 11 Suppl G:1–7
11. Lohler J, Timpl R, Jaenisch R (1984) Embryonic lethal mutation in mouse collagen I gene causes rupture of blood vessels and is associated with erythropoietic and mesenchymal cell death. Cell 38(2):597–607
12. Brenner DA, Kisseleva T, Scholten D et al (2012) Origin of myofibroblasts in liver fibrosis. Fibrogenesis Tissue Repair 5(Suppl 1):S17
13. Hinz B (2012) Mechanical aspects of lung fibrosis: a spotlight on the myofibroblast. Proc Am Thorac Soc 9(3):137–147
14. Hinz B, Phan SH, Thannickal VJ et al (2012) Recent developments in myofibroblast biology: paradigms for connective tissue remodeling. Am J Pathol 180(4):1340–1355
15. Park S, Park M, Kim BH et al (2015) Acute suppression of TGF-ss with local, sustained release of tranilast against the formation of fibrous capsules around silicone implants. J Control Release 200:125–137
16. Wolfram D, Rainer C, Niederegger H et al (2004) Cellular and molecular composition of fibrous capsules formed around silicone breast implants with special focus on local immune reactions. J Autoimmun 23(1):81–91
17. Bowen T, Jenkins RH, Fraser DJ (2013) MicroRNAs, transforming growth factor beta-1, and tissue fibrosis. J Pathol 229(2):274–285
18. Mammoto T, Ingber DE (2010) Mechanical control of tissue and organ development. Development 137(9):1407–1420
19. Tomeno W, Yoneda M, Imajo K et al (2013) Evaluation of the Liver Fibrosis Index calculated by using real-time tissue elastography for the non-invasive assessment of liver fibrosis in chronic liver diseases. Hepatol Res 43(7):735–742

20. Yin MF, Lian LH, Piao DM et al (2007) Tetrandrine stimulates the apoptosis of hepatic stellate cells and ameliorates development of fibrosis in a thioacetamide rat model. World J Gastroenterol 13(8):1214–1220
21. Lam WA, Cao L, Umesh V et al (2010) Extracellular matrix rigidity modulates neuroblastoma cell differentiation and N-myc expression. Mol Cancer 9:35
22. Paszek MJ, Zahir N, Johnson KR et al (2005) Tensional homeostasis and the malignant phenotype. Cancer Cell 8(3):241–254
23. Pathak A, Kumar S (2012) Independent regulation of tumor cell migration by matrix stiffness and confinement. Proc Natl Acad Sci U S A 109(26):10334–10339
24. Schaller MD, Borgman CA, Cobb BS et al (1992) pp125FAK a structurally distinctive protein-tyrosine kinase associated with focal adhesions. Proc Natl Acad Sci U S A 89(11):5192–5196
25. Schedin P, Keely PJ (2011) Mammary gland ECM remodeling, stiffness, and mechanosignaling in normal development and tumor progression. Cold Spring Harb Perspect Biol 3(1):a003228
26. Ulrich TA, de Juan Pardo EM, Kumar S (2009) The mechanical rigidity of the extracellular matrix regulates the structure, motility, and proliferation of glioma cells. Cancer Res 69(10):4167–4174
27. Choi TY, Ahmadi N, Sourayanezhad S et al (2013) Relation of vascular stiffness with epicardial and pericardial adipose tissues, and coronary atherosclerosis. Atherosclerosis 229(1):118–123
28. Wells RG (2013) Tissue mechanics and fibrosis. Biochim Biophys Acta 1832(7):884–890
29. Ho YY, Lagares D, Tager AM et al (2014) Fibrosis—a lethal component of systemic sclerosis. Nat Rev Rheumatol 10(7):390–402
30. Clark RA, Ashcroft GS, Spencer MJ et al (1996) Re-epithelialization of normal human excisional wounds is associated with a switch from alpha v beta 5 to alpha v beta 6 integrins. Br J Dermatol 135(1):46–51
31. Hinz B (2009) Tissue stiffness, latent TGF-beta1 activation, and mechanical signal transduction: implications for the pathogenesis and treatment of fibrosis. Curr Rheumatol Rep 11(2):120–126
32. Wang HB, Dembo M, Wang YL (2000) Substrate flexibility regulates growth and apoptosis of normal but not transformed cells. Am J Physiol Cell Physiol 279(5):C1345–C1350
33. Yeung T, Georges PC, Flanagan LA et al (2005) Effects of substrate stiffness on cell morphology, cytoskeletal structure, and adhesion. Cell Motil Cytoskeleton 60(1):24–34
34. Peyton SR, Putnam AJ (2005) Extracellular matrix rigidity governs smooth muscle cell motility in a biphasic fashion. J Cell Physiol 204(1):198–209
35. Engler AJ, Rehfeldt F, Sen S et al (2007) Microtissue elasticity: measurements by atomic force microscopy and its influence on cell differentiation. Methods Cell Biol 83:521–545
36. Georges PC, Hui JJ, Gombos Z et al (2007) Increased stiffness of the rat liver precedes matrix deposition: implications for fibrosis. Am J Physiol Gastrointest Liver Physiol 293(6):G1147–G1154
37. Mauch C, Adelmann-Grill B, Hatamochi A et al (1989) Collagenase gene expression in fibroblasts is regulated by a three-dimensional contact with collagen. FEBS Lett 250(2):301–305
38. Johnson LA, Rodansky ES, Sauder KL et al (2013) Matrix stiffness corresponding to strictured bowel induces a fibrogenic response in human colonic fibroblasts. Inflamm Bowel Dis 19(5):891–903
39. Liu F, Mih JD, Shea BS et al (2010) Feedback amplification of fibrosis through matrix stiffening and COX-2 suppression. J Cell Biol 190(4):693–706
40. Arora PD, Narani N, McCulloch CA (1999) The compliance of collagen gels regulates transforming growth factor-beta induction of alpha-smooth muscle actin in fibroblasts. Am J Pathol 154(3):871–882
41. Galie PA, Westfall MV, Stegemann JP (2011) Reduced serum content and increased matrix stiffness promote the cardiac myofibroblast transition in 3D collagen matrices. Cardiovasc Pathol 20(6):325–333

42. Huang X, Yang N, Fiore VF et al (2012) Matrix stiffness-induced myofibroblast differentiation is mediated by intrinsic mechanotransduction. Am J Respir Cell Mol Biol 47(3):340–348
43. Olsen AL, Bloomer SA, Chan EP et al (2011) Hepatic stellate cells require a stiff environment for myofibroblastic differentiation. Am J Physiol Gastrointest Liver Physiol 301(1):G110–G118
44. Shi Y, Dong Y, Duan Y et al (2013) Substrate stiffness influences TGF-beta1-induced differentiation of bronchial fibroblasts into myofibroblasts in airway remodeling. Mol Med Rep 7(2):419–424
45. Friedman SL (2008) Hepatic stellate cells: protean, multifunctional, and enigmatic cells of the liver. Physiol Rev 88(1):125–172
46. Friedman SL, Roll FJ, Boyles J et al (1989) Maintenance of differentiated phenotype of cultured rat hepatic lipocytes by basement membrane matrix. J Biol Chem 264(18):10756–10762
47. Gaca MD, Zhou X, Issa R et al (2003) Basement membrane-like matrix inhibits proliferation and collagen synthesis by activated rat hepatic stellate cells: evidence for matrix-dependent deactivation of stellate cells. Matrix Biol 22(3):229–239
48. Wang H, Haeger SM, Kloxin AM et al (2012) Redirecting valvular myofibroblasts into dormant fibroblasts through light-mediated reduction in substrate modulus. PLoS One 7(7), e39969
49. Balestrini JL, Chaudhry S, Sarrazy V et al (2012) The mechanical memory of lung myofibroblasts. Integr Biol (Camb) 4(4):410–421
50. Wilson CG, Stone JW, Fowlkes V et al (2011) Age-dependent expression of collagen receptors and deformation of type I collagen substrates by rat cardiac fibroblasts. Microsc Microanal 17(4):555–562
51. Wang S, Cukierman E, Swaim WD et al (1999) Extracellular matrix protein-induced changes in human salivary epithelial cell organization and proliferation on a model biological substratum. Biomaterials 20(11):1043–1049
52. Zhang YH, Zhao CQ, Jiang LS et al (2011) Substrate stiffness regulates apoptosis and the mRNA expression of extracellular matrix regulatory genes in the rat annular cells. Matrix Biol 30(2):135–144
53. Vandenburgh HH (1982) Dynamic mechanical orientation of skeletal myofibers in vitro. Dev Biol 93(2):438–443
54. Vandenburgh H, Kaufman S (1979) In vitro model for stretch-induced hypertrophy of skeletal muscle. Science 203(4377):265–268
55. Weber KT, Janicki JS, Shroff SG et al (1988) Collagen remodeling of the pressure-overloaded, hypertrophied nonhuman primate myocardium. Circ Res 62(4):757–765
56. Jalil JE, Doering CW, Janicki JS et al (1989) Fibrillar collagen and myocardial stiffness in the intact hypertrophied rat left ventricle. Circ Res 64(6):1041–1050
57. Kollros PR, Bates SR, Mathews MB et al (1987) Cyclic AMP inhibits increased collagen production by cyclically stretched smooth muscle cells. Lab Invest 56(4):410–417
58. Butt RP, Bishop JE (1997) Mechanical load enhances the stimulatory effect of serum growth factors on cardiac fibroblast procollagen synthesis. J Mol Cell Cardiol 29(4):1141–1151
59. Carver W, Nagpal ML, Nachtigal M et al (1991) Collagen expression in mechanically stimulated cardiac fibroblasts. Circ Res 69(1):116–122
60. Lee AA, Delhaas T, Waldman LK et al (1996) An equibiaxial strain system for cultured cells. Am J Physiol 271(4 Pt 1):C1400–C1408
61. Wang Z, Kuang R, Xu Q et al (2015) Reaction of human fibroblasts from different sites to the mechanical stress. Zhongguo Xiu Fu Chong Jian Wai Ke Za Zhi 29(4):467–471
62. Auluck A, Mudera V, Hunt NP et al (2005) A three-dimensional in vitro model system to study the adaptation of craniofacial skeletal muscle following mechanostimulation. Eur J Oral Sci 113(3):218–224
63. Birla RK, Huang YC, Dennis RG (2007) Development of a novel bioreactor for the mechanical loading of tissue-engineered heart muscle. Tissue Eng 13(9):2239–2248

64. Masoumi N, Howell MC, Johnson KL et al (2014) Design and testing of a cyclic stretch and flexure bioreactor for evaluating engineered heart valve tissues based on poly(glycerol sebacate) scaffolds. Proc Inst Mech Eng H 228(6):576–586
65. Tokuyama E, Nagai Y, Takahashi K et al (2015) Mechanical stretch on human skin equivalents increases the epidermal thickness and develops the basement membrane. PLoS One 10(11), e0141989
66. Imsirovic J, Derricks K, Buczek-Thomas JA et al (2013) A novel device to stretch multiple tissue samples with variable patterns: application for mRNA regulation in tissue-engineered constructs. Biomatter 3(3):pii: e24650
67. Obi S, Yamamoto K, Ando J (2014) Effects of shear stress on endothelial progenitor cells. J Biomed Nanotechnol 10(10):2586–2597
68. Dunn J, Simmons R, Thabet S et al (2015) The role of epigenetics in the endothelial cell shear stress response and atherosclerosis. Int J Biochem Cell Biol 67:167–176
69. Rodriguez I, Gonzalez M (2014) Physiological mechanisms of vascular response induced by shear stress and effect of exercise in systemic and placental circulation. Front Pharmacol 5:209
70. Sedmera D, Pexieder T, Rychterova V et al (1999) Remodeling of chick embryonic ventricular myoarchitecture under experimentally changed loading conditions. Anat Rec 254(2):238–252
71. Hove JR, Koster RW, Forouhar AS et al (2003) Intracardiac fluid forces are an essential epigenetic factor for embryonic cardiogenesis. Nature 421(6919):172–177
72. Tan H, Biechler S, Junor L et al (2013) Fluid flow forces and rhoA regulate fibrous development of the atrioventricular valves. Dev Biol 374(2):345–356
73. Egorova AD, Khedoe PP, Goumans MJ et al (2011) Lack of primary cilia primes shear-induced endothelial-to-mesenchymal transition. Circ Res 108(9):1093–1101
74. Misra S, Fu AA, Puggioni A et al (2008) Increased shear stress with upregulation of VEGF-A and its receptors and MMP-2, MMP-9, and TIMP-1 in venous stenosis of hemodialysis grafts. Am J Physiol Heart Circ Physiol 294(5):H2219–H2230
75. Markwald RR, Fitzharris TP, Manasek FJ (1977) Structural development of endocardial cushions. Am J Anat 148(1):85–119
76. Zeisberg EM, Tarnavski O, Zeisberg M et al (2007) Endothelial-to-mesenchymal transition contributes to cardiac fibrosis. Nat Med 13(8):952–961
77. Zeisberg EM, Potenta SE, Sugimoto H et al (2008) Fibroblasts in kidney fibrosis emerge via endothelial-to-mesenchymal transition. J Am Soc Nephrol 19(12):2282–2287
78. Moonen JR, Lee ES, Schmidt M et al (2015) Endothelial-to-mesenchymal transition contributes to fibro-proliferative vascular disease and is modulated by fluid shear stress. Cardiovasc Res 108(3):377–386
79. Grabias BM, Konstantopoulos K (2014) The physical basis of renal fibrosis: effects of altered hydrodynamic forces on kidney homeostasis. Am J Physiol Renal Physiol 306(5):F473–F485
80. Chao YH, Tsuang YH, Sun JS et al (2012) Centrifugal force induces human ligamentum flavum fibroblasts inflammation through activation of JNK and p38 pathways. Connect Tissue Res 53(5):422–429
81. Nogueira AV, Nokhbehsaim M, Eick S et al (2014) Biomechanical loading modulates proinflammatory and bone resorptive mediators in bacterial-stimulated PDL cells. Mediators Inflamm 2014:425421
82. Jacobs C, Walter C, Ziebart T et al (2014) Induction of IL-6 and MMP-8 in human periodontal fibroblasts by static tensile strain. Clin Oral Investig 18(3):901–908
83. D'Angelo E, Koutsoukou A, Della Valle P et al (2008) Cytokine release, small airway injury, and parenchymal damage during mechanical ventilation in normal open-chest rats. J Appl Physiol (1985) 104(1):41–49
84. Plataki M, Hubmayr RD (2010) The physical basis of ventilator-induced lung injury. Expert Rev Respir Med 4(3):373–385
85. Li G, Luna C, Qiu J et al (2010) Modulation of inflammatory markers by miR-146a during replicative senescence in trabecular meshwork cells. Invest Ophthalmol Vis Sci 51(6):2976–2985

86. Huang Y, Crawford M, Higuita-Castro N et al (2012) miR-146a regulates mechanotransduction and pressure-induced inflammation in small airway epithelium. FASEB J 26(8):3351–3364
87. Hartupee J, Mann DL (2016) Role of inflammatory cells in fibroblast activation. J Mol Cell Cardiol 93:143–148
88. Madjene LC, Pons M, Danelli L et al (2015) Mast cells in renal inflammation and fibrosis: lessons learnt from animal studies. Mol Immunol 63(1):86–93
89. Overed-Sayer C, Rapley L, Mustelin T et al (2013) Are mast cells instrumental for fibrotic diseases? Front Pharmacol 4:174
90. Fowlkes V, Wilson CG, Carver W et al (2013) Mechanical loading promotes mast cell degranulation via RGD-integrin dependent pathways. J Biomech 46(4):788–795
91. Komiyama H, Miyake K, Asai K et al (2014) Cyclical mechanical stretch enhances degranulation and IL-4 secretion in RBL-2H3 mast cells. Cell Biochem Funct 32(1):70–76
92. DuFort CC, Paszek MJ, Weaver VM (2011) Balancing forces: architectural control of mechanotransduction. Nat Rev Mol Cell Biol 12(5):308–319
93. Zhang H, Labouesse M (2012) Signalling through mechanical inputs: a coordinated process. J Cell Sci 125(Pt 13):3039–3049
94. MacKenna DA, Dolfi F, Vuori K et al (1998) Extracellular signal-regulated kinase and c-Jun NH2-terminal kinase activation by mechanical stretch is integrin-dependent and matrix-specific in rat cardiac fibroblasts. J Clin Invest 101(2):301–310
95. Buck CA, Horwitz AF (1987) Cell surface receptors for extracellular matrix molecules. Annu Rev Cell Biol 3:179–205
96. Humphries MJ, Yasuda Y, Olden K et al (1988) The cell interaction sites of fibronectin in tumour metastasis. Ciba Found Symp 141:75–93
97. Atance J, Yost MJ, Carver W (2004) Influence of the extracellular matrix on the regulation of cardiac fibroblast behavior by mechanical stretch. J Cell Physiol 200(3):377–386
98. Roca-Cusachs P, Iskratsch T, Sheetz MP (2012) Finding the weakest link: exploring integrin-mediated mechanical molecular pathways. J Cell Sci 125(Pt 13):3025–3038
99. Zaidel-Bar R, Itzkovitz S, Ma'ayan A et al (2007) Functional atlas of the integrin adhesome. Nat Cell Biol 9(8):858–867
100. Pasapera AM, Schneider IC, Rericha E et al (2010) Myosin II activity regulates vinculin recruitment to focal adhesions through FAK-mediated paxillin phosphorylation. J Cell Biol 188(6):877–890
101. Ingber DE (1991) Control of capillary growth and differentiation by extracellular matrix. Use of a tensegrity (tensional integrity) mechanism for signal processing. Chest 99(3 Suppl):34S–40S
102. Ingber DE (1997) Integrins, tensegrity, and mechanotransduction. Gravit Space Biol Bull 10(2):49–55
103. Burridge K, Mangeat P (1984) An interaction between vinculin and talin. Nature 308(5961):744–746
104. Critchley DR (2009) Biochemical and structural properties of the integrin-associated cytoskeletal protein talin. Annu Rev Biophys 38:235–254
105. Weiner TM, Liu ET, Craven RJ et al (1993) Expression of focal adhesion kinase gene and invasive cancer. Lancet 342(8878):1024–1025
106. Chen HC, Appeddu PA, Isoda H et al (1996) Phosphorylation of tyrosine 397 in focal adhesion kinase is required for binding phosphatidylinositol 3-kinase. J Biol Chem 271(42):26329–26334
107. Michael KE, Dumbauld DW, Burns KL et al (2009) Focal adhesion kinase modulates cell adhesion strengthening via integrin activation. Mol Biol Cell 20(9):2508–2519
108. Mammoto A, Mammoto T, Ingber DE (2012) Mechanosensitive mechanisms in transcriptional regulation. J Cell Sci 125(Pt 13):3061–3073
109. Carter DR, Beaupre GS, Giori NJ et al (1998) Mechanobiology of skeletal regeneration. Clin Orthop Relat Res (355 Suppl):S41–S55

110. Zheng W, Christensen LP, Tomanek RJ (2008) Differential effects of cyclic and static stretch on coronary microvascular endothelial cell receptors and vasculogenic/angiogenic responses. Am J Physiol Heart Circ Physiol 295(2):H794–H800
111. Lehoux S, Esposito B, Merval R et al (2005) Differential regulation of vascular focal adhesion kinase by steady stretch and pulsatility. Circulation 111(5):643–649
112. Discher DE, Mooney DJ, Zandstra PW (2009) Growth factors, matrices, and forces combine and control stem cells. Science 324(5935):1673–1677
113. Bishop JE, Butt R, Dawes K et al (1998) Mechanical load enhances the stimulatory effect of PDGF on pulmonary artery fibroblast procollagen synthesis. Chest 114(1 Suppl):25S
114. Wipff PJ, Rifkin DB, Meister JJ et al (2007) Myofibroblast contraction activates latent TGF-beta1 from the extracellular matrix. J Cell Biol 179(6):1311–1323
115. Tomasek JJ, Gabbiani G, Hinz B et al (2002) Myofibroblasts and mechano-regulation of connective tissue remodelling. Nat Rev Mol Cell Biol 3(5):349–363
116. Goffin JM, Pittet P, Csucs G et al (2006) Focal adhesion size controls tension-dependent recruitment of alpha-smooth muscle actin to stress fibers. J Cell Biol 172(2):259–268
117. Wipff PJ, Hinz B (2008) Integrins and the activation of latent transforming growth factor beta1—an intimate relationship. Eur J Cell Biol 87(8-9):601–615
118. Sarrazy V, Koehler A, Chow ML et al (2014) Integrins alphavbeta5 and alphavbeta3 promote latent TGF-beta1 activation by human cardiac fibroblast contraction. Cardiovasc Res 102(3): 407–417

Chapter 10
Advancing Islet Transplantation: From Donor to Engraftment

Omaima M. Sabek

Abstract Over the past few decades, tremendous efforts have been made to establish pancreatic islet transplantation as a standard therapy for the treatment of diabetes. Nevertheless, long-term efficacy has been limited to a marginal number of patients. Outcomes have been restricted, in part, by challenges associated with the transplant site, poor vascularization, and disruption of the native islet architecture during the isolation process. This chapter reviews possible solutions for the challenges encountered in the islet transplantation field, which include islet source limitation, suboptimal engraftment of islets, and lack of oxygen and blood supply for transplanted islets.

Keywords Human islet transplant • Metabolic function • Engraftment • Diabetes • Pancreatic islet

10.1 Introduction

Islet cell transplantation holds great promise for treating patients with type 1 diabetes mellitus (T1DM), and for preventing unstable metabolic state commonly referred to as brittle diabetes in patients that undergo pancreatic resection. Despite advances in monitoring and therapeutics, acute and chronic complications [1–3] remain a major source of morbidity and mortality in diabetic patients. Islet transplant is a relatively noninvasive procedure and an attractive alternative to pancreas transplantation for restoring endogenous insulin secretion. Studies have shown that

O.M. Sabek (✉)
Department Cell and Molecular Biology, Weill Cornell Medical College,
New York, NY, USA

Department of Surgery, Houston Methodist Hospital,
6550 Fannin Street, Smith Tower 1661, Houston, TX 77030, USA
e-mail: OMSabek@houstonmethodist.org

B. Corradetti (ed.), *The Immune Response to Implanted Materials and Devices*,
DOI 10.1007/978-3-319-45433-7_10

Table 10.1 Factors influencing islet transplant outcome

Donor factors	Pancreas procurement	Islet cell processing	Islet cell viability and function assay	Clinical transplant
Autotransplant Pancreatitis patients • Extent of the disease • Size of the pancreas • Metabolic function *Allotransplant* • Age • Gender • BMI • Pancreas morphology (fat and fibrosis) • Islet morphology • Metabolic function • Cause of death • ICU	• Procurement technique • Warm ischemia • Cold ischemia • Preservation	• Isolation technique • Enzyme • Reagents • Purification • Culture	*In vitro assay* • Static incubation • DNA content • O_2 consumption • ATP response • Apoptosis • Cell content (α,β,δ) • Insulin extraction *In vivo assay* • Animal diabetic and nondiabetic models	*Recipient* • BMI • Insulin requirement • Diabetes microvascular complications • Life style *Site of implantation* • Liver • Muscle *Monitoring function* • Metabolic function • Recurrence of autoimmunity • Rejection • Efficacy of immunotherapy • Tolerance

even partial graft function after islet transplantation is remarkably effective in protecting against severe hypoglycemic events, and reduced progression of diabetic nephropathy and retinopathy compared with the progression with intensive medical therapy [4–7]. Published data suggest that long-term insulin independence (>5 years) can be achieved in 45–50 % of recipients given T cell-depleting induction immunotherapy, matching insulin independence rates of solitary pancreas transplantation [8]. The success of recent clinical trials for allogeneic islet transplantation as well as the increasing centers that perform autotransplantation is showing that the beta-cell replacement therapy for the treatment of patients with diabetes after total pancreatectomy is promising. It needs only to be improved and made more widely available to the millions of desperate patients who search for alternatives to a life of insulin injections, hypoglycemia and the risks of end-organ damage. Steady progress has been achieved in recent years in different areas in the pancreatic islet transplantation process including islet cell processing [9], preservation [10], engraftment [11–16] and immune therapies that justify optimism. To implement this therapeutic approach to larger cohorts of patients that would benefit from the restoration of beta-cell function requires multiple interventions and the standardization of the different stages of islet transplant process. This chapter presents a review of the possible areas of intervention and the ongoing research toward this important goal (Table 10.1, Fig. 10.1).

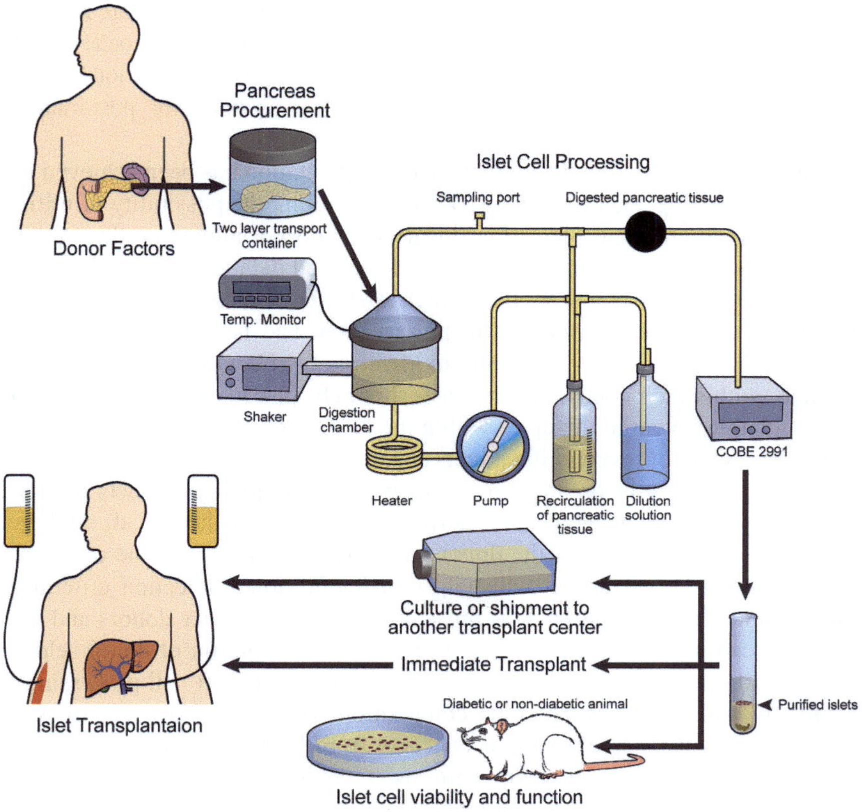

Fig. 10.1 Islet cell transplant process from donor to recipient

10.2 Donor Factors

Islet transplantation now faces significant challenges related to donor supply for clinical grade islets [17]. Despite evidence that the procedure can cease the demand for exogenous insulin therapy for ~3 years, and demonstrate functionality by the sustained presence of c-peptide synthesis for at least double that time frame, only ~750 patients received this treatment between 2004 and 2014 worldwide [18], with a dramatic decrease after 2009. Reasons for the limited transplant include a lack of available (cadaveric) islet cells. Current underuse of deceased donor pancreata still represents an area for intervention, a major problem in the islet field is the selectivity exercised in accepting cadaveric pancreas for islet isolation. Enlarging the donor pool is severely limited, mainly because of the poor function and high incidence of primary non-function of islets from less than ideal donors. Protocol modifications to improve the rate of islet function have included using the double layer method for pancreatic preservation [19] and the introduction of short-term culture [20–23]. Still, even with strict donor selection, the majority of patients require repeated transplants and there are no reliable methods that can predict which islet grafts are likely to function. We believe that to achieve successful islet transplantation, islet non-function and poor function have to be eliminated by identification, pre-transplant, of islets destined for post-transplant failure.

It is noteworthy that while clinical islet transplantation has been achieved from a non-heart beating cadaver donor [24], demonstrating the possibility of utilizing marginal organs for transplantation, ideal donor do not allows produce an adequate number of viable and function islet. Published data identifying donor variables vary from one center to another, which can be the result of donor demographic, management, pancreas procurement, preservation, and isolation technique [25–29]. As it is important for each center to identify the effect of donor and isolation variables within their center, the emphasis should be placed on donor management and evaluating potential therapies within the context of clinical trials. An aggressive management of severely ill patients and deceased donors has shown a positive effect of the quality of organ recovery as well as the quality of islets recovered [30]. Moreover, identifying the morphological, histopathological, and gene expression characteristics of the donor pancreas as predictor for the outcome of human islet in vivo function will help broadening donor selection criteria that include "marginal" organs such as those obtained from elderly donors and donation after cardiac death [28, 31–35]. The extensive definition of donor selection criteria will not only improve the success rate of human islet cell processing but eventually will overcome potential competition with whole pancreas transplant programs. For instance, the use of pancreas from high body mass index (BMI) donors is generally associated with better islet yields [26, 28, 32, 36–40], but they are commonly considered "not optimal" for whole pancreas. However, the relation between donor BMI and islet function is still controversial [28, 32]. While Matsumoto et al. [41] using an in vivo nude mouse transplant bioassay showed that human islet function was the same between two BMI groups (with group 1 being BMI is of 30 or more and group 2, BMI of less than 30), Benhamou et al. [42] showed that a lower donor BMI correlated with a better islet recovery and viability. A possible explanation by Larsson et al. [43–45] were they showed that obese subjects with impaired glucose tolerance have altered glucose modulation of islet function, yet insulin sensitivity is not different than in equally obese subjects with normal glucose tolerance. Their data suggest that the donor glucose tolerance level is an important factor in correlating BMI and islet function. Moreover, inflammation provokes significant abnormalities in host metabolism that result from the systemic release of cytokines. We, and others, have reported on the expression of pro-inflammatory genes and the increase of hypoxia and oxidative stress genes in donor pancreatic as well as isolated islets [32, 34]. Hypoxia has multiple detrimental effects on pancreatic tissues [46]. In the intact pancreas, chronic hypoxia [47] results in histologic changes in the pancreas including vascular congestion, edema of pancreatic lobules, and nesidioblastosis of the islets [48]. Oxygen consumption in pancreatic beta-cells is linked to the transduction mechanisms that mediate glucose-stimulated insulin secretion [49]. Hypoxia has been reported to induce hypoxia-induced factor 1 (HIF-1) and lead to development of islet apoptosis [50].

Recent reports indicate that interventions aimed at increasing oxygen tension result in improved outcome of islet transplantation by stabilizing post transplant mass [51–53]. Our preliminary data points to increased expression of hypoxia-induced genes in nonfunctional islet. These data are the first demonstration of a difference in the gene profile associated with cultured islet graft failure [54]. The overexpression of hypoxia and pro-inflammatory genes may result in reduced insulin secretion and lead to islet destruction post-transplantation. Increases in inflammation, such as activation of monocytes and increased levels of inflammatory markers, e.g., C-reactive protein, plasminogen activator inhibitor-1, and other cytokines, were reported in insulin-resistant states without diabetes. Hence, it is important to assess the metabolic function of the donor, but the inability to perform detailed metabolic evaluations on the cadaver donor is a major problem. However, performing a serious of histological and genetic tests on resected pancreas of auto-transplant recipients, and correlating the results with their metabolic function pre-pancreatic resection and post-islet transplantation would be beneficial in establishing a model which can be used to predict a posteriori the donor metabolic function in the allotransplant situations when donor data is not available. Studies aimed at Morphological and Gene expression study of donor pancreas can be very instrumental in predicting in vivo function in optimal and nonoptimal donor selection. Data from our laboratory shows that overexpression of pro-inflammatory such as interleukin-1, toll-like receptor 4, T-cell activation NFKB-like protein1, as well as genes that are associated with insulin resistance such as SOCS-1 and 6 and NFAT, calcineurin-dependent 1 and 2 may result in reduced insulin secretion and lead to islet destruction post-transplantation [34, 54]. Moreover, morphological characteristics of human donor pancreata such as dedifferentiation of beta-cell, the content of acinar and interstitial adipose cells, islet hyperplasia, and islet hypertrophy per defined area were found to be associated with reduced in vivo function [55, 56]. Interestingly, that observed histological difference were not related to demographic characteristics such as donor age, weight, BMI, cause of death, history of alcohol use and donor medications [34].

Morphological and gene expression studies of pancreas obtained from patients awaiting autotransplant or brain dead donors may shed more light on the mechanistic injury of the islet and can be very instrumental in predicting in vivo function in optimal and nonoptimal donor selection before islet isolation.

10.3 Metabolic Factors

The phenomenon of beta-cell function progressively declining in patients with type 2 diabetes [57–59] has also been observed in those with pre-diabetes [60, 61]. Cross-sectional and prospective clinic studies confirm that deterioration in beta-cell function that precedes hyperglycemia [62]. It is estimated that beta-cell function has declined

50 % before hyperglycemia occurs [63]. Several mechanisms have been put forward for the study of beta-cell function decline and the topic has been reviewed in some detail [64, 65]. Concurrent progressive beta-cell failure and insulin resistance are involved in most cases. The pathophysiology includes both metabolic and inflammatory mediators. Hyperglycemia, lipid deposition, high levels of fatty acids, pro-inflammatory cytokines, leptin, and islet-cell amyloid has been implicated. Insulin resistance in the tissues is an early feature of type II diabetes, and the beta-cell mass produces increased amounts of insulin in an attempt to overcome the resistance. This can result in beta-cell exhaustion and subsequent failure [56, 57]. In more than half of the cases, exogenous insulin injections are necessary. White et al. has shown that in recently diagnosed patients with diabetes, there is evidence of beta-cell dedifferentiation and its subsequent reprogramming to alpha-cell, underlying beta-cell failure [56].

In our center, patients awaiting pancreatic resection and islet autotransplantation (IAT) undergo a metabolic evaluation to identify insulin resistance and beta-cell functional abnormalities. The metabolic factors adversely affecting beta-cell function stem from glucotoxicity and lipotoxicity and both may be pertinent when considering IAT. Testing to look for evidence of either includes baseline chemistries, hemoglobin A1c, and lipid studies. We then look for insulin resistance and beta-cell functional abnormalities. Once identified, we can take advantage of the pre-surgical period to address the pathophysiology. Testing, for example, for relative insulin resistance and beta-cell dysfunction can be determined from fasting concentrations of insulin and glucose. Beta-cell stress can be assessed by calculating the ratio of proinsulin to immunoreactive insulin (PI/IRI). An increased ratio indicates altered processing of proinsulin; it correlates with beta-cell dysfunction [66, 67]. Increased insulin biosynthesis furthermore results in endoplasmic reticulum (ER) stress, a finding associated with the development of diabetes [68]. The resultant secretory defects lead to transient and chronic hyperglycemia, which in turn might culminate in either reversible beta-cell exhaustion or potentially irreversible cellular dysfunction [50]. When hyperglycemia is pronounced or persistent, pro-apoptotic signals appear [69] for which ER stress appears as signal [70]. Chronic hyperglycemia can lead to increased cytosolic Ca^{2+}, which can also be pro-apoptotic [71]. In particular, glucotoxicity results in the generation of reactive oxygen (ROS) and nitrogen species that in turn activates pre-apoptotic and inflammatory pathways leading to irreversible damage [72]. Lipotoxicity also contributes to beta-cell dysfunction. Saturated fatty acids, such a palmitate, are toxic to the beta cell [73]. They too can induce apoptosis and impair beta-cell function through the ceramide-mitochondrial apoptotic pathway [74]. The "malonyl-CoA/Long chain AcylCoa" hypothesis has been proposed as a basis for lipotoxic effects on the cell implicates the relationship between glucotoxicity and lipotoxicity as a mechanism of cell damage [75, 76].

Investigators and practitioners must therefore consider the status of beta-cell function in these patients who serve as their own donors. Since donor for islet autotransplantation (IAT) is predetermined, autotransplantation protocols allow for donor assessments that are not possible in cadaveric donors programs and afford the opportunity to prospectively track islet function. Several of the donor factors shown to be critical with regard to cadaveric transplants are now not applicable [77, 78]. The assessment can address not only the extent to which metabolic and inflammatory processes have damaged the donor's beta-cells but also the sensitivity to insulin's action. These factors as noted above can influence the successful outcome of autotransplanted islets. In particular, the presence or absence of metabolic syndrome components such as central obesity, hypertension, combined dyslipidemia, and glucose abnormalities and family history of diabetes are associated with enhanced risk. These clinical features are associated with insulin resistance, some degree of beta-cell dysfunction and systemic inflammation. Some of these can be quantified and addressed prior to the IAT procedure. Techniques developed to assess an individual's diabetes risk apply to IAT candidates. Simple clinical tools are therefore available [79]. Their use however in IAT protocols has not been validated, and as such these metabolic tests that estimate beta-cell function and insulin resistance might play an important but undefined roll in the decision for IAT.

Several such tests and indices are available. The measurement of fasting glucose (FG), insulin, C-peptide, and proinsulin can be quite informative [66, 67, 80]. The results can be used to calculate indices of insulin resistance, beta-cell function and reserve. Elevated FG identifies impaired fasting glucose tolerance (IFG), defined as FG between >100 mg/dL (5.56 mmol/L) and < 126 mg/dL (7.0 mmol/L), and diabetes when the glucose is ≥126 mg/dL (7.0 mmol/L) or higher. The FG and insulin results can then be used to calculate the HOMA-IR, a measure of insulin resistance, and beta-cell function index, a measure of beta-cell reserve, indices [81]. Additionally, assuming the subject is not taking insulin, the proinsulin to insulin ratio (PI/IRI) provides a measure of beta-cell stress. The relative simplicity lies in the fact that these findings are based on results obtained from a baseline fasting blood sample. More detailed dynamic tests are also available.

The standard 75 g 2-h oral glucose tolerance test (OGTT) when completed to insulin levels can confirm and expands the information obtained from the fasting sample alone. The OGTT can be used to assess insulin release and sensitivity [82]. Insulin sensitivity indices have been compared favorably with more cumbersome euglycemic insulin clamp test technique [83]. Moreover, the rapid intravenous injection of glucose tolerance test (IVGTT) using 20 g and the intravenous arginine stimulation test (AST) using arginine 5 g each followed by frequent blood sampling allows for determination of measures of insulin secretory capacity (the acute insulin release in response to glucose AIR_{gluc} and arginine AIR_{arg}) and glucose disposal. The glucose disappearance constant (K_g) is obtained from measurements made during the IVGTT. K_g is a measure of how rapidly the glucose returns toward baseline. In subjects with insulin resistance, glucose intolerance, and diabetes the K_g level is reduced [84].

Assessment of the metabolic function pre-pancreatic resection and post-islet transplantation would be beneficial not only in establishing a model which can be used to predict a posteriori the donor metabolic function in the allotransplant but also to explain the variability of the autotransplant outcome. Islet autotransplantation (IAT) is a unique mode for examining islet function in human due to the absence of immune rejection. The status of autotransplant describes wide variation of results of the procedure that has been done in a single center [85, 86], where some patients achieve insulin independent receiving less than 1000 IE/kg and remains insulin independence for 4 years while another receiving 20,385 IEQ/kg did not achieve insulin independence

10.4 Pancreas Procurement and Preservation

Success of human pancreatic islet isolation depends largely on the techniques used during pancreas procurement with careful excision of the gland before the liver, and on the maintenance of a low core pancreas temperature by adequate surface cooling of the pancreas. Pancreatic preservation in a double layer of oxygenated perfluorocarbons and University of Wisconsin (UW) solution has been shown to provide higher oxygen availability and reduce cold ischemic damage to the gland [19, 38, 87–91]. Hence, it has a positive impact on the yield and post-transplant function even from marginal donors with extended cold ischemia and after cardiac death [19, 38, 88–96] by protecting islets from apoptosis through the mitochondrial pathway.

Studies by Edmonton and others [36, 97, 98], showed that combining 2-layer storage with trypsin inhibition using Pefabloc allows prolongation of the interval between recovery and isolation and improves yield, and in vitro function in marginal pancreata. Standardizing the pancreas procurement technique for pancreatic islet transplantation has resulted in a 67 % islet isolation success rate, despite the use of a remote islet isolation center [99].

10.5 Islet Cell Processing

10.5.1 Islet Isolation

The overall cost of islets transplantation is slightly higher than of pancreas transplantation with the cell isolation process being critical to the overall cost of islet transplantation [100]. An optimization of islet isolation process would permit an increased number of successfully transplant islet from one donor pancreas to achieve insulin independence, therefore represents a major step toward cost reduction, hence securing reimbursement for this procedure from health insurance providers and national health systems. The average number of islets in a pancreas is estimated between 300,000 to 1.5 million; however, only 30–50 % of islets are being recovered from pancreata [78], with an average of 65 % of the human islet preparations being viable [101, 102]. Over the last three decades since the introduction of the automated method by Ricordi [103], there has been a steady improvement in the processing of human pancreata for isolating islet cell. The efficiency of this process highly depends on the quality of the organ, the protocols used for isolation and purification, and the experience of the center. The process of islet isolation from the pancreas essentially involves dissociation of islets from the exocrine pancreas by enzymatic digestion combined with mechanical agitation followed by purification on density gradients [104, 105]. Pancreatic islet cells have a unique capillary network and high blood perfusion, which is necessary for a high delivery of oxygen and nutrients to the islets cells and for optimizing the dispersal of the secreted hormones to their target organs [106]. During pancreatic islet isolations, islet vasculature disrupts and degenerates leading to loss of islet integrity and viability [107]. The ability to digest the extracellular matrix within the islet–exocrine interface, without inflicting significant damage on the pancreatic islet cells is the key to successful islet transplantation. The finding by Edmonton group that using the low-pressure perfusion instead the injection method presents an advantage improves islet yield, yet using the perfusion technique still allows the collagenase to penetrate the islet interior [108]. The presence of the collagenase in the islet interior may result in more digestion of the internal structure framework of the islet and that can lead to lower islet yield and/or islet dysfunction. Further research to determine the optimal methods of collagenase delivery that can digest pancreatic tissue to free viable intact islet without the disruption of islet vasculature and structure is needed.

Availability of improved enzyme blend with reduce the lot-to-lot variability of enzyme effectiveness is another major step in standardizing islet cell process. A review of the recent literature shows that Liberase HI is the most widely used commercially available enzyme for clinical islet isolation. However, this enzyme blend still exhibits significant lot-to-lot and even intra-lot variability [109–111]. Moreover, recent studies have shown that Liberase is no more effective than crude collagenase and induces functional damage to human islets. A newly developed Liberase Collagenase Blend (Roche Applied Science) contains only class I and class II collagenase [112], and Collagenase NB1 (Serva Electrophoresis GmbH, Heidelberg, Germany) in conjunction with separate Neutral Protease NB (Serva Electrophoresis GmbH) has been developed and used for clinical islet isolation [110]. In our center, we examined the effect of using different enzymes combined with different isolation methods (mechanical or hand shaking) on human islet recovery, purity, and in vivo function. Our data demonstrate a number of points: first, when using the same isolation method whether it is mechanical or hand shaking; there is no significant difference between Liberase, Serva, and Collagenase P in islet yield or purity. Second, the total yield (islet equivalent) and the integrity of the islet is significantly higher using the hand shaking method thereby resulting in better viability and in vivo function. These results indicate the importance of considering different aspects of the isolation when a new enzyme is developed and tested by different centers.

10.5.2 Islet Purification

Another important aspect of islet cell processing and standardization is islet purification. Islet purification is another important step in extracting the islet from the digested pancreatic tissue. Human islet purification is a process that based on density gradient centrifugation using COBE 2991 [104, 105, 113]. Islet purification has proven particularly difficult because both the density and the diameter of acinar tissue, and of the islets, changes from one preparation to another [114, 115]. Islets normally vary in diameter from 15 to 500 μm and, in addition, the diameter of islets and acinar tissue is critically dependent upon the collagenase digestion stage of the isolation process. Moreover, the density of the acinar tissue varies as a result of cellular swelling and edema, size of the aggregates formed [116]. Acinar cell swelling can be provoked by a number of insults, including mechanical trauma [117] and hypothermia, and cell membrane permeability as a result of collagenase digestion [118]. The challenge is to produce solutions that will minimize acinar tissue swelling without compromising islet yield and viability. One approach is washing the pancreatic digest and storing in UW solution at 4 °C for 1 h prior to density gradient centrifugation [113]. Implementation of an additional purification step for low purity fractions after standard purification has also been shown to increase the efficiency of the process and the recovery of more islets.

10.5.3 Islet Culture

Culturing human islets offers a window of opportunity for potential interventions aiming at preserving islet mass and at conferring cytoprotection to the graft. Islet culture prior to transplantation can be considered in order to allow sufficient time for extensive viability and functionality testing of islets, screening human islet preparations for additional pathogens and, achieve therapeutic levels of immunosuppressive drugs in the recipients, and transportation of islets to distant centers [100, 119]. Optimization of culture protocols for human islets with improved oxygen availability and customized media formulations to prevent oxidative stress while providing proper nutrients and extracellular matrices may allow for improved quality of islet cell products for transplantation, and reduce immunogenicity [120–122]. Our laboratory has been interested in the potential development of islet repositories that would allow pooling of cultured islets and their maintenance for relatively long periods. To achieve this, we have worked extensively on methods that could allow prolonged in vitro culture of viable islets [20]. We have shown that recovery of viable islets approaches 85 % of the original number at 1 month and between 65 and 75 % at 2 months of culture.

Importantly, we have demonstrated that the function of cultured islets following transplantation into nonimmune NOD-SCID mice was significantly improved following 1-month culture [21]. In fact, C-peptide production was doubled that which was seen in mice transplanted with the same amount of fresh islets. Recently, our laboratory has extended the culture time to 2 and 6 months in an attempt to test the robustness of the culture technique. We demonstrated that 6 month cultured islets in our Memphis Serum Free Media (MSFM) not only function well following transplantation into the mice but also cure streptozotocin-induced diabetes, which is the ultimate test for establishing functional viability [23].

In an attempt to characterize the morphological changes seen in islets maintained in culture, we performed a series of evaluations on our cultured islets. Our data indicates that the degree islet apoptosis immediately following isolation and during the culture period correlates with subsequent islet yield [122]. Thus, apoptosis appears to be a major reason why islet mass is lost on culture, islet apoptosis can be caused by a long list offending stimuli including hypoxia following isolation and during cultures, disruption of the islet matrix [123], and by cytokines and endotoxin produced during the isolation procedure and carried through to the culture media [124]. In addition to apoptosis, central necrosis of pancreatic islets has been described and may contribute to islet death on culture. The degree of necrosis is determined by the density of the islet culture, the amount of islet clumping, the diameter of the islets, and the degree of apoptosis induced during pre-culture islet processing and during the culture itself. Loss of necrotic matter during prolonged culture may act as delayed purification yielding a higher percentage of functional islets for transplantation. In essence, the culture may act as a "biological screen" allowing removal of marginally viable, apoptotic or necrotic islets that not only affect performance of the islet preparation, but may also be harmful to the implantation and engraftment process by inducing inflammatory responses to the islet infusion. Cell death may be a beneficial effect of culture, however, since contaminating exocrine cells and dendritic type cells are seen to decline in number during the period of islet culture, particularly the first 10 days. To demonstrate this, we carried out staining experiments that demonstrated almost complete disappearance of CD45+ve cells by day 10 and 80 % reduction in amylase positive cells and in the first 2 weeks of culture.

Our culture results are unique in the literature because of the serum free media we use and the duration of time, we have maintained islets. The mechanism by which islet culture results in improved function is not yet understood. Work by Lacy and Scharp described decreased immunogenicity of islets following short-term (~7 day) culture related to the elimination of passenger leukocytes [121, 125]. Although the NOD-scid mouse does not mount a specific immune response to the human islet graft because of its lack of T and B-lymphocytes, the mouse has an intact macrophage/monocyte system that is capable of mounting nonspecific cytokine based responses. The increased purification of the islets by prolonged culture may decrease the occurrence of these inflammatory responses by reducing the number of passenger leukocytes or other cells that can stimulate a macrophage response. In fact, some of our preliminary data demonstrate almost complete disappearance of CD45 positive cells. We have also analyzed different human islet preparation for gene expression over 2 weeks of culture in MSFM. The data demonstrate a down regulation of genes related to inflammation and exocrine cells while genes levels related to insulin production were not changed [122]. The morphological changes that islets undergo following prolonged cultures have been variable depending on the medium, in which the cultures are grown and the technique of culture. The work of Schmied et al. demonstrated morphological "trans-differentiation" of human islets to exocrine type elements [126, 127]. In that report Schmied demonstrated the ability to maintain human islet tissue in culture for up to 12 months, but the culture technique used resulted in loss of all endocrine function after 60 days in culture [126, 128]. Their culture technique utilized bovine serum, growth factors and supplements. Our media is serum free and contains only insulin as a growth factor. Insulin may help preserve the islet tissue by reducing metabolic demand through negative feedback inhibition of insulin secretion and preventing metabolic burnout. Unlike Schmied et al., we do not attempt to purify our culture by hand picking islets or placing the culture on a rocker. It is possible that acinar tissue contaminants produce factors that contribute to the stability of the islet's structure, therefore helping preserve their functionality. In addition our islets are cultured in suspension rather than allowing the islets to attach to the bottom of the plate. Earlier literature indicated that free floating long term cultures had improved insulin biosynthesis and maintained insulin response to glucose stimulation when grown in serum supplemented media [129]. We believe that maintaining islets in suspension is important in maintaining the islets' three-dimensional structure. This is supported by the fact that when allowed to attach to the bottom of a culture plate islets quickly loss their rounded structural orientation [130]. The structural change is then followed by eventual loss of the islet's endocrine function [130]. Insight into this phenomenon is offered by the work of Yuan et al. Using human islet tissue embedded in type 1 collagen matrix, the group observed trans-differentiation of the endocrine tissue to duct-like tissue after only 96 h in culture [130].

Recently, there has been interest in the integrin–matrix interactions [131, 132] and the interrelationship of islets driven. Novel approaches to islet cultures aimed at maintaining the three-dimensional structure of islets and the islet interface. One of these approaches included culturing the islets under microgravity conditions using rotating bioreactor culture equipment. Under these microgravity conditions, islet culture appeared to loose dendritic cells to increase in size and to develop multiple nutritional channels with improvements of over functional characteristics [133]. In addition, other techniques such as short term culture in fibrin clots [134] or the addition of various hormones [135] and or growth factors [136–141] all have resulted in either trans-differentiation of the islets or in their improved functionality and survival.

In summary, we described the maintenance of human islets for extended periods on serum-free culture and their ability to function and cure diabetes after this prolonged culture. Our data indicates that significant loss of mass occurs beyond the first 30 days of culture and suggests that despite its success in maintaining functional islets the culture technique can still be substantially improved.

10.5.4 Islet Cell Viability and Functional Assay

A major impediment to the optimization of islet isolation process and the increased number of clinical transplant is the lack of methods that adequately assess the quality of islets. Pancreatic islets harvested for autotransplants or allotransplants are susceptible to multiple insults that occur because of chronic or acute inflammation, donor brain death, organ procurement and preservation, isolation, and transplantation. The isolation procedure and disruption of the integrin-related islet extracellular matrix connection leads to islet apoptosis [123]. After the isolation insult, islets are exposed to additional post-transplant stresses that compromise function. These include hypoxia before secondary revascularization, exposure to blood and pro-inflammatory cytokines, hyperglycemia, immune rejection, and diabetogenic effects of immunosuppressive drugs [142]. The outcome is both early and late loss of functional islet mass, which is the major constraint in achieving an optimal glycemic state.

Assessment of human islet preparations prior to transplantation is a fundamental step to characterize the quality of the final cell product for clinical use. Besides exclusion of adventitious potential (e.g., endotoxin, mycoplasma, and bacterial contamination), it is required to characterize islet cell viability and potency. The limitation of the methods used for product release in recent years has been well recognized, and has led to the quest for more sensitive tests predictive of clinical islet outcome.

Standard tests of islet quality in clinical studies include cell composition, total cellular insulin, DNA and the extent of insulin secretion in vitro in response to secretagogues; due to the lack of success of insulin secretion to provide meaningful quality criteria, the latter three parameters have not proved to accurately predict in vivo islet function. Other quality assessment parameters have been proposed such as membrane integrity, conversion of proinsulin, adenylate cyclase activity, metabolic activity and nitrite release, ATP, and nuclear magnetic resonance, but none have been successfully transferred into the clinic. To date the use of animal models still present the most accurate methods to predict in vivo human islet function [143–145].

In our laboratory, we have extensively characterized a NOD-scid mouse as a model that allows the evaluation of in vivo function of islets prior to human transplantation [21, 146, 147]. The NOD-scid mouse contains a homozygous, spontaneous mutation (Prkdscid) which produces an immunodeficiency characterized by an absence of functional T and B cells, and defective NK cell function. This eliminates specific immune-mediated rejection from the in vivo model and allows us to investigate other causes of graft failure (e.g., those related to isolation, enzymes, donor factors, etc.), which may otherwise be overlooked in the context of immune responses. Although induction of diabetes by chemotherapeutics is difficult in the NOD-scid model because of the inability of this mouse to repair DNA breaks, we were able to titrate different doses of streptozotocin (STZ) and achieve a model with 85 % induction of diabetes (blood glucose > 300 mg/dl in two consecutive measurements), and low animal mortality of 10 % [145]. Low mortality allowed us to maintain these mice for 4–6 weeks or until islet isolation [145]. Different preparations of pancreatic islets were tested for in vivo function in our diabetic NOD-scid mice. Most animals were cured (blood glucose < 200 mg/dl in two consecutive measurements) of diabetes when viable and functioning islets were used, although some preparations produced higher rates of animal cure than others. The overall cure rate was 85 %. Cured mice reverted to the diabetic state when the kidney containing the islet graft was surgically removed. These data confirm that we have a successful diabetic NOD-scid model. However, use of STZ has found to concentrates in the liver and kidney, with as much as 20 % of the drug metabolized and/or excreted by the kidney. In addition, intraperitoneal STZ administration has been associated with increased incidence of kidney, lung, and uterine tumors in mice. Another drawback to the use of diabetic model is the ability to maintain normal insulin levels by insulin injection to avoid glucose toxicity effect the islet post transplantation. As part of our characterization of the NOD-scid mouse model, we compared the use of nondiabetic NOD-scid mouse model to its diabetic counterpart in terms of predicting islet viability [145]. Transplantation of 2000 human islet equivalents was performed in both models using the same technique. Islet function was determined in the diabetic mice by return to normoglycemia (blood glucose < 200 mg/dl in two consecutive measurements) and measurement of fasting human C-peptide on days 7 and 14 post-transplant. For nondiabetic NOD-scid mice, function was tested 1 week after transplantation and following an overnight fast and intraperitoneal glucose challenge (2 g/kg of body weight). Thirty minutes after glucose injection, blood samples for human plasma insulin and C-peptide levels were drawn. The 30-min time point was determined because of a series of complete 2-h intraperitoneal glucose challenge tests that demonstrated a peak in glucose, insulin, and C-peptide in the NOD-scid mouse at that time. The glucose stimulation test was repeated on day 14 and 30 post-transplant. Human insulin levels were determined by an ELISA assay (Alpco Diagnostics, Windham, NH) that has minimal cross-reactivity to mouse insulin (<1.0 %). Human C-peptide levels were determined by radioimmunoassay (Diagnostic Products Corporation, Los Angeles, CA) with <1.0 % cross-reactivity to mouse C-peptide. Since assay cross-reactivity with murine insulin and C-peptide can be of concern in experiments employing a nondiabetic mouse model, we continually validate these assays in our laboratory. The human insulin ELISA and C-peptide radioimmunoassay have been tested in over 2000 NOD-scid mice prior to islet transplantation. Background levels of 1.62 ± 1.21 μU/ml and 0.35 to ±0.32 ng/ml for human insulin and C-peptide, respectively, is measured in the NOD-scid mice. Values >2 standard deviations above background levels for each assay (insulin levels and C-peptide of 5.0 μU/ml and >1.5 ng/ml, respectively) are thus used as cutoff for determination of islet function. Transplanting 23 different islet preparations in both diabetic and nondiabetic models showed 100 % concordance between the two models in terms of testing for islet function [145]. We had used the nondiabetic NOD-scid model to evaluate the in vivo function of over 200 human islet preparations [23, 34, 143, 145, 147–150]. This model is now being used by other centers to validate new in vitro assay such as oxygen consumption measurement [147] as well as the efficacy of different gene transfection experiments to improve human islet model [149–151]. Moreover, the nondiabetic mouse model has been used to predict clinical islet transplant outcome and to correlate with [143, 145].

10.5.5 Islet Vascularization After Transplant

Patients require the equivalent islet mass from between 2 and 4 donors, as the majority of islets die in the first days after transplant [152]. Most of this cell death results from an instant blood mediated immune reaction (IBMIR) [153], and some research has focussed on mitigating the factors responsible for this process, such as modulating platelet–monocyte interactions [154] or by blocking complement activation [155]. However, the remaining islet death is due to the loss of blood supply, causing hypoxia, ischemia–reperfusion injury [156], and amyloid deposition as time progresses. Islet vascularization is a critical determinant of cell survival in the long term [157]. While most of the large blood vessels are sheared off during isolation, smaller capillaries remain. The revascularization process starts within 2 days of transplant, and is mostly complete within 2 weeks [158]. The new blood vessels were surprisingly shown to be derived from both host and donor endothelial cells [158], demonstrating chimerism. While highly vascularized islets ostensibly demonstrate higher functional capacity, these same islets are also highly susceptible to various cell stressors that access the islet via the vasculature, and in fact cause preferential death early after transplant [159]. It may be stated that islet transplant success could be tied to site-dependent factors. While the liver is the predominant site for transplantation currently, there is evidence that the pancreas itself, or even muscle, is preferred for their improved metabolic profile [160], and the liver has been shown to be poorer for revascularization post-transplant than the pancreas [161]. Some moderate success has been shown by culturing islets prior to transplant in materials previously shown to improve vascularization, such as vascular endothelial growth factor (VEGF) via endothelial progenitor cells (EPCs) [162], or human umbilical vein endothelial cells (HUVECs) [163], neural crest cells [164, 165], or polymerized hemoglobin [162]. Furthermore, human VEGF gene delivery has been effective [149, 150, 166]. Coculture approaches have been used in upstream applications, specifically by the addition of bone-marrow derived mesenchymal stem cells (MSCs) which promote angiogenesis [167–171]. Some of this success can be attributed to the use of small islets over larger ones, which allow enhanced vascularization and nutrient exchange [172, 173]. Means to physically separate the islets from the host defenses, such as cassettes, pouches, or other encapsulation implant devices [174–177] have also gained favor for their ability to mitigate cell death. Other materials have been used to protect the transplanted islets, such as alginate-based ECM coatings [176–178]. It is important when designing implanting materials or devices to focus on the engraftment, which includes vascularization and immune response.

References

1. Realsen J, Goettle H, Chase HP (2012) Morbidity and mortality of diabetic ketoacidosis with and without insulin pump care. Diabetes Technol Ther 14(12):1149–1154
2. Giannini C, Mohn A, Chiarelli F (2009) Technology and the issue of cost/benefit in diabetes. Diabetes Metab Res Rev 25(Suppl 1):S34–S44
3. Gaba MK, Gaba S, Clark LT (1999) Cardiovascular disease in patients with diabetes: clinical considerations. J Assoc Acad Minor Phys 10(1):15–22
4. Thompson DM, Meloche M, Ao Z et al (2011) Reduced progression of diabetic microvascular complications with islet cell transplantation compared with intensive medical therapy. Transplantation 91(3):373–378
5. Noguchi H (2009) Pancreatic islet transplantation. World J Gastrointest Surg 1(1):16–20

6. Warnock GL, Thompson DM, Meloche RM et al (2008) A multi-year analysis of islet transplantation compared with intensive medical therapy on progression of complications in type 1 diabetes. Transplantation 86(12):1762–1766
7. Thompson DM, Begg IS, Harris C et al (2008) Reduced progression of diabetic retinopathy after islet cell transplantation compared with intensive medical therapy. Transplantation 85(10):1400–1405
8. Lehmann R, Graziano J, Brockmann J et al (2015) Glycemic control in simultaneous islet-kidney versus pancreas-kidney transplantation in type 1 diabetes: a prospective 13-year follow-up. Diabetes Care 38(5):752–759
9. Hilling DE, Bouwman E, Terpstra OT et al (2014) Effects of donor-, pancreas-, and isolation-related variables on human islet isolation outcome: a systematic review. Cell Transplant 23(8):921–928
10. Oniscu GC, Randle LV, Muiesan P et al (2014) In situ normothermic regional perfusion for controlled donation after circulatory death—the United Kingdom experience. Am J Transplant 14(12):2846–2854
11. Farney AC, Sutherland DE, Opara EC (2016) Evolution of islet transplantation for the last 30 years. Pancreas 45(1):8–20
12. Liljeback H, Grapensparr L, Olerud J et al (2016) Extensive loss of islet mass beyond the first day after intraportal human islet transplantation in a mouse model. Cell Transplant 25(3):481–489
13. Grapensparr L, Vasylovska S, Li Z et al (2015) Co-transplantation of human pancreatic islets with post-migratory neural crest stem cells increases beta-cell proliferation and vascular and neural regrowth. J Clin Endocrinol Metab 100(4):E583–E590
14. Lau J, Vasylovska S, Kozlova EN et al (2015) Surface coating of pancreatic islets with neural crest stem cells improves engraftment and function after intraportal transplantation. Cell Transplant 24(11):2263–2272
15. Naziruddin B, Iwahashi S, Kanak MA et al (2014) Evidence for instant blood-mediated inflammatory reaction in clinical autologous islet transplantation. Am J Transplant 14(2):428–437
16. Pepper AR, Gala-Lopez B, Ziff O et al (2013) Revascularization of transplanted pancreatic islets and role of the transplantation site. Clin Dev Immunol 2013:352315
17. Robertson RP (2002) Islet transplantation: travels up the learning curve. Curr Diab Rep 2(4):365–370
18. Bruni A, Gala-Lopez B, Pepper AR et al (2014) Islet cell transplantation for the treatment of type 1 diabetes: recent advances and future challenges. Diabetes Metab Syndr Obes 7:211–223
19. Matsumoto S, Rigley TH, Qualley SA et al (2002) Efficacy of the oxygen-charged static two-layer method for short-term pancreas preservation and islet isolation from nonhuman primate and human pancreata. Cell Transplant 11(8):769–777
20. Fraga DW, Sabek O, Hathaway DK et al (1998) A comparison of media supplement methods for the extended culture of human islet tissue. Transplantation 65(8):1060–6
21. Gaber AO, Fraga DW, Callicutt CS et al (2001) Improved in vivo pancreatic islet function after prolonged in vitro islet culture. Transplantation 72(11):1730–1736
22. Odorico JS, Heisey DM, Voss BJ et al (1998) Donor factors affecting outcome after pancreas transplantation. Transplant Proc 30(2):276–277
23. Rush BT, Fraga DW, Kotb MY et al (2004) Preservation of human pancreatic islet in vivo function after 6-month culture in serum-free media. Transplantation 77(8):1147–1154
24. Markmann JF, Deng S, Desai NM et al (2003) The use of non-heart-beating donors for isolated pancreatic islet transplantation. Transplantation 75(9):1423–1429
25. Iwanaga Y, Sutherland DE, Harmon JV et al (2008) Pancreas preservation for pancreas and islet transplantation. Curr Opin Organ Transplant 13(2):135–141
26. Liu X, Matsumoto S, Okitsu T et al (2008) Analysis of donor- and isolation-related variables from non-heart-beating donors (NHBDs) using the Kyoto islet isolation method. Cell Transplant 17(6):649–656

27. Andrades P, Asiedu CK, Gansuvd B et al (2008) Pancreatic islet isolation variables in non-human primates (rhesus macaques). Diabetologia 51(7):1236–1244
28. Sabek O, Cowan P, Fraga D et al (2006) The effect of donor factors on human islet yield and their in vivo function. Progess Transplant 16(4):350–354
29. Ponte GM, Pileggi A, Messinger S et al (2007) Toward maximizing the success rates of human islet isolation: influence of donor and isolation factors. Cell Transplant 16(6):595–607
30. Rosendale JD, Kauffman HM, McBride MA et al (2003) Aggressive pharmacologic donor management results in more transplanted organs. Transplantation 75(4):482–487
31. Mahler R, Franke FE, Hering BJ et al (1999) Evidence for a significant correlation of donor pancreas morphology and the yield of isolated purified human islets. J Mol Med 77(1):87–89
32. Sabek OM, Cowan P, Fraga DW et al (2008) The effect of isolation methods and the use of different enzymes on islet yield and in vivo function. Cell Transplant 17(7):785–792
33. Sabek O, Hathaway D, Fraga D et al (1998) Influence of human donor factors on pancreatic collagenase digestion. Transplant Proc 30(2):353
34. Sabek OM, Marshall DR, Minoru O et al (2005) Gene expression profile of nonfunctional human pancreatic islets: predictors of transplant failure? Transplant Proc 37(8):3441–3443
35. Sabek OM, Nishimoto SK, Fraga D et al (2015) Osteocalcin effect on human beta-cells mass and function. Endocrinology 156(9):3137–3146
36. Lakey JR, Helms LM, Kin T et al (2001) Serine-protease inhibition during islet isolation increases islet yield from human pancreases with prolonged ischemia. Transplantation 72(4):565–570
37. Brandhorst D, Brandhorst H, Hering BJ et al (1995) Islet isolation from the pancreas of large mammals and humans: 10 years of experience. Exp Clin Endocrinol Diabetes 103(Suppl 2):3–14
38. Matsumoto S, Kuroda Y (2002) Perfluorocarbon for organ preservation before transplantation. Transplantation 74(12):1804–1809
39. Kim SC, Han DJ, Kang CH et al (2005) Analysis on donor and isolation-related factors of successful isolation of human islet of Langerhans from human cadaveric donors. Transplant Proc 37(8):3402–3403
40. Hanley SC, Paraskevas S, Rosenberg L (2008) Donor and isolation variables predicting human islet isolation success. Transplantation 85(7):950–955
41. Matsumoto S, Zhang G, Qualley S et al (2004) Analysis of donor factors affecting human islet isolation with current isolation protocol. Transplant Proc 36(4):1034–1036
42. Benhamou PY, Watt PC, Mullen Y et al (1994) Human islet isolation in 104 consecutive cases. Factors affecting isolation success. Transplantation 57(12):1804–1810
43. Larsson H, Ahren B (1996) Failure to adequately adapt reduced insulin sensitivity with increased insulin secretion in women with impaired glucose tolerance. Diabetologia 39(9):1099–1107
44. Larsson H, Ahren B (1996) Islet dysfunction in obese women with impaired glucose tolerance. Metabolism 45(4):502–509
45. Larsson H, Berglund G, Ahren B (1995) Glucose modulation of insulin and glucagon secretion is altered in impaired glucose tolerance. J Clin Endocrinol Metab 80(6):1778–1782
46. Dionne KE, Colton CK, Yarmush ML (1989) Effect of oxygen on isolated pancreatic tissue. ASAIO Trans 35(3):739–741
47. Rosso D, Carnazzo G, Giarelli L et al (2001) Atherosclerosis and pancreatic damage. Arch Gerontol Geriatr 32(2):95–100
48. Culberson DE, Manci EA, Shah AK et al (2001) Nesidioblastosis in sickle cell disease. Pediatr Pathol Mol Med 20(2):155–165
49. Sweet IR, Khalil G, Wallen AR et al (2002) Continuous measurement of oxygen consumption by pancreatic islets. Diabetes Technol Ther 4(5):661–672
50. Moritz W, Meier F, Stroka DM et al (2002) Apoptosis in hypoxic human pancreatic islets correlates with HIF-1alpha expression. FASEB J 16(7):745–747

51. Juang JH, Hsu BR, Kuo CH et al (2002) Beneficial effects of hyperbaric oxygen therapy on islet transplantation. Cell Transplant 11(2):95–101
52. Hughes SJ, Davies SE, Powis SH et al (2003) Hyperoxia improves the survival of intraportally transplanted syngeneic pancreatic islets. Transplantation 75(12):1954–1959
53. Schrezenmeir J, Velten F, Beyer J (1994) Immobilized hemoglobin improves islet function and viability in the bioartificial pancreas in vitro and in vivo. Transplant Proc 26(2):792–800
54. Marshall D, Sabek O, Fraga D et al (2005) Examination of the molecular signature associated with islet dysfunction. Transplant Proc 37(2):1311–1312
55. Sabek OM, Hamilton DJ, Gaber AO (2009) Prospects for future advancements in islet cell transplantation. Minerva Chir 64(1):59–73
56. White MG, Marshall HL, Rigby R et al (2013) Expression of mesenchymal and alpha-cell phenotypic markers in islet beta-cells in recently diagnosed diabetes. Diabetes Care 36(11):3818–3820
57. Guo S, Dai C, Guo M et al (2013) Inactivation of specific beta cell transcription factors in type 2 diabetes. J Clin Invest 123(8):3305–3316
58. Qi M, McFadden B, Valiente L et al (2015) Human pancreatic islets isolated from donors with elevated HbA1c levels: islet yield and graft efficacy. Cell Transplant 24(9):1879–1886
59. White MF (2003) Insulin signaling in health and disease. Science 302(5651):1710–1711
60. Butler AE, Janson J, Bonner-Weir S et al (2003) Beta-cell deficit and increased beta-cell apoptosis in humans with type 2 diabetes. Diabetes 52(1):102–110
61. Butler AE, Janson J, Soeller WC et al (2003) Increased beta-cell apoptosis prevents adaptive increase in beta-cell mass in mouse model of type 2 diabetes: evidence for role of islet amyloid formation rather than direct action of amyloid. Diabetes 52(9):2304–2314
62. Bogardus C, Lillioja S, Howard BV et al (1984) Relationships between insulin secretion, insulin action, and fasting plasma glucose concentration in nondiabetic and noninsulin-dependent diabetic subjects. J Clin Invest 74(4):1238–1246
63. Weyer C, Bogardus C, Mott DM et al (1999) The natural history of insulin secretory dysfunction and insulin resistance in the pathogenesis of type 2 diabetes mellitus. J Clin Invest 104(6):787–794
64. Wajchenberg BL (2007) Beta-cell failure in diabetes and preservation by clinical treatment. Endocr Rev 28(2):187–218
65. Donath MY, Schumann DM, Faulenbach M et al (2008) Islet inflammation in type 2 diabetes: from metabolic stress to therapy. Diabetes Care 31(Suppl 2):S161–S164
66. Kahn BB (1996) Lilly lecture 1995. Glucose transport: pivotal step in insulin action. Diabetes 45(11):1644–1654
67. Kahn SE, Leonetti DL, Prigeon RL et al (1996) Proinsulin levels predict the development of non-insulin-dependent diabetes mellitus (NIDDM) in Japanese-American men. Diabet Med 13(9 Suppl 6):S63–S66
68. Eizirik DL, Cardozo AK, Cnop M (2008) The role for endoplasmic reticulum stress in diabetes mellitus. Endocr Rev 29(1):42–61
69. Robertson RP, Harmon J, Tran PO et al (2003) Glucose toxicity in beta-cells: type 2 diabetes, good radicals gone bad, and the glutathione connection. Diabetes 52(3):581–587
70. Donath MY, Halban PA (2004) Decreased beta-cell mass in diabetes: significance, mechanisms and therapeutic implications. Diabetologia 47(3):581–589
71. Harding HP, Ron D (2002) Endoplasmic reticulum stress and the development of diabetes: a review. Diabetes 51(Suppl 3):S455–S461
72. Grill V, Bjorklund A (2001) Overstimulation and beta-cell function. Diabetes 50(Suppl 1):S122–S124
73. Newsholme P, Haber EP, Hirabara SM et al (2007) Diabetes associated cell stress and dysfunction: role of mitochondrial and non-mitochondrial ROS production and activity. J Physiol 583(Pt 1):9–24

74. Maedler K, Oberholzer J, Bucher P et al (2003) Monounsaturated fatty acids prevent the deleterious effects of palmitate and high glucose on human pancreatic beta-cell turnover and function. Diabetes 52(3):726–733
75. Summers SA (2006) Ceramides in insulin resistance and lipotoxicity. Prog Lipid Res 45(1):42–72
76. Poitout V (2008) Glucolipotoxicity of the pancreatic beta-cell: myth or reality? Biochem Soc Trans 36(Pt 5):901–904
77. Poitout V, Robertson RP (2008) Glucolipotoxicity: fuel excess and beta-cell dysfunction. Endocr Rev 29(3):351–366
78. Lakey JR, Warnock GL, Rajotte RV et al (1996) Variables in organ donors that affect the recovery of human islets of Langerhans. Transplantation 61(7):1047–1053
79. Toso C, Oberholzer J, Ris F et al (2002) Factors affecting human islet of Langerhans isolation yields. Transplant Proc 34(3):826–827
80. Heikes KE, Eddy DM, Arondekar B et al (2008) Diabetes risk calculator: a simple tool for detecting undiagnosed diabetes and pre-diabetes. Diabetes Care 31(5):1040–1045
81. Abdul-Ghani MA, Matsuda M, Jani R et al (2008) The relationship between fasting hyperglycemia and insulin secretion in subjects with normal or impaired glucose tolerance. Am J Physiol Endocrinol Metab 295(2):E401–E406
82. Matthews DR, Hosker JP, Rudenski AS et al (1985) Homeostasis model assessment: insulin resistance and beta-cell function from fasting plasma glucose and insulin concentrations in man. Diabetologia 28(7):412–419
83. Shungin D, Winkler TW, Croteau-Chonka DC et al (2015) New genetic loci link adipose and insulin biology to body fat distribution. Nature 518(7538):187–196
84. Matsuda M, DeFronzo RA (1999) Insulin sensitivity indices obtained from oral glucose tolerance testing: comparison with the euglycemic insulin clamp. Diabetes Care 22(9):1462–1470
85. Teuscher AU, Kendall DM, Smets YF et al (1998) Successful islet autotransplantation in humans: functional insulin secretory reserve as an estimate of surviving islet cell mass. Diabetes 47(3):324–330
86. Nath DS, Kellogg TA, Sutherland DE (2004) Total pancreatectomy with intraportal auto-islet transplantation using a temporarily exteriorized omental vein. J Am Coll Surg 199(6):994–995
87. Webb MA, Illouz SC, Pollard CA et al (2008) Islet auto transplantation following total pancreatectomy: a long-term assessment of graft function. Pancreas 37(3):282–287
88. Kuroda Y, Fujino Y, Morita A et al (1992) Oxygenation of the human pancreas during preservation by a two-layer (University of Wisconsin solution/perfluorochemical) cold-storage method. Transplantation 54(3):561–562
89. Kuroda Y, Fujino Y, Kawamura T et al (1990) Mechanism of oxygenation of pancreas during preservation by a two-layer (Euro-Collins' solution/perfluorochemical) cold-storage method. Transplantation 49(4):694–696
90. Matsumoto S, Qualley SA, Goel S et al (2002) Effect of the two-layer (University of Wisconsin solution-perfluorochemical plus O2) method of pancreas preservation on human islet isolation, as assessed by the Edmonton Isolation Protocol. Transplantation 74(10):1414–1419
91. Lakey JR, Kneteman NM, Rajotte RV et al (2002) Effect of core pancreas temperature during cadaveric procurement on human islet isolation and functional viability. Transplantation 73(7):1106–1110
92. Lakey JR, Tsujimura T, Shapiro AM et al (2002) Preservation of the human pancreas before islet isolation using a two-layer (UW solution-perfluorochemical) cold storage method. Transplantation 74(12):1809–1811
93. Ricordi C, Fraker C, Szust J et al (2003) Improved human islet isolation outcome from marginal donors following addition of oxygenated perfluorocarbon to the cold-storage solution. Transplantation 75(9):1524–1527

94. Tsujimura T, Kuroda Y, Avila JG et al (2004) Influence of pancreas preservation on human islet isolation outcomes: impact of the two-layer method. Transplantation 78(1):96–100
95. Brandhorst H, Muehling B, Yamaya H et al (2008) New class of oxygen carriers improves islet isolation from long-term stored rat pancreata. Transplant Proc 40(2):393–394
96. Saito T, Gotoh M, Satomi S et al (2010) Islet transplantation using donors after cardiac death: report of the Japan Islet Transplantation Registry. Transplantation 90(7):740–747
97. Matsumoto S, Rigley TH, Reems JA et al (2003) Improved islet yields from Macaca nemestrina and marginal human pancreata after two-layer method preservation and endogenous trypsin inhibition. Am J Transplant 3(1):53–63
98. Al-Abdullah IH, Bentsi-Barnes K, Valiente L et al (2008) Testing combinations of protease inhibitor and preservation solution to improve islet quality and yield. Transplant Proc 40(2):390–392
99. Rose NL, Palcic MM, Helms LM et al (2003) Evaluation of Pefabloc as a serine protease inhibitor during human-islet isolation. Transplantation 75(4):462–466
100. Lee TC, Barshes NR, Brunicardi FC et al (2004) Procurement of the human pancreas for pancreatic islet transplantation. Transplantation 78(3):481–483
101. Guignard AP, Oberholzer J, Benhamou PY et al (2004) Cost analysis of human islet transplantation for the treatment of type 1 diabetes in the Swiss-French Consortium GRAGIL. Diabetes Care 27(4):895–900
102. Huang GC, Zhao M, Jones P et al (2004) The development of new density gradient media for purifying human islets and islet-quality assessments. Transplantation 77(1):143–145
103. Nano R, Clissi B, Melzi R et al (2005) Islet isolation for allotransplantation: variables associated with successful islet yield and graft function. Diabetologia 48(5):906–912
104. Ricordi C, Lacy PE, Finke EH et al (1988) Automated method for isolation of human pancreatic islets. Diabetes 37(4):413–420
105. Robertson GS, Chadwick DR, Contractor H et al (1993) The use of continuous density gradients for the assessment of islet and exocrine tissue densities and islet purification. Acta Diabetol 30(3):175–180
106. Robertson GS, Chadwick DR, Contractor H et al (1993) The optimization of large-scale density gradient isolation of human islets. Acta Diabetol 30(2):93–98
107. Jansson L (1994) The regulation of pancreatic islet blood flow. Diabetes Metab Rev 10(4):407–416
108. Parr EL, Bowen KM, Lafferty KJ (1980) Cellular changes in cultured mouse thyroid glands and islets of Langerhans. Transplantation 30(2):135–141
109. Cross SE, Hughes SJ, Partridge CJ et al (2008) Collagenase penetrates human pancreatic islets following standard intraductal administration. Transplantation 86(7):907–911
110. O'Gorman D, Kin T, Murdoch T et al (2005) The standardization of pancreatic donors for islet isolations. Transplantation 80(6):801–806
111. Bucher P, Mathe Z, Morel P et al (2005) Assessment of a novel two-component enzyme preparation for human islet isolation and transplantation. Transplantation 79(1):91–97
112. Bucher P, Bosco D, Mathe Z et al (2004) Optimization of neutral protease to collagenase activity ratio for islet of Langerhans isolation. Transplant Proc 36(4):1145–1146
113. Antonioli B, Fermo I, Cainarca S et al (2007) Characterization of collagenase blend enzymes for human islet transplantation. Transplantation 84(12):1568–1575
114. van der Burg MP, Guicherit OR, Frolich M et al (1994) Impact of donor-related variables on islet isolation outcome in dogs. Diabetologia 37(1):111–114
115. London NJ, James R, Bell PR et al (1992) Islet purification. In: Pancreatic islet transplantation. Ed Landes:113
116. London NJM, James RFL, Bell PRF (1992) Pancreatic islet transplantation, islet purification. In: Ricordi C (ed) Vol. Chapter 11. Landes, Austin, pp 113–131
117. Pretlow TG II (1975) Disaggregation of prostates and purification of epithelial cells from normal and cancerous prostates using sedimentation in an isokinetic density gradient of Ficoll in tissue culture medium. Cancer Chemother Rep 1 59(1):143–145

118. Schwartz BD, Traverso LW (1984) Morphological changes in pancreatic fragments prepared for transplantation by collagenase treatment. Transplantation 38(3):273–280
119. Burg MB, Orloff J (1964) Active cation transport by kidney tubules at OC. Am J Physiol 207:983–988
120. Goss JA, Schock AP, Brunicardi FC et al (2002) Achievement of insulin independence in three consecutive type-1 diabetic patients via pancreatic islet transplantation using islets isolated at a remote islet isolation center. Transplantation 74(12):1761–1766
121. MacKenzie DA, Sollinger HW, Hullett DA (2003) Removal of CD45+ cells from human fetal pancreas alters immunogenicity in vitro. Transplant Proc 35(4):1506–1507
122. Scharp DW, Lacy PE, Finke E et al (1987) Low-temperature culture of human islets isolated by the distention method and purified with Ficoll or Percoll gradients. Surgery 102(5):869–879
123. Sabek OM, Marshall DR, Penmetsa R et al (2006) Examination of gene expression profile of functional human pancreatic islets after 2-week culture. Transplant Proc 38(10):3678–3679
124. Cattan P, Berney T, Schena S et al (2001) Early assessment of apoptosis in isolated islets of Langerhans. Transplantation 71(7):857–862
125. Caulin-Glaser T, Watson CA, Pardi R et al (1996) Effects of 17beta-estradiol on cytokine-induced endothelial cell adhesion molecule expression. J Clin Invest 98(1):36–42
126. Lacy PE, Davie JM, Finke EH (1979) Prolongation of islet allograft survival following in vitro culture (24 degrees C) and a single injection of ALS. Science 204(4390):312–313
127. Schmied BM, Ulrich A, Matsuzaki H et al (2000) Maintenance of human islets in long-term culture. Differentiation 66(4-5):173–180
128. Schmied BM, Ulrich A, Matsuzaki H et al (2001) Transdifferentiation of human islet cells in a long-term culture. Pancreas 23(2):157–171
129. Schmied BM, Liu G, Matsuzaki H et al (2000) Differentiation of islet cells in long-term culture. Pancreas 20(4):337–347
130. Buitrago A, Gylfe E (1983) Significance of serum for the preservation of insulin secretion during culture. Med Biol 61(2):133–138
131. Yuan S, Rosenberg L, Paraskevas S et al (1996) Transdifferentiation of human islets to pancreatic ductal cells in collagen matrix culture. Differentiation 61(1):67–75
132. Wang J, Zhang X (2001) Needle-type dual microsensor for the simultaneous monitoring of glucose and insulin. Anal Chem 73(4):844–847
133. Thomas FT, Contreras JL, Bilbao G et al (1999) Anoikis, extracellular matrix, and apoptosis factors in isolated cell transplantation. Surgery 126(2):299–304
134. Rutzky LP, Bilinski S, Kloc M et al (2002) Microgravity culture condition reduces immunogenicity and improves function of pancreatic islets. Transplantation 74(1):13–21
135. Beattie GM, Montgomery AM, Lopez AD et al (2002) A novel approach to increase human islet cell mass while preserving beta-cell function. Diabetes 51(12):3435–3439
136. Contreras JL, Smyth CA, Bilbao G et al (2002) 17beta-Estradiol protects isolated human pancreatic islets against proinflammatory cytokine-induced cell death: molecular mechanisms and islet functionality. Transplantation 74(9):1252–1259
137. Sabek OM, Fraga DW, Henry J et al (2007) Expression of transforming growth factor-beta by human islets: impact on islet viability and function. Cell Transplant 16(8):775–785
138. Beattie GM, Cirulli V, Lopez AD et al (1997) Ex vivo expansion of human pancreatic endocrine cells. J Clin Endocrinol Metab 82(6):1852–1856
139. Beattie GM, Otonkoski T, Lopez AD et al (1997) Functional beta-cell mass after transplantation of human fetal pancreatic cells: differentiation or proliferation? Diabetes 46(2):244–248
140. Beattie GM, Rubin JS, Mally MI et al (1996) Regulation of proliferation and differentiation of human fetal pancreatic islet cells by extracellular matrix, hepatocyte growth factor, and cell-cell contact. Diabetes 45(9):1223–1228
141. Otonkoski T, Beattie GM, Rubin JS et al (1994) Hepatocyte growth factor/scatter factor has insulinotropic activity in human fetal pancreatic cells. Diabetes 43(7):947–953
142. Otonkoski T, Mally MI, Hayek A (1994) Opposite effects of beta-cell differentiation and growth on Reg expression in human fetal pancreatic cells. Diabetes 43(9):1164–1166

143. Dionne KE, Colton CK, Yarmush ML (1993) Effect of hypoxia on insulin secretion by isolated rat and canine islets of Langerhans. Diabetes 42(1):12–21
144. Gaber AO, Fraga D, Kotb M et al (2004) Human islet graft function in NOD-SCID mice predicts clinical response in islet transplant recipients. Transplant Proc 36(4):1108–1110
145. Hubert T, Strecker G, Gmyr V et al (2008) Acute insulin response to arginine in deceased donors predicts the outcome of human islet isolation. Am J Transplant 8(4):872–876
146. Sabek OM, Fraga DW, Minoru O et al (2005) Assessment of human islet viability using various mouse models. Transplant Proc 37(8):3415–3416
147. Gerling IC, Kotb M, Fraga D et al (1998) No correlation between in vitro and in vivo function of human islets. Transplant Proc 30(2):587–588
148. Sweet IR, Gilbert M, Jensen R et al (2005) Glucose stimulation of cytochrome C reduction and oxygen consumption as assessment of human islet quality. Transplantation 80(8):1003–1011
149. Dobson T, Fraga D, Saba C et al (2000) Human pancreatic islets transfected to produce an inhibitor of TNF are protected against destruction by human leukocytes. Cell Transplant 9(6):857–865
150. Narang AS, Sabek O, Gaber AO et al (2006) Co-expression of vascular endothelial growth factor and interleukin-1 receptor antagonist improves human islet survival and function. Pharm Res 23(9):1970–1982
151. Narang AS, Cheng K, Henry J et al (2004) Vascular endothelial growth factor gene delivery for revascularization in transplanted human islets. Pharm Res 21(1):15–25
152. Hardstedt M, Lindblom S, Karlsson-Parra A et al (2016) Characterization of innate immunity in an extended whole blood model of human islet allotransplantation. Cell Transplant 25(3):503–515
153. Kourtzelis I, Kotlabova K, Lim JH et al (2016) Developmental endothelial locus-1 modulates platelet-monocyte interactions and instant blood-mediated inflammatory reaction in islet transplantation. Thromb Haemost 115(4):781–788
154. Xiao F, Ma L, Zhao M et al (2016) APT070 (mirococept), a membrane-localizing C3 convertase inhibitor, attenuates early human islet allograft damage in vitro and in vivo in a humanized mouse model. Br J Pharmacol 173(3):575–587
155. Chhabra P, Linden J, Lobo P et al (2012) The immunosuppressive role of adenosine A2A receptors in ischemia reperfusion injury and islet transplantation. Curr Diabetes Rev 8(6):419–433
156. Brissova M, Aamodt K, Brahmachary P et al (2014) Islet microenvironment, modulated by vascular endothelial growth factor-A signaling, promotes beta cell regeneration. Cell Metab 19(3):498–511
157. Brissova M, Fowler M, Wiebe P et al (2004) Intraislet endothelial cells contribute to revascularization of transplanted pancreatic islets. Diabetes 53(5):1318–1325
158. Ullsten S, Lau J, Carlsson PO (2015) Vascular heterogeneity between native rat pancreatic islets is responsible for differences in survival and revascularisation post transplantation. Diabetologia 58(1):132–139
159. Lau J, Mattsson G, Carlsson C et al (2007) Implantation site-dependent dysfunction of transplanted pancreatic islets. Diabetes 56(6):1544–1550
160. Lau J, Carlsson PO (2009) Low revascularization of human islets when experimentally transplanted into the liver. Transplantation 87(3):322–325
161. Kang S, Park HS, Jo A et al (2012) Endothelial progenitor cell cotransplantation enhances islet engraftment by rapid revascularization. Diabetes 61(4):866–876
162. Paget MB, Murray HE, Bailey CJ et al (2011) Rotational co-culture of clonal beta-cells with endothelial cells: effect of PPAR-gamma agonism in vitro on insulin and VEGF secretion. Diabetes Obes Metab 13(7):662–668
163. Espes D, Lau J, Quach M et al (2015) Cotransplantation of polymerized hemoglobin reduces beta-cell hypoxia and improves beta-cell function in intramuscular islet grafts. Transplantation 99(10):2077–2082
164. Zhang N, Richter A, Suriawinata J et al (2004) Elevated vascular endothelial growth factor production in islets improves islet graft vascularization. Diabetes 53(4):963–970

165. Borg DJ, Weigelt M, Wilhelm C et al (2014) Mesenchymal stromal cells improve transplanted islet survival and islet function in a syngeneic mouse model. Diabetologia 57(3):522–531
166. Buitinga M, Janeczek Portalska K, Cornelissen DJ et al (2016) Coculturing human islets with proangiogenic support cells to improve islet revascularization at the subcutaneous transplantation site. Tissue Eng Part A 22(3-4):375–385
167. Cao XK, Li R, Sun W et al (2016) Co-combination of islets with bone marrow mesenchymal stem cells promotes angiogenesis. Biomed Pharmacother 78:156–164
168. Fransson M, Brannstrom J, Duprez I et al (2015) Mesenchymal stromal cells support endothelial cell interactions in an intramuscular islet transplantation model. Regen Med Res 3:1
169. Park KS, Kim YS, Kim JH et al (2010) Trophic molecules derived from human mesenchymal stem cells enhance survival, function, and angiogenesis of isolated islets after transplantation. Transplantation 89(5):509–517
170. Lehmann R, Zuellig RA, Kugelmeier P et al (2007) Superiority of small islets in human islet transplantation. Diabetes 56(3):594–603
171. Li W, Zhao R, Liu J et al (2014) Small islets transplantation superiority to large ones: implications from islet microcirculation and revascularization. J Diabetes Res 2014:192093
172. Motte E, Szepessy E, Suenens K et al (2014) Composition and function of macroencapsulated human embryonic stem cell-derived implants: comparison with clinical human islet cell grafts. Am J Physiol Endocrinol Metab 307(9):E838–E846
173. Pepper AR, Pawlick R, Gala-Lopez B et al (2015) Diabetes is reversed in a murine model by marginal mass syngeneic islet transplantation using a subcutaneous cell pouch device. Transplantation 99(11):2294–2300
174. Sabek OM, Ferrati S, Fraga DW et al (2013) Characterization of a nanogland for the autotransplantation of human pancreatic islets. Lab Chip 13(18):3675–3688
175. Veiseh O, Doloff JC, Ma M et al (2015) Size- and shape-dependent foreign body immune response to materials implanted in rodents and non-human primates. Nat Mater 14(6):643–651
176. Duvivier-Kali VF, Omer A, Parent RJ et al (2001) Complete protection of islets against allorejection and autoimmunity by a simple barium-alginate membrane. Diabetes 50(8):1698–1705
177. Llacua A, de Haan BJ, Smink SA et al (2016) Extracellular matrix components supporting human islet function in alginate-based immunoprotective microcapsules for treatment of diabetes. J Biomed Mater Res A 104(7):1788–96
178. Omer A, Duvivier-Kali V, Fernandes J et al (2005) Long-term normoglycemia in rats receiving transplants with encapsulated islets. Transplantation 79(1):52–58

Chapter 11
Targeted Modulation of Macrophage Functionality by Nanotheranostics in Inflammatory Liver Disease and Cancer

Matthias Bartneck and Frank Tacke

Abstract Liver diseases characteristically progress from chronic inflammation to aberrant wound-healing with excessive scarring, termed fibrosis, and eventually to liver cancer. Since hepatic macrophages are critical regulators of these inflammatory processes, it appears promising to target these cells with novel nanomedicine-based therapeutics. Nanomedicine bears a large potential for the design of novel drugs by site-specific delivery and controlled release. Nanotheranostics allow for additional in vivo tracing of the therapeutics. Therapeutic nanoparticles are, in most cases, composed of biodegradable compounds such as phospholipids, which are an essential part of biological membranes. Nanodrugs may interact with soluble parts of the immune system (humoral immunity), specifically with components that help immune cells in pathogen recognition such as antibodies or complement factors. Macrophages are a heterogeneous cell type being composed of pro- or anti-inflammatory subtypes that can either heal or worsen inflammatory diseases as well as combat or support cancer growth. Due to their inherent capability of foreign material uptake, macrophages are relatively easy to target, but may also hinder particles from reaching other target cells. A variety of receptors attractive for targeting was found to be useful in more specific strategies for selectively modulating macrophages to overcome effects on other cell types. In this chapter, current strategies to target macrophages in liver diseases and cancer are reviewed.

Keywords Chemokines • Liver fibrosis • Liver inflammation • Macrophage • Monocyte • Nanotheranostics

---

M. Bartneck • F. Tacke (✉)
Department of Medicine III, RWTH University Hospital Aachen, Aachen, Germany
e-mail: frank.tacke@gmx.net

B. Corradetti (ed.), *The Immune Response to Implanted Materials and Devices*,
DOI 10.1007/978-3-319-45433-7_11

11.1 Introduction

Theranostics, diagnostic tracing of therapeutics, comprises interventions that are oftentimes imaging-guided administrations of drugs. Nanotheranostics additionally bring up novel features of nanomedicine with a large potential for the design of novel drugs by optimizing site-specific delivery and controlled release of drugs [1]. Immune system elements represent important interactors for nanotheranostics both on a cellular and humoral (body fluids) level. The noncellular components belong to humoral immunity and typically consist of antibodies, antimicrobial peptides, and complement factors. The normal function of antibodies and complement molecules is to improve recognition by immune cells or to facilitate cell killing by complement factors. The process of molecules binding to pathogenic targets such as bacteria or to foreign material is called opsonization. Opsonins potentially also bind to nanoparticles, and their coverage is an important issue for circulating nanoconstructs, since it may affect the recognition by other cells, including the targeting capability of the formulations [2]. Similar to other nano-sized constructs, nanotheranostics may attract binding of antithetically charged serum proteins to their surface and thereby change their load, as reflected by the zeta potential [3]. In vitro characterizations of nanomaterials should therefore also include assays in physiological media such as phosphate-buffered saline and biological media such as plasma and serum from the species of interest.

Antibodies further constitute large amounts of serum proteins and are generated by cells of the adaptive immune system (lymphocytes). De novo generation of antibodies by the adaptive lymphocytes requires the recognition, uptake, and processing of antigen by professional antigen-presenting cells, leading to antigen presentation to lymphocytes, which are then able to produce antibodies. In addition to their recognition of novel pathogenic threats, phagocytes, and especially macrophages, are the most decisive cells for interactions with foreign material. This is important for applications, where macrophages are intended to be targeted by nanotheranostics, but also to reach other cell types, since any nanoformulation may in part act on macrophages, due to their inherent phagocytic activity [4]. The fact that macrophages can also initiate the production of antibodies is relevant for many nanotherapeutics, since there are reports that macrophages may trigger B cells to produce antibodies directed against poly ethylene glycol (PEG) [5]. This mechanism could potentially contribute to the accelerated blood clearance effect of PEGylated liposomes, which relies on the enhanced liposome uptake by macrophages, most likely due to the opsonization of liposomes with IgM antibodies [5]. PEGylation of drugs for liver targeting has been successfully done for many years, one example given by Pegasys® (Roche Inc.), which is PEGylated interferon α-2a, and has been used for the treatment of viral hepatitis for many years [6]. In the past decade, PEG has also been deployed on nanosystems to prolong half-life of circulating drugs and delay nanoparticle recognition by human phagocytes [7, 8].

11.2 Role of Hepatic Macrophages in Inflammation and Cancer

Due to their inherent uptake of any sort of foreign material, macrophages are probably the most important immune cell type for injectable nanotheranostics. Macrophages are a heterogeneous cell type composed of mixtures of cells with varying portions of proinflammatory (M1) and alternatively activated macrophages (M2), and stimuli-dependent subpopulations thereof [9]. M1 macrophages secrete proinflammatory mediators such as the tumor necrosis factor (TNF) and interleukin 1β (IL1β) and exhibit antitumoral activity, whereas M2 have a stronger phagocytic activity, release antiinflammatory and angiogenic factors, and thereby support tumor growth, but inhibit inflammation [10]. The heterogeneity of macrophages is especially prominent in the liver. During homeostasis, liver macrophages rather promote immunological tolerance and prevent immune activation, while they are critical orchestrators of inflammatory and wound-healing responses in conditions of liver disease [11]. Thus, the phenotype of macrophages is highly dependent on environmental signals and can switch from pro- to anti-inflammatory functions and vice versa [12].

In addition to the M1-M2 dichotomy, the ancestry of hepatic macrophages is an important factor for understanding their function. They originate from at least two different types of progenitors, which differentially contribute to nanoparticle clearance: the monocyte-derived macrophages (MoMF), oftentimes also referred to as inflammatory macrophages [13], which are $CD45^{+}Ly6G^{-}F4/80^{+}CD11b^{+}$ in mice and originate from circulating monocytes (and thus are of hematopoietic origin), and those derived from progenitor cells that are resident to the liver and are termed Kupffer cells (KC). KC in mice are typically characterized and isolated via fluorescence-activated cell sorting (FACS) by their surface expression pattern $CD45^{low}$ $Ly6G^{-}F4/80^{++}CD11b^{-}$ [13]. Earlier studies of our group indicate that the capabilities for nanoparticle uptake by both subtypes of hepatic macrophages can strongly differ: we discovered a 30-fold enhanced amount of gold nanorods in liver MoMF compared to the KC [13] after intravenous administration in mice—it thus appears that small metal-based nanocarriers may be more efficiently cleared by the MoMF compared to the KC. In contrast, uptake of 100 nm-sized liposomes was similar between both macrophage subtypes [14]. However, many studies in the past have not differentiated hepatic macrophage subtypes and simply designated all hepatic macrophages as KC [15]. As the functions of MoMF and KC can be different or even partially opposing in health and disease, it is important to relate effects of nanotherapeutics in the liver to the affected type of macrophages.

In liver disease, macrophages sense initiating signals that lead to inflammation [11]. These signals include molecules from damaged cells (such as hepatocytes), bacteria, or fatty acids, and probably many more stimuli to be identified in future. These inflammation-initiating stimuli activate the inflammasome of immune cells, specifically in hepatic macrophages, which for their part activate other cell types and thereby exacerbate liver inflammation [16]. In carbon tetrachloride

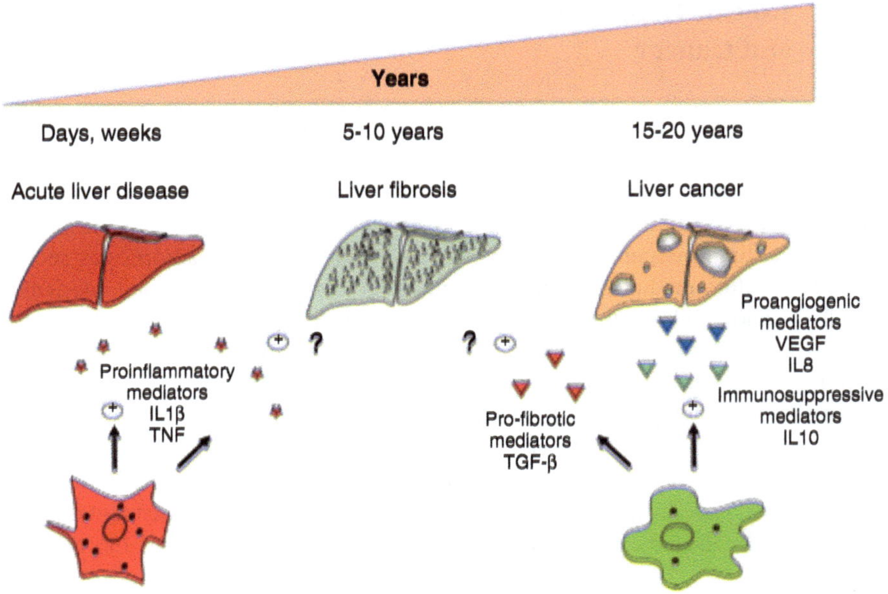

Fig. 11.1 Role of macrophage subpopulations during the course of liver disease progression. Macrophages exert manifold actions during liver diseases, ranging from primarily inflammatory activation in the initiation of disease to fibrogenic and proangiogenic functions that provide an environment for liver cancer development

(CCl_4)-induced liver injury in mice, an influx of monocyte-derived macrophages occurs [17], due to signals from CCl_4-injured hepatocytes (Fig. 11.1). During the course of this initiation, macrophages stimulate a further infiltration of monocytes by production of the chemokine CCL2 and also trigger activation of a key cell type in fibrogenesis, the hepatic stellate cells (HSC). HSC become activated, subsequently proliferate and produce increased amounts of collagen, which leads to organ fibrosis and ultimately loss of liver function [18]. In many cases, a sequence of events can be noted, which finally leads from liver inflammation, accompanied by fibrosis and cirrhosis, to the development of hepatocellular carcinoma [19] (Fig. 11.2).

Despite the underlying origin (e.g., viral hepatitis, alcohol abuse, cholestasis, metabolic or hereditary disorders), liver diseases characteristically progress from chronic inflammation ("hepatitis") to fibrosis to cirrhosis and eventually to hepatocellular carcinoma [20]. Macrophages accompany and fuel these processes, starting with the initiation of inflammation by means of proinflammatory cytokines, which among other cells, activate hepatic stellate cells (Fig. 11.2). The role of macrophages in fibrosis is ambiguous, because not only do they activate fibrogenesis, but it appears that an M2-polarization could also be anti-fibrotic in liver. Anti-fibrotic properties of M2 polarized hepatic macrophages were suggested in one of our own recent studies in histidine-rich glycoprotein (HRG) deficient mice, which exhibited significantly reduced levels of liver fibrosis due to the lack of an inflammatory M1

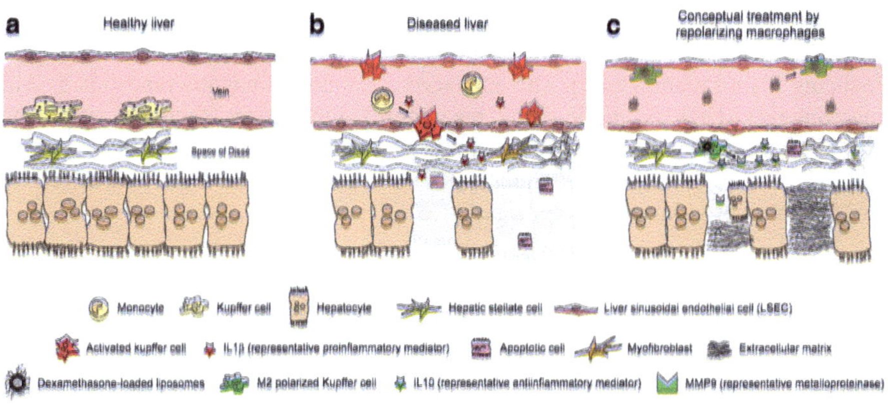

Fig. 11.2 Macrophage phenotype in the healthy (**a**) and diseased liver (**b**) as well as after therapeutic interventions with macrophage-specific nanotheranostics (**c**). Please note that there are different cell types, from which macrophages originate such as local progenitors for Kupffer cells and monocytes for monocyte-derived macrophages (MoMF), which is simplified here for clarity reasons

and a pronounced M2 polarization [21]. Nevertheless, M2-polarized macrophages can also release transforming growth factor β (TGF-β), which is a key activator of fibrogenesis [22]. These findings illustrate that the concept of "M2 cells" is too superficial and does not reflect the full spectrum of macrophage subsets in the liver, especially in conditions of fibrosis [10]. Profibrogenic activities of tissue macrophages were also identified in other organs such as in pulmonary [23] or cardiac [24] fibrosis. However, macrophages can also promote the resolution of hepatic fibrosis, by means of anti-inflammatory cytokines or secretion of matrix-degrading enzymes [25]. Macrophages, especially tumor-associated macrophages (TAM), also affect the development of cancer [26]. M2-polarized TAM support tumor growth by the release of proangiogenic cytokines such as CXCL8 (IL8) and growth factors such as the vascular endothelial growth factor (Vegf) [10, 27].

Given the key role of macrophages for the different steps of liver disease progression, these cells might represent an interesting target for therapeutic approaches, including new nanotheranostics, at different stages of disease [4]. Since inflammation leads, via fibrosis, to cancer, it might be advisable to block disease development already at early stages (Fig. 11.2c). An idealized conceptual treatment of liver inflammation by nanotheranostics makes use of nanoconstructs, which trigger M2 polarization such as dexamethasone-containing liposomes that were successfully used to treat acute liver inflammation in mice [28]. Macrophages internalize nanotheranostics, and are subsequently switched to M2 polarization. Ideally, they would deactivate activated myofibroblasts and degrade extracellular matrix (ECM) by means of their matrix metalloproteinases such as MMP9 [11] (Fig. 11.2c). However, such a repolarization principally carries the risk of favoring a tumor-promoting function of macrophages, and thus, timing and treatment duration would have to be optimized before clinical application.

Generally, novel nanotheranostics would allow for modifying the macrophage phenotype, for instance, either increasing inflammatory activation (as in cancer), or reducing it (in inflammation). The enhanced clearance of nanotheranostics by M2 macrophages [29] is particularly meaningful for cancer therapy, since the M2-polarized TAM clear large amounts of nanotheranostics [29]. Vice versa, targeting M1 cells might require modifications in the compounds or increased doses for the treatment of inflammatory disease [11, 29].

11.3 Classifications of Nanomaterials and Nanotheranostics

Nanotheranostic tracing strategies strongly rely on the possibilities for detecting particles in cells and organisms. Nonfluorescent nanodrugs such as lipid-based carriers or polymers can get detectable in vivo by using additional fluorescent tags such as 1,2-dioleoyl-sn-glycero-3-phosphoethanolamine-N-(7-nitro-2-1,3-benzoxadiazol-4-yl) (NBD-PE), therefore enabling studies on their biodistribution in living organisms or cells [14]. Many inorganic nanoparticles, specifically metals such as gold nanoparticles, can be detected in computerized tomography techniques due to their electron density and allow for an assessment of in vivo biodistribution [13]. However, their persistence in the body [13] remains a critical issue for using metal-based nanotheranostics in the clinics. The biodistribution of nanoparticles relies on the material, size, and additional functionalizations. Nanoparticles sizing around 10–250 nm were found to be preferentially present in liver and spleen, whereas those below 10 nm were found throughout the body in many organs, including kidney, testis, and brain [30]. Though many nanoconstructs accumulate in both spleen and liver to a similar extent, hepatic accumulation is considered far more relevant, since the dry weight of liver is about 50 times larger than that of spleen [13]. The large portion of hepatic compared to all macrophages, which makes up about 90 % of tissue macrophages in the body, is assumed to be the major reason for hepatic accumulation [31]. Cytotoxicity, as demonstrated by Pan and Jahnen-Dechent, occurs particularly for particles below 1.4 nm [32]. In detail, it is probably the hydrodynamic diameter that decides for particle clearance [33]. Nanoparticle charge, which depends on the physicochemical composition of the constructs, remains a critical issue for their uptake by cells and especially by phagocytic cells. Zwitterionic and neutral charges were shown to reduce, while positive charges evoked increased uptake in a study facilitating quantum dots [33].

11.4 Targeting of Hepatic Macrophages Using Nanotheranostics

Many different research groups currently test the manipulation of macrophage behavior using nanomedicine as a therapeutic strategy against liver diseases. Due to the inherent phagocytic capabilities of macrophages, reaching hepatic macrophages should be, at least in part, possible by simply using either positively or negatively charged particles without the need for a specific "active" targeting strategy. However, this most clearly holds true for solid nanoparticles like those made of gold, whereas liposomal formulations are more intensively also internalized by lymphocytes [13, 28]. Earlier studies have demonstrated that surface chemistry, charge, functional end-groups and biomimetic molecules such as the RGD tripeptide affect macrophage polarization [7, 34]. Despite the promising M2 polarization in vitro [7, 34], RGD-capped gold nanorods corroborated acute liver injury in mouse models in vivo [13]. Pharmacologically active substances such as corticosteroids may represent an advance for hepatic therapy based on macrophage polarization, since their application abrogates hepatic injury in the same model in case of liposomal dexamethasone [14].

Nanotechnological alterations in drug pharmacokinetics may help to significantly reduce side effects due to a targeted delivery of compounds (Fig. 11.3). The most frequently used strategy for active targeting of macrophages is probably the decoration of the formulations with mannose, which targets the mannose receptor of macrophages (CD206) [35–38]. The delivery of siRNA with mannose-modified nanocarriers against tumor necrosis factor represents an innovative strategy for curing hepatitis [37]. As an example of targeting macrophages based on positive charge,

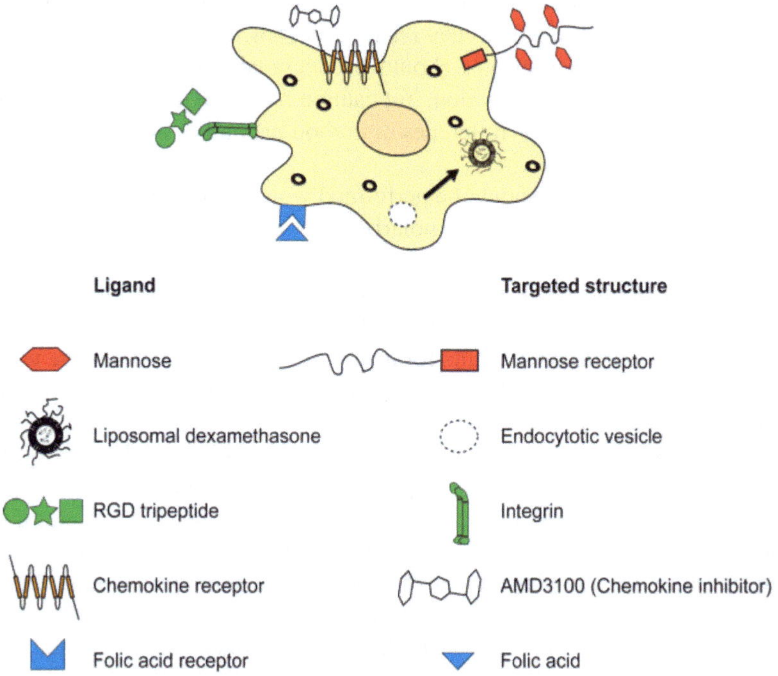

Fig. 11.3 Ligands and the corresponding receptors useful for selective targeting of hepatic macrophages

ritonavir, which targets the liver enzyme CYP 3A4 activity in the liver [39], has recently been delivered using a novel formulation of 200 nm sizing cationic lipid–polymer hybrid nanoparticles. The strongly positive charge of +30 mV evoked an efficient delivery of cargo to hepatic macrophages [40]. This effect is similar to our own studies using CTAB-stabilized gold nanorods, which exhibit a similarly positive surface charge [13].

Recently, tumor-associated macrophages were unintentionally targeted based on their expression of chemokine receptor CXCR4, because CXCR4-directed nanoparticles were designed to target hepatocellular carcinoma cells [41]. Additionally, Plerixafor (AMD3100), the ligand for CXCR4, exhibits antiangiogenic properties intrinsically which acts as an additional therapeutic property that can be incorporated into a single particle [41].

11.5 Conclusions

Macrophages appear as a promising target for liver diseases and cancer. The selection of suitable targeting nanosystems has to cautiously consider all properties of a carrier such as particle material, potential spacers, and targeting moieties. One major issue is to assure biodegradability for all single compounds but also for the combined structure—therefore, organic carriers appear most feasible compared to the metal-based particles, which may reside in organs for weeks [13]. Ideal targeting compounds for macrophages should be degradable in body fluids after the desired time for pharmaceutic action. Yet caution has to be taken with organic particles, because materials like fullerenes or carbon nanotubes are not degradable at all in aqueous environments [42].

Targeting macrophages should be performed with caution, since there is a sensitive balance between differentially polarized subsets. An overstimulation of M1, which might be a goal in cancer therapy to overcome M2-related immune suppression, could potentially result in proinflammatory over-activation of macrophages, which in turn might lead to a systemic inflammatory response like in sepsis [43]. Conversely, M1-suppressing agents for acute inflammatory liver disease may support cancer growth, since M2 potentially fuel cancer progression when reaching non-appropriate sites in the body.

The administration route is a major issue in nanomaterial design—among patients, oral administration is clearly preferred. However, most novel nanomaterials require parenteral administration such as subcutaneous or intravenous administration. In order to evoke oral administration, special requirements for the carrier systems arise allowing enteral absorption and enterohepatic redistribution, in addition to the requirement for cellular targeting n. Preclinical studies should provide a solid concept for targeting macrophages, which may then be adapted to nanocarriers with appropriate oral delivery capabilities.

References

1. Lammers T, Aime S, Hennink WE et al (2011) Theranostic nanomedicine. Acc Chem Res 44(10):1029–1038
2. Petros RA, DeSimone JM (2010) Strategies in the design of nanoparticles for therapeutic applications. Nat Rev Drug Discov 9(8):615–627
3. Bartneck M, Keul HA, Zwadlo-Klarwasser G et al (2010) Phagocytosis independent extracellular nanoparticle clearance by human immune cells. Nano Lett 10(1):59–63
4. Bartneck M, Warzecha KT, Tacke F (2014) Therapeutic targeting of liver inflammation and fibrosis by nanomedicine. Hepatobiliary Surg Nutr 3(6):364–376
5. Abu Lila AS, Ichihara M, Shimizu T et al (2013) Ex-vivo/in-vitro anti-polyethylene glycol (PEG) immunoglobulin M production from murine splenic B cells stimulated by PEGylated liposome. Biol Pharm Bull 36(11):1842–1848
6. Pawlotsky JM, Feld JJ, Zeuzem S et al (2015) From non-A, non-B hepatitis to hepatitis C virus cure. J Hepatol 62(1 Suppl):S87–S99
7. Bartneck M, Keul HA, Singh S et al (2010) Rapid uptake of gold nanorods by primary human blood phagocytes and immunomodulatory effects of surface chemistry. ACS Nano 4(6):3073–3086
8. Moghimi SM, Szebeni J (2003) Stealth liposomes and long circulating nanoparticles: critical issues in pharmacokinetics, opsonization and protein-binding properties. Prog Lipid Res 42(6):463–478
9. Murray PJ, Allen JE, Biswas SK et al (2014) Macrophage activation and polarization: nomenclature and experimental guidelines. Immunity 41(1):14–20
10. Martinez FO, Sica A, Mantovani A et al (2008) Macrophage activation and polarization. Front Biosci 13:453–461
11. Tacke HW, Zimmermann F (2014) Macrophage heterogeneity in liver injury and fibrosis. J Hepatol 60(5):1090–1096
12. Dal-Secco D, Wang J, Zeng Z et al (2015) A dynamic spectrum of monocytes arising from the in situ reprogramming of CCR2+ monocytes at a site of sterile injury. J Exp Med 212(4):447–456
13. Bartneck M, Ritz T, Keul HA et al (2012) Peptide-functionalized gold nanorods increase liver injury in hepatitis. ACS Nano 6(10):8767–8777
14. Bartneck M, Peters FM, Warzecha KT et al (2014) Liposomal encapsulation of dexamethasone modulates cytotoxicity, inflammatory cytokine response, and migratory properties of primary human macrophages. Nanomedicine 10(6):1209–1220
15. Sadauskas E, Wallin H, Stoltenberg M et al (2007) Kupffer cells are central in the removal of nanoparticles from the organism. Part Fibre Toxicol 4:10
16. Csak T, Ganz M, Pespisa J et al (2011) Fatty acid and endotoxin activate inflammasomes in mouse hepatocytes that release danger signals to stimulate immune cells. Hepatology 54(1):133–144
17. Karlmark KR, Weiskirchen R, Zimmermann HW et al (2009) Hepatic recruitment of the inflammatory Gr1+ monocyte subset upon liver injury promotes hepatic fibrosis. Hepatology 50(1):261–274
18. Bataller DA, Brenner R (2005) Liver fibrosis. J Clin Invest 115(2):209–218
19. Vucur M, Roderburg C, Bettermann K et al (2010) Mouse models of hepatocarcinogenesis: what can we learn for the prevention of human hepatocellular carcinoma? Oncotarget 1(5):373–378
20. Ehling J, Tacke F (2016) Role of chemokine pathways in hepatobiliary cancer. Cancer Lett 379(2):173–183
21. Bartneck M, Fech V, Ehling J et al (2015) Histidine-rich glycoprotein promotes macrophage activation and inflammation in chronic liver disease. Hepatology 63(4):1310–1324
22. Wynn TA, Barron L (2010) Macrophages: master regulators of inflammation and fibrosis. Semin Liver Dis 30(3):245–257

23. He C, Ryan AJ, Murthy S et al (2013) Accelerated development of pulmonary fibrosis via Cu, Zn-sod-induced alternative activation of macrophages. J Biol Chem 288(28):20745–20757
24. Meznarich J, Malchodi L, Helterline D et al (2013) Urokinase plasminogen activator induces pro-fibrotic/m2 phenotype in murine cardiac macrophages. PLoS One 8(3), e57837
25. Tacke C, Trautwein F (2015) Mechanisms of liver fibrosis resolution. J Hepatol 63(4):1038–1039
26. Mossanen JC, Tacke F (2013) Role of lymphocytes in liver cancer. Oncoimmunology 2(11), e26468
27. Movahedi K, Laoui D, Gysemans C et al (2010) Different tumor microenvironments contain functionally distinct subsets of macrophages derived from Ly6C(high) monocytes. Cancer Res 70(14):5728–5739
28. Bartneck M, Scheyda KM, Warzecha KT et al (2015) Fluorescent cell-traceable dexamethasone-loaded liposomes for the treatment of inflammatory liver diseases. Biomaterials 37:367–382
29. Kai MP, Brighton HE, Fromen CA et al (2016) Tumor presence induces global immune changes and enhances nanoparticle clearance. ACS Nano 10(1):861–70
30. Muller RH, Maassen S, Weyhers H et al (1996) Phagocytic uptake and cytotoxicity of solid lipid nanoparticles (SLN) sterically stabilized with poloxamine 908 and poloxamer 407. J Drug Target 4(3):161–170
31. Bilzer M, Roggel AL, Gerbes F (2006) Role of Kupffer cells in host defense and liver disease. Liver Int 26(10):1175–1186
32. Pan Y, Neuss S, Leifert A et al (2007) Size-dependent cytotoxicity of gold nanoparticles. Small 3(11):1941–1949
33. Choi HS, Liu W, Misra P et al (2007) Renal clearance of quantum dots. Nat Biotechnol 25(10):1165–1170
34. Bartneck M, Keul HA, Wambach M et al (2012) Effects of nanoparticle surface coupled peptides, functional endgroups and charge on intracellular distribution and functionality of human primary reticuloendothelial cells. Nanomedicine 8(8):1282–1292
35. Beljaars L, Molema G, Weert B et al (1999) Albumin modified with mannose 6-phosphate: a potential carrier for selective delivery of antifibrotic drugs to rat and human hepatic stellate cells. Hepatology 29(5):1486–1493
36. Blykers A, Schoonooghe S, Xavier C et al (2015) PET imaging of macrophage mannose receptor-expressing macrophages in tumor stroma using 18F-radiolabeled camelid single-domain antibody fragments. J Nucl Med 56(8):1265–1271
37. He C, Yin L, Tang C et al (2013) Multifunctional polymeric nanoparticles for oral delivery of TNF-alpha siRNA to macrophages. Biomaterials 34(11):2843–2854
38. Melgert BN, Olinga P, Van Der Laan JM et al (2001) Targeting dexamethasone to Kupffer cells: effects on liver inflammation and fibrosis in rats. Hepatology 34(4 Pt 1):719–728
39. Dresser GK, Spence JD, Bailey DG (2000) Pharmacokinetic-pharmacodynamic consequences and clinical relevance of cytochrome P450 3A4 inhibition. Clin Pharmacokinet 38(1):41–57
40. Asthana S, Jaiswal AK, Gupta PK et al (2015) Th-1 biased immunomodulation and synergistic antileishmanial activity of stable cationic lipid-polymer hybrid nanoparticle: biodistribution and toxicity assessment of encapsulated amphotericin B. Eur J Pharm Biopharm 89:62–73
41. Liu JY, Chiang T, Liu CH et al (2015) Delivery of siRNA using CXCR4-targeted nanoparticles modulates tumor microenvironment and achieves a potent antitumor response in liver cancer. Mol Ther 23(11):1772–1782
42. Kummerer K, Menz J, Schubert T et al (2011) Biodegradability of organic nanoparticles in the aqueous environment. Chemosphere 82(10):1387–1392
43. Schulte W, Bernhagen R, Bucala J (2013) Cytokines in sepsis: potent immunoregulators and potential therapeutic targets—an updated view. Mediators Inflamm 2013:165974

Chapter 12
T-Cell Mediated Immunomodulation and Transplant Optimization

Sandra Grass, Sara Khalid Al-Ageel, and Martin Hildebrandt

Abstract T-cell mediated immune responses are decisive for the success or failure of transplantation. As a consequence, T-cell mediated events have been the first and oldest major target for therapeutic efforts to optimize transplant survival. With cells, tissues, scaffolds, and devices more and more often merged to become medicinal products, i.e., tissue-engineered products or combination of advanced therapy medicinal product (ATMPs), understanding the T-cell response may be valuable for the purpose of this book to pave the way for novel, smart strategies to therapeutically modulate infiltrating immune cells, reducing side effects and improving the therapeutic outcome, i.e., healing, tissue restoration, and transplant survival. In this chapter, current and novel concepts of T-cell immunomodulation and their clinical translation will be presented and discussed to allow the transfer of the knowledge gained to implanted materials and devices as well as to combination ATMPs. Starting from a hypothesis as to the similarities and differences between classical T-cell immune responses and those directed against scaffolds and devices, the mechanisms behind T-cell responses in tolerance and rejection are unraveled and therapeutic strategies to modulate and control T-cells in front of implanted materials and devices are proposed.

Keywords T-cells • Costimulatory response • GvHD • Immunotolerance • Transplantation • Allograft • Rejection

S. Grass
Medizinische Klinik III, Klinikum rechts der Isar, Technische Universität München, München, Germany

S.K. Al-Ageel
Department of Pathology and Laboratory Medicine, King Faisal Specialist Hospital & Research Center, Riyadh, Saudi Arabia

TUMCells Interdisciplinary Center for Cellular Therapies, TUM School of Medicine, Technische Universität München, München, Germany

M. Hildebrandt (✉)
TUMCells Interdisciplinary Center for Cellular Therapies, TUM School of Medicine, Technische Universität München, München, Germany
e-mail: martin.hildebrandt@tum.de

B. Corradetti (ed.), *The Immune Response to Implanted Materials and Devices*,
DOI 10.1007/978-3-319-45433-7_12

12.1 Introduction

The classic model of T-cell mediated immune response has been elucidated in the recent decades to a large extent [1]. Major improvements in control of T-cell responses have been achieved, allowing for success rates in stem cell transplantation, organ transplantation and quality of life that meanwhile render transplantation a long-term option that, especially in solid organs, should be considered earlier, more seriously, and more often [2]. However, enhanced understanding does not match yet with therapeutic control, and rates of rejection, graft vs host disease (GvHD), and loss of transplant function are still unsatisfactory [3]. More recent developments in immunosuppressive drugs, including mTOR inhibitors, biologicals such as monoclonal antibodies directed against key elements of immune responses, and combinations thereof, still lack the specificity needed to circumvent major toxicity, avoid exaggerated immunosuppression, and improve transplant survival.

The activation of the immune response can be summarized briefly to require two distinct signals: the interaction of the T-cell receptor (first signal) and costimulatory molecules, surface molecules that bind to specific receptors on T cells [1, 4, 5]. Upon ligation, these costimulatory signals provide the mitogenic signals necessary for subsequent T-cell activation, clonal expansion, and maintenance of mature T cells. Several receptor-ligand pairs have been shown to exert a critical role in providing the costimulatory signal 2, including CD28 [6], CD40, ICOS [7], PD1 (for review cf. [4]), and CD26 (for review cf. [8]). CD28, by interacting with B7-molecules B7.1 and B7.2 (CD80, CD86), has defined our classical understanding of costimulation in T-cell mediated immune responses [9]; however, the authors of multiple studies have demonstrated that T-cell activation and allograft rejection can occur in the absence of both CD40 and CD28 signals [10–12]. The inducible costimulatory (ICOS) molecule is expressed on activated T cells [13], is constitutively expressed on APCs, and regulates both Th1 and Th2 cell differentiation [7, 14, 15]. The addition of anti-ICOS to the CD40 and CD28/B7 blockade (triple costimulatory blockade) has been shown to have synergistic effects [16]. CD26, a highly glycosylated surface molecule known to occur on many cell types, organs and as a soluble form in serum, exerts a highly specific ectopeptidase activity and occupies a complex immunomodulatory role in T-cell activation and costimulation [8, 17].

Physiological termination of immune responses may occur by competition with costimulatory events, as shown for CTLA-4 (CD154) in terminating CD28-mediated costimulatory events (for review cf. [18]). However, only facets of the mechanisms behind the induction of immune tolerance have been elucidated [19, 20], and the complex interplay of ligands competing at the immune synapses remains understood incompletely to date [4]. Although several animal studies have demonstrated allograft acceptance with the CD40 L blockade [21–23], the induced hyporesponsiveness caused by this blockade appears to be transient, and in several studies has failed to prevent chronic rejection [24, 25]. Clinical side effects, namely thromboembolic events, led to a cessation of clinical trials testing therapeutic antibodies inhibiting CD40/ CD40L ligation (for review cf. [26]). More recently, however, resurgence in the therapeutic potential inherent in CD40/CD40L blockade has been reported [27].

12.2 T-Cell Immune Responses and Age

Depending on the developmental stage, the immune system matures and changes, and the immune response to the implant and/or foreign material may vary during life span. In comparison to adults, neonates, preterm babies, and newborns mount a different cytokine response when exposed to pathogens, with a preponderance of IL-1alpha, IL-23, IL-10, and IL-6, and a diminished expression of IFN-gamma, INF1, and IL-12 [28]. The newborn's innate immune response tends to exhibit hyperactivity [29], while the T-cell activation is deficient [30–32] or inclined towards Foxp3+ CD25+ regulatory T cells [33, 34]. At birth, around 3 % of total CD4+ cells are peripheral Tregs resulting in an anti-inflammatory early-life immune profile [35, 36].

As the human advances in age, so does the immune system. Profound remodeling, reduction in polyclonality and decline in the immune system are associated with advanced aging [35, 37]. It begins to fail in maintaining full tolerance to self-antigens, due to lymphopenia occurring with age [38] and a decrease in regulatory T-cell function resulting in an increased incidence of autoimmune diseases among geriatric patients [39]. In some aspects the immune system of an aging organism resembles that of the newborn, with reduced antimicrobial activity by neutrophils and macrophages, reduced antigen presentation by DCs and decreased NK killing, and somewhat compromised adaptive lymphocyte responses making it more susceptible to infection.

12.3 Allogeneic HSCT: The Paradigm and Spearhead of T-Cell Immunotherapy

Allogeneic hematopoietic stem cell transplantation (allo-HSCT) is the paradigm of immunotherapy acting through a graft-versus-tumor reaction in the treatment of hematological malignancies. Still today, this curative allogeneic response can be associated with severe drawbacks, such as frequent and severe graft-versus-host disease (GvHD). More recently, other approaches of cancer immunotherapy such as cell therapy and immune checkpoint blockade provide promising data in various types of cancers including hematologic malignancies, and may develop into a valid option when allogeneic HSCT has not proven successful, is not applicable or insufficient [40–43]. For the purpose of this chapter, it may be useful to outline key points in allogeneic HSCT to project similarities and differences for tissues and scaffolds.

In allogeneic HSCT, we traditionally recognize two forms of GvHD: acute (aGvHD) and chronic (cGvHD). aGvHD remains, directly or indirectly, the major cause of short-term (day 100) mortality after allogenic HSCT. The pathology of aGvHD is a process comprising initial tissue damage from the conditioning regimen which in turn leads to activation of host antigen-presenting cells and activation and proliferation of donor T cells and finally to the release of inflammatory cytokines such as interleukin-1 and tissue necrosis factor (TNF-α) that eventually produce tissue necrosis. The action of this pathogenetic process in the induction of a GvHD is modulated in part by the presence of cells capable of inhibiting immune responses, most notably T-regulatory cells (T-regs).

As aGvHD is a result of an alloimmune effect the major risk for occurrence is the presence of human leukocyte antigen (HLA) disparity, and increasing degrees of HLA-mismatching increase the probability of more severe disease. Further risk factors include older patient age, the use of female donors for male recipients, prior alloimmunization of the donor, the nature of GvHD prophylaxis, increasing donor age, the use of peripheral blood stem cells as opposed to bone marrow, and recipient seropositivity for cytomegalovirus. More recently, we understand the importance of non-HLA genetic factors in the development of GvHD. Examples include polymorphisms in the genes encoding cytokines such as the tumor necrosis factors, the interleukins (IL-1, IL-6, and IL-10), interferon, and transforming growth factor (TGF)-β and the expression of the killer cell immunoglobulin-like receptors (KIR). Interestingly, one of the common features of the organs involved in aGvHD is that they are all exposed to microbial pathogens through the intestinal mucosa, epidermis and portal circulation, and early murine studies confirmed a reduction in the severity and incidence of GvHD in animals that received antibiotic prophylaxis to decontaminate the gastrointestinal tract or those kept in germ-free environments. This has led to the speculation that potential differences within individuals in the interactions of antigens derived from infective organisms and pattern recognition receptors (PRR) might protect or predispose to the occurrence of GvHD [44].

The first-line therapy of GvHD is single agent methylprednisolone, which is effective in 40–50 % of cases. But second line treatment for steroid refractory disease is largely unsatisfactory and therefore major efforts are exerted to prevent the occurrence of aGvHD. The most effective method of prophylaxis is T-cell depletion but in good risk transplants is accompanied by an unacceptable level of infection and relapse of the original disease. Still today, the most frequently used regimen for aGvHD prevention remains methotrexate and a calcineurin inhibitor. Manipulation of cellular subpopulations with immunosuppressive properties are promising new strategies for both prevention and treatment.

The chronic GvHD is an immunoregulatory disorder occurring after allogeneic HSCT and shares features of autoimmunity and immunodeficiency. Features of cGvHD resemble other autoimmune diseases such as Sjögren syndrome, scleroderma, primary biliary cirrhosis, and immuncytopenias. Similarly to aGvHD, cGvHD is also thought to be induced by the immune cells of the donor but the pathophysiology is even less well understood. Although autoreactive T-lymphocytes are considered to play the key role, recent data revealed the importance of B-cells

[44–46]. There is no standard salvage therapy for cGvHD. cGvHD is the main cause of late non-relapse mortality and morbidity after allogeneic HSCT. Mortality is primarily caused by infections either due to the immunodeficiency of cGvHD or due to its treatment.

Genetic disparity, in particular at the HLA loci, between patient and donor is a critical factor influencing transplantation outcome. Nevertheless, several issues remain to be explored, such as the impact of stem cell source and conditioning, T-cell depletion, non-HLA genetic factors, other donor factors (age, CMV, gender), delays in accessing the donor (time of transplantation) and the impact of HLA matching. The importance of large collaborative studies should be emphasized.

Novel approaches are needed to further improve the quality of the stem cell graft and to control side effects such as GvHD. Maybe CD26/Dipeptidyl Peptidase 4 (DPP4/DPPIV) is an appropriate target for attempts to further improve the outcome. *CD26/DPP4* is a surface T cell activation antigen and has been shown to have DPP4 enzymatic activity with critical impact on the turnover of various chemokines and bioactive peptides (ADA, CD45, Caveolin-1, CXCR-4, CARMA-1, M6P/IGFII). A large number of studies demonstrate that CD26/DPP4 plays an important role in the immune system, particularly in T cell activation and co-stimulation, memory T cell generation, in maintaining lymphocyte composition and function, and thymic emigration patterns during immune-senescence [8, 17]. Evidence is mounting that DPP4 expression and enzymatic activity contribute to the emergence of autoimmune disease and allograft rejection. Inhibitors of DPP4 activity have received marketing authorization as a treatment option for type II diabetes. However, the immunomodulatory facet of this therapeutic approach maybe even more important.

Over the past decade, mounting evidence has supported the therapeutic utility of T-cell-centered immunotherapy targeting check-point inhibitors, which, in various iterations, has been shown capable of eliciting highly precise preclinical and clinical responses with relevance for immune-oncology, transplantation immunology, and other fields of immunotherapy [47]. Nevertheless, further research is needed, such as the establishment of biomarkers for targeted therapy, duration of therapeutic activity and compatibility of combined strategies.

12.4 T-Cell Responses to Implants and Scaffolds: Challenges and Options

Decellularized (acellular) scaffolds, composed of natural extracellular matrix, form the basis of an emerging generation of tissue-engineered organ and tissue replacements capable of transforming healthcare. Prime requirements for allogeneic, or even xenogeneic, decellularized scaffolds are biocompatibility and absence of rejection.

The transfer of an implant or scaffold will inevitably trigger an inflammatory response towards the surgical trauma, the materials used and their interaction with the site of insertion. The intensity, the course, and the extent of the immune response will depend on the individual immune responsiveness and on factors related to the toxicity, corrosion and stability of the materials used [48]. Given the scarcity of data, the type of immune response will be nonspecific to a large extent; however, changes in material composition and the introduction of novel materials and technologies, including nanoparticles, will inevitably bear the risk of more specific, aberrant and unexpected immune responses that may threaten the function and survival of the implant and, more importantly, endanger the patient.

When comparing T-cell immune responses against scaffolds, implant materials and devices to those directed against organs and cells, differences are conceivable or evident and will have to be taken into account:

- A lack of HLA ("signal 1");
- A lack of immune cells of donor or graft origin that could mount a potent immune response against the host;
- A lack of tolerogenic immune cells of donor or graft origin that could alleviate a pathologic immune response that could lead to rejection, GvHD, transplant failure and death.

Irrespectively of the absence of donor-derived T cells or classical T cell partners, the following potential problems may mount a similarly strong and undesired effect:

- Cross-reactivity and molecular mimicry may initiate a T cell response from the side of the recipient,
- Superagonistic effects may initiate a cascade of immune effects in the absence of HLA and/ or costimulatory signals,
- A lack of biocompatibility, especially when materials of nonhuman biological origin are used, may unleash a cascade of responses that are by no means less potent than classical mechanisms of rejection, because the final routes are similar to a large extent, albeit more potent [49].

Sensitization against allergens derived from implants, scaffolds, and devices has been shown to occur, bearing features of a classical T-cell-mediated Type IV hypersensitivity response [50]. Allergens include known allergens such as nickel, cobalt, chromium, and also acrylate and antibiotics contained in bone cement. Interestingly, the numbers of patients reported to be allergic against such implant components are surprisingly low when compared with sensitization rates of up to 12 % in a broader community (for review cf. [48]). More recently, lymphoid aggregates found as histological lesions in adverse reactions to metal debris, were shown to have striking similarity with *tertiary lymphoid organs* [51] seen in chronic inflammatory diseases such as rheumatoid arthritis [52–54], Sjogren's syndrome [55], and Hashimoto's threoiditis [56]. Although these structures affect local immune responses, the contribution of these lymphoid aggregates to the underlining pathology is highly context dependent and can elicit either protective or deleterious outcomes [57].

The impact of such ectopic lymphoid organs, involving a plethora of immune cell populations, will reach beyond local effects and exert systemic responses. In fact, systemic reactions have been observed in patients with implants of various kinds [58]. In patients supplied with left ventricular assist devices (LVADs), several canonical signaling pathways of were shown to be affected 7 days and six months after implant, including NF-kB, HLA class II-mediated antigen presentation, OX40, IL-15, 14-3-3, and cytotoxic T-lymphocyte mediated apoptosis [59].

Fishman and colleagues [60] studied the cellular immune response against a decellularized skeletal muscle scaffold in a xenotransplant model. T-cell proliferation and cytokines served as surrogate markers for the in vivo host immune response toward the scaffold. Decellularized scaffolds were found to exert anti-inflammatory and immunosuppressive effects, as evidenced by delayed biodegradation time in vivo; reduced sensitized T-cell proliferative activity in vitro; reduced IL-2, IFN-gamma, and raised IL-10 levels; polarization of the macrophage response in vivo toward an M2 phenotype; and improved survival of donor-derived xenogeneic cells at 2 and 4 weeks in vivo. Decellularized scaffolds were shown to polarize host responses away from a classical Th1-proinflammatory profile and appear to down-regulate T-cell xenoresponses and Th1 effector function by inducing a state of peripheral T-cell hyporesponsiveness.

Common techniques utilized to process biomaterials include decellularization, chemical cross-linking to stabilize the matrices and/or remove or mask antigenic epitopes, DNA, and damage-associated molecular pattern (DAMP) molecules [61–63]. However, chemical cross-linking can impact the immune response and the process of decellularization does not fully remove DNA. In porcine-derived scaffolds for orthopedic applications, remnant porcine DNA within decellularized biological scaffold materials has been implicated as the cause of inflammatory immune reactions post transplantation [61–65]. In addition, studies have indicated that coating with serum substrates induces IL-10 production and Th-2 response, while adhesive substrates such as vitronectin or collagen have been implicated in the support of Th1 DC responses [66].

12.5 Perspectives for Transplant Optimization

For the topic addressed here, it is still difficult to draw conclusions as to potential strategies to optimize immune tolerance for biomaterials, implants and scaffolds.

Some contours, however, become apparent:

1. The *choice of material* can influence the immune response to the implant through the adsorption of proteins, which in turn facilitates the interaction with cells of the immune system [67, 68]. Monocyte adhesion to the surface of the implant is generally enhanced when the implant is made of hydrophobic materials in comparison to hydrophilic materials [68, 69]. Therefore, the utilization of neutral or hydrophilic materials decreases the formation of foreign-body giant cells through the reduction of monocyte/macrophage adhesion [70]. Conversely, a greater amount of inflammatory cytokines is produced by the adherent monocytes/macrophages on neutral or hydrophilic biomaterials [70, 71].

2. The *association of tertiary lymphoid organs with T-cell mediated inflammatory responses against implants*, devices and scaffolds may provide new perspectives for therapies to overcome transplant rejection, loss of function and implant failure.

   However, to date the number of clinical studies targeting tertiary lymphoid organs remains limited. One potential strategy is to disrupt the spatial arrangement of T and B cells. In chronic renal allograft rejection, B-cell depletion with rituximab (anti-CD20 monoclonal antibody) had a limited impact on the maintenance of ELFs and biological intervention promoted expression of the B-cell survival factor, B-cell-activating factor [72]. Other licensed biological therapies including IL-6 or IL-6 receptor-specific monoclonal antibodies (e.g., tocilizumab), T-cell activation antagonists (e.g., abatacept), and Janus kinase inhibitors (e.g., tofacitinib) are likely to target pathways linked with tertiary lymphoid organs, albeit untested [73]. Given their prominent roles in lymphoid neogenesis, drugs targeting lymphotoxin beta (e.g., baminercept and pateclizumab) may prove effective in blocking tertiary lymphoid organ activity [74–77]. Similarly, CXCL13 blockade has shown some promise in preclinical studies for the treatment of experimental inflammatory arthritis, diabetes, and Sjögren's syndrome [78–80]. Finally, preclinical data provide encouraging evidence for amelioration of disease following blockade of IL-21 involvement in germinal center reactions [81–83].
3. *Decellularized scaffolds exert an immunomodulatory effect in favor of immune tolerance*: First evidence suggests that such tissue preparations may be better suited than artificial compounds, at least in terms of immune tolerance. The immune response pattern observed here might, in fact, be seen as similar to an induction of tolerance and is reminiscent of studies in models of pregnancy, Th1-mediated abortions and their prevention by a shift of T cell responses to a Th2 hyporesponsive state [84, 85]. The fact that immunomodulatory effector mechanisms involving CD26/DPP IV inhibition have been shown to exert the shift towards a state of tolerance [8, 17] could pave the way for a further enhanced tolerance of such scaffolds in T-cell hyporesponsiveness as discussed above; in solid organ transplantation, the inhibition of CD26/DPP IV activity has at least contributed to improved transplant survival [86]. In line with reports on systemic effects of LVAD implants described above, the very same canonical signaling pathways [47] may be expected to reflect changes corresponding to an altered, tolerogenic systemic pattern. The *absence of HLA molecules* can be assumed to be beneficial, matching favorably with elegant data by Figueiredo et al. [87] who virtually neutralized cells by downregulation of both HLA class I and class II molecules.
4. *Scaffolds of xenogeneic origin*, however, will continue to be confronted with immune barriers that restrict their use in transplantation: for xenogeneic nonviable xenografts, increasing evidence suggests that considerable immune reactions, mediated by both innate and adaptive immunity, take place and influence the long-term outcome of xenogeneic materials in patients, possibly precluding the use of bioprosthetic heart valves in young individuals [59]. Solutions may

include (a) classical [88] and innovative concepts of immunosuppression as described for tertiary lymphoid organs above, (b) immunomodulatory approaches such as DPP IV inhibition, and (c) genetic modifications that could also contribute to bridging cross-species barriers.

Various factors lead to transplant optimization and immunomodulation, addressing both, the graft itself and the host's immune system. Age and/or state of the recipient must be taken into consideration when modulating therapy or preparing a graft for implantation/transplantation, as the humans differ in their immune profile depending on their age. In transplantation, many strategies have been designed and tested to induce clinically relevant immune tolerance, targeting different cell types including T cells and the interface between T cells and APCs (for review cf. [8, 17, 71]). Many important lessons have been learned especially in hematopoietic stem cell transplantation as outlined above. The comparison of known concepts in transplantation immunology with the emerging field of scaffolds used in regenerative medicine remains limited; however, some parallels pointed out in this chapter are hoped to provide useful hints for further development.

References

1. Barrett AJ, Rezvani K, Solomon S et al (2003) New developments in allotransplant immunology. Hematology Am Soc Hematol Educ Program:350–371
2. Rana A, Gruessner A, Agopian VG et al (2015) Survival benefit of solid-organ transplant in the United States. JAMA Surg 150(3):252–259
3. Blaise D, Castagna L (2012) Do different conditioning regimens really make a difference? Hematology Am Soc Hematol Educ Program 2012:237–245
4. Dilek N, Poirier N, Hulin P et al (2013) Targeting CD28, CTLA-4 and PD-L1 costimulation differentially controls immune synapses and function of human regulatory and conventional T-cells. PLoS One 8(12), e83139
5. Ray WZ, Kasukurthi R, Papp EM et al (2010) The role of T helper cell differentiation in promoting nerve allograft survival with costimulation blockade. J Neurosurg 112(2):386–393
6. Alegre ML, Frauwirth KA, Thompson CB (2001) T-cell regulation by CD28 and CTLA-4. Nat Rev Immunol 1(3):220–228
7. Dong C, Juedes AE, Temann UA et al (2001) ICOS co-stimulatory receptor is essential for T-cell activation and function. Nature 409(6816):97–101
8. Klemann C, Wagner L, Stephan M et al (2016) Cut to the chase: a review of CD26/dipeptidyl peptidase-4's (DPP4) entanglement in the immune system. Clin Exp Immunol 185(1):1–21
9. Miller SD, Vanderlugt CL, Lenschow DJ et al (1995) Blockade of CD28/B7-1 interaction prevents epitope spreading and clinical relapses of murine EAE. Immunity 3(6):739–745
10. Lin H, Rathmell JC, Gray GS et al (1998) Cytotoxic T lymphocyte antigen 4 (CTLA4) blockade accelerates the acute rejection of cardiac allografts in CD28-deficient mice: CTLA4 can function independently of CD28. J Exp Med 188(1):199–204
11. Mandelbrot DA, Oosterwegel MA, Shimizu K et al (2001) B7-dependent T-cell costimulation in mice lacking CD28 and CTLA4. J Clin Invest 107(7):881–887
12. Yamada A, Kishimoto K, Dong VM et al (2001) CD28-independent costimulation of T cells in alloimmune responses. J Immunol 167(1):140–146
13. Hutloff A, Dittrich AM, Beier KC et al (1999) ICOS is an inducible T-cell co-stimulator structurally and functionally related to CD28. Nature 397(6716):263–266

14. McAdam AJ, Chang TT, Lumelsky AE et al (2000) Mouse inducible costimulatory molecule (ICOS) expression is enhanced by CD28 costimulation and regulates differentiation of CD4+ T cells. J Immunol 165(9):5035–5040
15. Ozkaynak E, Gao W, Shemmeri N et al (2001) Importance of ICOS-B7RP-1 costimulation in acute and chronic allograft rejection. Nat Immunol 2(7):591–596
16. Tai CY, Weber RV, Mackinnon SE et al (2010) Multiple costimulatory blockade in the peripheral nerve allograft. Neurol Res 32(3):332–336
17. Hildebrandt M, Reutter W, Arck P et al (2000) A guardian angel: the involvement of dipeptidyl peptidase IV in psychoneuroendocrine function, nutrition and immune defence. Clin Sci (Lond) 99(2):93–104
18. Gardner D, Jeffery LE, Sansom DM (2014) Understanding the CD28/CTLA-4 (CD152) pathway and its implications for costimulatory blockade. Am J Transplant 14(9):1985–1991
19. Graca L, Honey K, Adams E et al (2000) Cutting edge: anti-CD154 therapeutic antibodies induce infectious transplantation tolerance. J Immunol 165(9):4783–4786
20. Li Y, Li XC, Zheng XX et al (1999) Blocking both signal 1 and signal 2 of T-cell activation prevents apoptosis of alloreactive T cells and induction of peripheral allograft tolerance. Nat Med 5(11):1298–1302
21. Kirk AD, Burkly LC, Batty DS et al (1999) Treatment with humanized monoclonal antibody against CD154 prevents acute renal allograft rejection in nonhuman primates. Nat Med 5(6):686–693
22. van Maurik A, Herber M, Wood KJ et al (2002) Cutting edge: CD4+CD25+ alloantigen-specific immunoregulatory cells that can prevent CD8+ T cell-mediated graft rejection: implications for anti-CD154 immunotherapy. J Immunol 169(10):5401–5404
23. Zhai Y, Meng L, Gao F et al (2006) CD4+ T regulatory cell induction and function in transplant recipients after CD154 blockade is TLR4 independent. J Immunol 176(10):5988–5994
24. Ensminger SM, Witzke O, Spriewald BM et al (2000) CD8+ T cells contribute to the development of transplant arteriosclerosis despite CD154 blockade. Transplantation 69(12):2609–2612
25. Guillot C, Guillonneau C, Mathieu P et al (2002) Prolonged blockade of CD40-CD40 ligand interactions by gene transfer of CD40Ig results in long-term heart allograft survival and donor-specific hyporesponsiveness, but does not prevent chronic rejection. J Immunol 168(4):1600–1609
26. Law CL, Grewal IS (2009) Therapeutic interventions targeting CD40L (CD154) and CD40: the opportunities and challenges. Adv Exp Med Biol 647:8–36
27. Pinelli DF, Ford ML (2015) Novel insights into anti-CD40/CD154 immunotherapy in transplant tolerance. Immunotherapy 7(4):399–410
28. Levy O (2007) Innate immunity of the newborn: basic mechanisms and clinical correlates. Nat Rev Immunol 7(5):379–390
29. Casey A, Dirks F, Liang OD et al (2014) Bone marrow-derived multipotent stromal cells attenuate inflammation in obliterative airway disease in mouse tracheal allografts. Stem Cells Int 2014:468927
30. Prescott SL, Macaubas C, Holt BJ et al (1998) Transplacental priming of the human immune system to environmental allergens: universal skewing of initial T cell responses toward the Th2 cytokine profile. J Immunol 160(10):4730–4737
31. Prescott SL, Macaubas C, Smallacombe T et al (1999) Development of allergen-specific T-cell memory in atopic and normal children. Lancet 353(9148):196–200
32. Wiles K, Fishman JM, De Coppi P et al (2016) The host immune response to tissue-engineered organs: current problems and future directions. Tissue Eng Part B Rev 22(3):208–219
33. Mold JE, Venkatasubrahmanyam S, Burt TD et al (2010) Fetal and adult hematopoietic stem cells give rise to distinct T cell lineages in humans. Science 330(6011):1695–1699
34. Simon AK, Hollander GA, McMichael A (2015) Evolution of the immune system in humans from infancy to old age. Proc Biol Sci 282(1821):20143085
35. Mackroth MS, Malhotra I, Mungai P et al (2011) Human cord blood CD4+CD25hi regulatory T cells suppress prenatally acquired T cell responses to Plasmodium falciparum antigens. J Immunol 186(5):2780–2791

36. Takahata Y, Nomura A, Takada H et al (2004) CD25+CD4+ T cells in human cord blood: an immunoregulatory subset with naive phenotype and specific expression of forkhead box p3 (Foxp3) gene. Exp Hematol 32(7):622–629
37. Weiskopf D, Weinberger B, Grubeck-Loebenstein B (2009) The aging of the immune system. Transpl Int 22(11):1041–1050
38. Sheu TT, Chiang BL, Yen JH et al (2014) Premature CD4+ T cell aging and its contribution to lymphopenia-induced proliferation of memory cells in autoimmune-prone non-obese diabetic mice. PLoS One 9(2), e89379
39. Goronzy JJ, Weyand CM (2003) Aging, autoimmunity and arthritis: T-cell senescence and contraction of T-cell repertoire diversity—catalysts of autoimmunity and chronic inflammation. Arthritis Res Ther 5(5):225–234
40. Hodi FS, O'Day SJ, McDermott DF et al (2010) Improved survival with ipilimumab in patients with metastatic melanoma. N Engl J Med 363(8):711–723
41. Pardoll DM (2012) The blockade of immune checkpoints in cancer immunotherapy. Nat Rev Cancer 12(4):252–264
42. Rizvi NA, Hellmann MD, Snyder A et al (2015) Cancer immunology. Mutational landscape determines sensitivity to PD-1 blockade in non-small cell lung cancer. Science 348(6230):124–128
43. Vanneman M, Dranoff G (2012) Combining immunotherapy and targeted therapies in cancer treatment. Nat Rev Cancer 12(4):237–251
44. Penack O, Holler E, van den Brink MR (2010) Graft-versus-host disease: regulation by microbe-associated molecules and innate immune receptors. Blood 115(10):1865–1872
45. Flowers ME, Martin PJ (2015) How we treat chronic graft-versus-host disease. Blood 125(4):606–615
46. Shimabukuro-Vornhagen A, Hallek MJ, Storb RF et al (2009) The role of B cells in the pathogenesis of graft-versus-host disease. Blood 114(24):4919–4927
47. Mellman I, Hubbard-Lucey VM, Tontonoz MJ et al (2016) De-risking immunotherapy: report of a consensus workshop of the Cancer Immunotherapy Consortium of the Cancer Research Institute. Cancer Immunol Res 4(4):279–288
48. Thomas P, Thomsen M (2010) Implant allergies. Hautarzt 61(3):255–262, quiz 263-254
49. Vadori M, Cozzi E (2015) The immunological barriers to xenotransplantation. Tissue Antigens 86(4):239–253
50. Hallab N, Merritt K, Jacobs JJ (2001) Metal sensitivity in patients with orthopaedic implants. J Bone Joint Surg Am 83-A(3):428–436
51. Mittal S, Revell M, Barone F et al (2013) Lymphoid aggregates that resemble tertiary lymphoid organs define a specific pathological subset in metal-on-metal hip replacements. PLoS One 8(5), e63470
52. Manzo A, Paoletti S, Carulli M et al (2005) Systematic microanatomical analysis of CXCL13 and CCL21 in situ production and progressive lymphoid organization in rheumatoid synovitis. Eur J Immunol 35(5):1347–1359
53. Takemura S, Braun A, Crowson C et al (2001) Lymphoid neogenesis in rheumatoid synovitis. J Immunol 167(2):1072–1080
54. Weyand CM, Goronzy JJ (2003) Ectopic germinal center formation in rheumatoid synovitis. Ann N Y Acad Sci 987:140–149
55. Barone F, Bombardieri M, Manzo A et al (2005) Association of CXCL13 and CCL21 expression with the progressive organization of lymphoid-like structures in Sjogren's syndrome. Arthritis Rheum 52(6):1773–1784
56. Armengol MP, Cardoso-Schmidt CB, Fernandez M et al (2003) Chemokines determine local lymphoneogenesis and a reduction of circulating CXCR4+ T and CCR7 B and T lymphocytes in thyroid autoimmune diseases. J Immunol 170(12):6320–6328
57. Jones GW, Jones SA (2016) Ectopic lymphoid follicles: inducible centres for generating antigen-specific immune responses within tissues. Immunology 147(2):141–151
58. Pearson MJ, Williams RL, Floyd H et al (2015) The effects of cobalt-chromium-molybdenum wear debris in vitro on serum cytokine profiles and T cell repertoire. Biomaterials 67:232–239

59. Mitchell A, Guan W, Staggs R et al (2013) Identification of differentially expressed transcripts and pathways in blood one week and six months following implant of left ventricular assist devices. PLoS One 8(10), e77951
60. Fishman JM, Lowdell MW, Urbani L et al (2013) Immunomodulatory effect of a decellularized skeletal muscle scaffold in a discordant xenotransplantation model. Proc Natl Acad Sci U S A 110(35):14360–14365
61. Gilbert TW, Sellaro TL, Badylak SF (2006) Decellularization of tissues and organs. Biomaterials 27(19):3675–3683
62. Lotze MT, Deisseroth A, Rubartelli A (2007) Damage associated molecular pattern molecules. Clin Immunol 124(1):1–4
63. Ma B, Wang X, Wu C et al (2014) Crosslinking strategies for preparation of extracellular matrix-derived cardiovascular scaffolds. Regen Biomater 1(1):81–89
64. Badylak SF, Gilbert TW (2008) Immune response to biologic scaffold materials. Semin Immunol 20(2):109–116
65. Zheng MH, Chen J, Kirilak Y et al (2005) Porcine small intestine submucosa (SIS) is not an acellular collagenous matrix and contains porcine DNA: possible implications in human implantation. J Biomed Mater Res B Appl Biomater 73(1):61–67
66. Acharya AP, Dolgova NV, Clare-Salzler MJ et al (2008) Adhesive substrate-modulation of adaptive immune responses. Biomaterials 29(36):4736–4750
67. Brodbeck WG, Anderson JM (2009) Giant cell formation and function. Curr Opin Hematol 16(1):53–57
68. Thevenot P, Hu W, Tang L (2008) Surface chemistry influences implant biocompatibility. Curr Top Med Chem 8(4):270–280
69. Hezi-Yamit A, Sullivan C, Wong J et al (2009) Impact of polymer hydrophilicity on biocompatibility: implication for DES polymer design. J Biomed Mater Res A 90(1):133–141
70. Jones JA, Chang DT, Meyerson H et al (2007) Proteomic analysis and quantification of cytokines and chemokines from biomaterial surface-adherent macrophages and foreign body giant cells. J Biomed Mater Res A 83(3):585–596
71. Kontos S, Grimm AJ, Hubbell JA (2015) Engineering antigen-specific immunological tolerance. Curr Opin Immunol 35:80–88
72. Thaunat O, Patey N, Gautreau C et al (2008) B cell survival in intragraft tertiary lymphoid organs after rituximab therapy. Transplantation 85(11):1648–1653
73. Hunter CA, Jones SA (2015) IL-6 as a keystone cytokine in health and disease. Nat Immunol 16(5):448–457
74. Fava RA, Kennedy SM, Wood SG et al (2011) Lymphotoxin-beta receptor blockade reduces CXCL13 in lacrimal glands and improves corneal integrity in the NOD model of Sjogren's syndrome. Arthritis Res Ther 13(6):R182
75. Fava RA, Notidis E, Hunt J et al (2003) A role for the lymphotoxin/LIGHT axis in the pathogenesis of murine collagen-induced arthritis. J Immunol 171(1):115–126
76. Gatumu MK, Skarstein K, Papandile A et al (2009) Blockade of lymphotoxin-beta receptor signaling reduces aspects of Sjogren's syndrome in salivary glands of non-obese diabetic mice. Arthritis Res Ther 11(1):R24
77. Lee Y, Chin RK, Christiansen P et al (2006) Recruitment and activation of naive T cells in the islets by lymphotoxin beta receptor-dependent tertiary lymphoid structure. Immunity 25(3):499–509
78. Henry RA, Kendall PL (2010) CXCL13 blockade disrupts B lymphocyte organization in tertiary lymphoid structures without altering B cell receptor bias or preventing diabetes in non-obese diabetic mice. J Immunol 185(3):1460–1465
79. Kramer JM, Klimatcheva TL, Rothstein E (2013) CXCL13 is elevated in Sjogren's syndrome in mice and humans and is implicated in disease pathogenesis. J Leukoc Biol 94(5):1079–1089
80. Zheng B, Ozen Z, Zhang X et al (2005) CXCL13 neutralization reduces the severity of collagen-induced arthritis. Arthritis Rheum 52(2):620–626
81. Bucher C, Koch L, Vogtenhuber C et al (2009) IL-21 blockade reduces graft-versus-host disease mortality by supporting inducible T regulatory cell generation. Blood 114(26):5375–5384

82. Herber D, Brown TP, Liang S et al (2007) IL-21 has a pathogenic role in a lupus-prone mouse model and its blockade with IL-21R.Fc reduces disease progression. J Immunol 178(6): 3822–3830
83. Young DA, Hegen M, Ma HL et al (2007) Blockade of the interleukin-21/interleukin-21 receptor pathway ameliorates disease in animal models of rheumatoid arthritis. Arthritis Rheum 56(4):1152–1163
84. Arck PC, Rose M, Hertwig K et al (2001) Stress and immune mediators in miscarriage. Hum Reprod 16(7):1505–1511
85. Ruter J, Hoffmann T, Heiser U et al (2002) The expression of T-cell surface antigens CTLA-4, CD26, and CD28 is modulated by inhibition of dipeptidylpeptidase IV (DPP IV, CD26) activity in murine stress-induced abortions. Cell Immunol 220(2):150–156
86. Jungraithmayr W, De Meester I, Matheeussen V et al (2010) Inhibition of CD26/DPP IV attenuates ischemia/reperfusion injury in orthotopic mouse lung transplants: the pivotal role of vasoactive intestinal peptide. Peptides 31(4):585–591
87. Figueiredo C, Blasczyk R (2015) A future with less HLA: potential clinical applications of HLA-universal cells. Tissue Antigens 85(6):443–449
88. Snanoudj R, Tinel C, Legendre C (2015) Immunological risks of minimization strategies. Transpl Int 28(8):901–910

Index

B. Corradetti (ed.), *The Immune Response to Implanted Materials and Devices*,
DOI 10.1007/978-3-319-45433-7

MIX
Papier aus verantwortungsvollen Quellen
Paper from responsible sources
FSC® C105338

If you have any concerns about our products,
you can contact us on
ProductSafety@springernature.com

In case Publisher is established outside the EU,
the EU authorized representative is:
Springer Nature Customer Service Center GmbH
Europaplatz 3, 69115 Heidelberg, Germany

Printed by Libri Plureos GmbH
in Hamburg, Germany